POPULATION PARAMETERS:
ESTIMATION FOR
ECOLOGICAL MODELS

METHODS IN ECOLOGY

Series Editors

J.H. LAWTON FRS
Imperial College at Silwood Park
Ascot, UK

G.E. LIKENS
Institute of Ecosystem Studies
Millbrook, USA

METHODS IN ECOLOGY

Population Parameters: Estimation for Ecological Models

HAMISH McCALLUM

Department of Zoology and Entomology and
Centre for Conservation Biology
University of Queensland
Brisbane, Australia

Blackwell
Science

© 2000 by
Blackwell Science Ltd
Editorial Offices:
Osney Mead, Oxford OX2 0EL
25 John Street, London WC1N 2BL
23 Ainslie Place, Edinburgh EH3 6AJ
350 Main Street, Malden
  MA 02148 5018, USA
54 University Street, Carlton
  Victoria 3053, Australia
10, rue Casimir Delavigne
  75006 Paris, France

Other Editorial Offices:
Blackwell Wissenschafts-Verlag GmbH
Kurfürstendamm 57
10707 Berlin, Germany

Blackwell Science KK
MG Kodenmacho Building
7–10 Kodenmacho Nihombashi
Chuo-ku, Tokyo 104, Japan

The right of the Author to be
identified as the Author of this Work
has been asserted in accordance
with the Copyright, Designs and
Patents Act 1988.

First published 2000

Set by Graphicraft Limited, Hong Kong
Printed and bound in Great Britain
at MPG Books Ltd, Bodmin, Cornwall

The Blackwell Science logo is a
trade mark of Blackwell Science Ltd,
registered at the United Kingdom
Trade Marks Registry

DISTRIBUTORS

Marston Book Services Ltd
PO Box 269
Abingdon, Oxon OX14 4YN
(*Orders*: Tel: 01235 465500
    Fax: 01235 465555)

USA
Blackwell Science, Inc.
Commerce Place
350 Main Street
Malden, MA 02148 5018
(*Orders*: Tel: 800 759 6102
    781 388 8250
  Fax: 781 388 8255)

Canada
Login Brothers Book Company
324 Saulteaux Crescent
Winnipeg, Manitoba R3J 3T2
(*Orders*: Tel: 204 837-2987)

Australia
Blackwell Science Pty Ltd
54 University Street
Carlton, Victoria 3053
(*Orders*: Tel: 3 9347 0300
    Fax: 3 9347 5001)

A catalogue record for this title
is available from the British Library
and the Library of Congress

ISBN   0-86542-740-2

For further information on
Blackwell Science, visit our website:
www.blackwell-science.com

Contents

The Methods in Ecology Series

The explosion of new technologies has created the need for a set of concise and authoritative books to guide researchers through the wide range of methods and approaches that are available to ecologists. The aim of this series is to help graduate students and established scientists choose and employ a methodology suited to a particular problem. Each volume is not simply a recipe book, but takes a critical look at different approaches to the solution of a problem, whether in the laboratory or in the field, and whether involving the collection or the analysis of data.

Rather than reiterate established methods, authors have been encouraged to feature new technologies, often borrowed from other disciplines, that ecologists can apply to their work. Innovative techniques, properly used, can offer particularly exciting opportunities for the advancement of ecology.

Each book guides the reader through the range of methods available, letting ecologists know what they could, and could not, hope to learn by using particular methods or approaches. The underlying principles are discussed, as well as the assumptions made in using the methodology, and the potential pitfalls that could occur – the type of information usually passed on by word of mouth or learned by experience. The books also provide a source of reference to further detailed information in the literature. There can be no substitute for working in the laboratory of a real expert on a subject, but we envisage the Methods in Ecology Series as being the 'next best thing'. We hope that, by consulting these books, ecologists will learn what technologies and techniques are available, what their main advantages and disadvantages are, when and where not to use a particular method, and how to interpret the results.

Much is now expected of the science of ecology, as humankind struggles with a growing environmental crisis. Good methodology alone never solved any problem, but bad or inappropriate methodology can only make matters worse. Ecologists now have a powerful and rapidly growing set of methods and tools with which to confront fundamental problems of a theoretical and applied nature. We see the Methods in Ecology Series as a major contribution towards making these techniques known to a much wider audience.

John H. Lawton
Gene E. Likens

Preface

One of the axioms of modelling is GIGO: garbage in, garbage out. The fundamental aim of this book is to have less garbage going into ecological models, and thus better predictions and improved contributions to ecological understanding coming out the other end.

Modelling has a central role in ecology, as it does in all sciences. Any attempt to generalize or understand how an ecological process works requires a model. Whilst such a model need not be in mathematical form, or involve quantitative parameter estimates, a mathematical model has the advantage that it will require an explicit statement of the underlying assumptions. Any attempt to manage an ecological system also involves a model. Management of a system entails making a choice from a range of possible actions. Making the correct choice depends upon a capacity to predict the likely response of the system to these different actions. Prediction necessarily requires models, and almost always it will be worthwhile to have quantitative predictions, which must be based on quantitative parameter estimates.

Ecological modelling is currently moving away from being largely undertaken by expert mathematicians and programmers and into a situation where models are being used by field ecologists and wildlife managers. Most of these users will rely upon prewritten models, into which they will put their own parameter estimates. Their main requirement is not for nuts and bolts details of how models are constructed, but for information on how to provide appropriate input to the models. A recurrent theme in this book is that the most appropriate way to estimate a given parameter depends on the nature of the model in which the parameter is used, and on the objective of the modelling process. It is also essential for model users to understand how to interpret model results, given the inevitable approximations and errors in the parameter estimation process.

A particular feature of this book is the emphasis on estimating parameters for ecological interactions. Charles Elton (1933) defined animal ecology as 'the inter-relations of animals, numbers, social organization, migration, food and many [other factors]'. Much theoretical ecology has also, of course, concentrated on interactions. However, most of the existing books concerned with statistical aspects of animal ecology have concentrated on estimating population size and density for single populations, and on estimating birth, death and growth rates, without taking biotic interactions into account. The literature on estimating parameters for interactions has remained scattered.

I hope that, by bringing it together, I will enable researchers and managers to include interactions in their models more easily, producing improved ecological understanding and predictions.

Hamish McCallum
Brisbane

Acknowledgements

I am very grateful to Tony Pople (University of Queensland) for reading several of the chapters, providing unpublished data, and performing some of the example calculations. I wrote part of the book whilst on Study Leave. I thank Ilkka Hanski (University of Helsinki), Bryan Grenfell (University of Cambridge) and Roy Anderson (Oxford University) for their hospitality, and for the productive discussions with them and their research groups. Support from the Australian Research Council also contributed to the book. Finally, I am very grateful for the patience of Blackwell Science, and the series editors, John Lawton and Gene Likens, throughout the gestation of this book.

I would also like to thank the following publishers and journals for allowing material from their publications to be used in this book: Academic Press: Fig. 10.9e (McCallum & Scott, 1994); *Biometrics*: Box 3.2 (Smith *et al.*, 1995); Blackwell Science: Fig. 7.7 (van den Bosch *et al.*, 1992); *Canadian Journal of Fisheries and Aquatic Sciences*: Fig. 6.2 (Walters & Ludwig, 1981), Table 6.3 (Getz & Swartzman, 1981); Chapman and Hall: Fig. 6.1 (Hilborn & Walters, 1992); Cambridge University Press: Fig. 5.3 (Bayliss, 1987), Box 9.1 (Peters, 1983); *Ecology*: Box 6.2 (Turchin & Taylor, 1992), Fig. 7.2 (Okubo & Levin, 1989), Fig. 8.5 and Table 8.2 (Pascual & Kareiva, 1996), Fig. 9.3c (Messier, 1994), Fig. 9.3d (Crawford & Jennings, 1989), Figs 9.8 and 9.9 (Carpenter *et al.*, 1994), Fig. 10.10 (Dobson & Meagher, 1996); *Epidemiology and Infection*: Fig. 10.4 (Hone *et al.*, 1992); *Journal of Animal Ecology*: Figs 10.7 and 10.8 (Hudson *et al.*, 1992); *Journal of Parasitology*: Fig. 10.1 (Hudson & Dobson, 1997); *Limnology and Oceanography*: Fig. 4.4 (Hairston & Twombly, 1985); MacMillan: Fig. 2.1 (Swartzman & Kaluzny, 1987); *Oecologia*: Fig. 9.3b (Pech *et al.*, 1992); *Oikos*: Figs 8.1 and 8.2 (Abramsky *et al.*, 1992, 1994), Fig. 8.3 (Abramsky *et al.*, 1994), Fig. 8.4 (Chase, 1996), Fig. 9.3a (Sinclair *et al.*, 1990), Figs 9.4 and 9.5 (Caughley & Gunn, 1993); Oxford University Press: Fig. 10.9a and b (Anderson & May, 1991); *Parasitology*: Fig. 10.2 (Woolhouse & Chandiwana, 1992), Fig. 10.3 (McCallum, 1982), Fig. 10.6 (Gulland & Fox, 1992), Fig. 10.9c and d (Gulland, 1992); Princeton University Press: Figs 8.6 and 8.7 (Tilman, 1982 © 1982, reprinted by permission of Princeton University Press); *Science*: Fig. 6.3 (reprinted with permission from Myers *et al.*, 1995 © 1995 American Association for the Advancement of Science), Box 7.1 (reprinted with permission from Wahlberg *et al.*, 1996 © 1996 American Association for the Advancement of Science); *Science of the Total Environment*: Fig. 10.5 (Grenfell *et al.*, 1992 © 1992, reprinted with permission from Elsevier Science); *Trends in Ecology and Evolution*: Figs 7.3 and 7.4 (Koenig *et al.*, 1996 © 1996, reprinted with permission from Elsevier Science), Fig. 7.6a (Waser & Strobeck, 1998 © 1998, reprinted with permission from Elsevier Science); University of Chicago Press: Table 8.1 (Pfister, 1995), Figs 9.6 and 9.7 (Turchin & Hanski, 1997); *Wildlife Monographs*: Fig. 3.5 (Pollock *et al.*, 1990).

CHAPTER 1

Introduction

Scope of modelling

Any attempt to draw generalizations in science involves a model: an abstracted version of reality. A standard dictionary definition of a model is:

A simplified description of a system, process, etc., put forward as a basis for theoretical or empirical understanding.

(New Shorter Oxford English Dictionary)

This is a fairly accurate summary of the objective of ecological modelling.

Models may be verbal or diagrammatic, but this book is concerned with mathematical models: the subset of ecological models that uses mathematical representation of ecological processes. All statistical methods implicitly involve mathematical models, which are necessary in order to produce parameter estimates from the raw data used as input. This book is particularly concerned with models that use parameter estimates as input, rather than those that produce them as output. Statistical models are included only to the extent that is necessary to understand the properties and limitations of the parameter estimates they generate.

Scope of this book

This book concentrates on parameter estimation for models in animal population ecology. It does not attempt to deal with models in plant biology, nor to cover processes at the community or ecosystem level in any detail. I have biased the contents somewhat towards 'wildlife' ecology. The term 'wildlife' is a rubbery one, but it usually means vertebrates, other than fish.

I have not aimed to provide a full coverage of the methods required for parameter estimation in fisheries biology. Some fisheries-based approaches are included, largely where they have relevance to other taxa. As the branch of population ecology with the longest history of the practical application of mathematical modelling (although not with unqualified success: Beddington & Basson, 1994; Hutchings & Myers, 1994), fisheries has much to offer other areas of ecology. It has a large literature of its own, and to attempt a coverage that would be useful for fisheries managers, whilst also covering other areas of ecology, would be impractical. The classic reference on parameterizing fisheries models is Ricker (1975). An excellent, more recent treatment can be found in Hilborn and Walters (1992).

I have written the book with two principal audiences in mind. First, wildlife managers in most countries are now required by legislation to ensure that populations, communities and ecosystems are managed on an ecologically sustainable basis. Any attempt to ensure sustainability entails predicting the state of the system into the future. Prediction of any sort requires models. Frequently, managers use packaged population viability analysis models (Boyce, 1992) in an attempt to predict the future viability of populations. These models require a rather bewildering number of parameter estimates. This book provides managers with clear recommendations on how to estimate parameters, and how to ensure that the parameters included in the model are the appropriate ones to answer particular management questions. Most population viability analysis models currently model a single species at a time. Parameters for single species are the topic of the first few chapters in the book. However, it is very clear that management decisions will increasingly require interactions between species to be taken into account. The final chapters deal with estimating parameters that describe interactions.

My second target audience is researchers in ecology, including postgraduate students. In addition to recipes and recommendations for parameter estimation in straightforward cases, I have reviewed some of the approaches that have been taken recently in specific, more complicated circumstances. These provide jumping-off points from which researchers can develop novel approaches for their own particular problems. For this audience, it was not possible to make the book entirely self-contained, but I hope it provides a good summary and entry into the current literature.

Almost the entire book is accessible to anyone who has completed an introductory ecology course, together with a biometrics course covering analysis of variance and regression. No computing skills beyond a familiarity with simple spreadsheets are assumed. Spreadsheets are an excellent tool for simple modelling and analysis, and are particularly good for working through examples to make 'black boxes' more transparent. I have used them extensively throughout the book. Beyond what can be done easily with a spreadsheet, there is little point in providing computer code or detailed instructions for applying particular numerical methods. There is now an enormous range of software available, both for general statistical analysis and for particular ecological estimation problems. Much of this software, particularly for specific ecological problems, is freely available on the World Wide Web. As the web addresses for software tend to move around, the Blackwell Science Web site (www. blackwell-science.com) should be consulted for current links.

Why model?

Before commencing any modelling exercise, it is absolutely essential to decide what the model is for. There is little point in constructing an ecological model if it is simply going to be placed on a shelf and admired, like a ship in a bottle. A model can be thought of as a logical proposition (Starfield, 1997). Given a set of assumptions (and these assumptions include estimates of the parameters used in the model), then certain consequences, or model outputs, follow. In this sense, all models make predictions. However, it has often been suggested that a distinction can be made between models developed for explanation and models developed for prediction (Holling, 1966; May, 1974b). Models for explanation are developed to better understand ecological processes. Their goal is to make qualitative predictions about how particular processes influence the behaviour of ecological systems. This means that they will usually be highly abstract and very general. As a result, they are likely to be quite poor at making accurate quantitative predictions about the outcome of any specific interaction. General models developed for understanding are often called 'strategic' models. A classic example of a strategic model is the model developed by May (1974a) to investigate the influence of time delays and overcompensation in populations with non-overlapping generations. This model has only one parameter, but it operates at such a high level of abstraction that there is little point in trying to estimate this parameter for any real population. To even begin to see which of the general properties of the model apply to real populations, it is necessary to use slightly more complex and realistic models (Hassell *et al.*, 1976).

In contrast, 'tactical' models may be developed in an attempt to forecast quantitatively the state of particular ecological systems in specific circumstances. This is usually what is meant by a 'predictive' model. Such models must typically be quite complex and will often require a large amount of ecological data to generate estimates of a considerable number of parameters. The highly specific and complex models developed for the northern spotted owl (Lande 1988; Lamberson *et al.*, 1992; Lahaye *et al.*, 1994), are examples of tactical models. For such models to be valuable, it is absolutely essential that parameter estimates should be as accurate as is possible. For reasons I discuss at the beginning of the next chapter, do not assume that a complex model with many parameters will necessarily produce more accurate quantitative predictions than a simpler model.

These two examples represent ends of a continuum; in between are strategic models intended to provide general predictions about particular classes of ecological interactions, and tactical models intended to provide approximate predictions in the absence of detailed ecological data. Locating a model along this continuum is very important from the point of view of parameterizing models. There is little value in attempting to derive highly accurate parameter estimates for a strategic model. As they operate on a high level of abstraction, many processes are frequently subsumed into one parameter, making the translation from real biological data into an appropriate parameter estimate quite difficult. Parameter estimates for models of this type are also context-dependent. An estimate of the intrinsic rate of increase r for a particular species in a particular place that is appropriate for one abstract model may well be quite inappropriate for another model of the same species, in the same place, if that model has a different purpose (see Chapter 5). For a strategic model, estimates of the appropriate order of magnitude are typically required, and the principal problem is to ensure that the translation from biological complexity to mathematical simplicity is carried out correctly. One major advantage of simple models is that at least some results can be expressed algebraically in terms of the parameters. This means that the dependence of the results on particular parameters or combinations of parameters is obvious.

Models for prediction will usually require estimates of more parameters than models for understanding, and these estimates usually must be more accurate if the model is to be useful. Stochastic models require information on the probability distribution of parameters: the variance, shape of the distribution and correlation between parameters.

It is very tempting to be prescriptive, and demand that all parameter estimates should be based on sound experimental designs and rigorous statistical methods. Indeed, it is highly desirable that they should be, and the more rigour that can be introduced at the parameter estimation stage, the more reliable will be the results of the resulting models, provided the model itself is appropriate. However, rigour is a luxury that may not be affordable in many practical contexts. All management decisions in ecology are based on models, even if those models are verbal or even less distinct 'gut feeling'. If it is decided not to use a mathematical model because the parameters cannot properly be estimated, the management decision will proceed regardless. We often are not able to postpone decisions until the appropriate experiments can be done. By that time, the species in question may have ceased to exist, or the ecosystem may have become irreversibly degraded.

It will usually be the case that a set of approximate parameter estimates in a formal mathematical model is preferable to an arm-waving argument. At least, in the former case, the assumptions on which the model is based will be transparent, and there is the possibility of investigating the robustness of the conclusions to the uncertainty in parameters and assumptions. Problems occur when the results of such modelling are presented to managers as coming from a black box, and no commitment is made to testing the assumptions and tightening parameter estimates so that the next version of the model is more reliable. It is also important to perform 'sensitivity analysis': to consider the range of model behaviour produced by the range of plausible values for the input parameters.

Successful applications of modelling in ecology are almost always iterative.

The initial versions of the model highlight areas of uncertainty and weakness, which can then be addressed in later versions.

Brief taxonomy of ecological models

Continuous and discrete time

One of the major distinctions between models is whether they deal with time continuously or in discrete steps. Continuous-time models are usually structured as one or more *differential equations*, in which the rate of change of one or more variables (usually population size) is represented as a function of the various parameters and variables. The simplest of these is the equation for exponential growth

$$\frac{dN}{dt} = rN, \tag{1.1}$$

which states simply that the rate at which the size of the population N increases with time is a constant r (the intrinsic rate of increase) times the current population size. This equation thus has a single parameter r to be estimated, but as will be seen in Chapter 5, this is not a straightforward task.

The principal advantage of differential equation models is that considerable progress can be made in understanding their behaviour analytically, using the standard tools of calculus. For an ecologist, the advantage of this is that the sensitivity of the model's behaviour to changes in parameter values can be understood by looking at the algebraic structure of the analytic results, without the need for exhaustive and necessarily incomplete numerical sensitivity analysis. This process will be illustrated with an example in the next chapter. Only rarely, however, is it possible to obtain an algebraic closed form for population size as a function of time. That is, it is usually impossible to represent population size as

$$N(t) = f(t,p), \tag{1.2}$$

where $N(t)$ is population size at time t, p is a set of parameters and f is an arbitrary function. Usually, it will be necessary to use a numerical algorithm to obtain an approximate result. All algorithms proceed by approximating the continuous process by a series of discrete steps. However, in contrast to discrete-time models, the step length is determined by the accuracy required, not the structure of the model itself; it can be made as small as necessary, and, using many algorithms, it can be altered as the solution is calculated. A good indication that the step length in a numerical solution of a continuous-time model is sufficiently short is if the solution does not change, depending on the step length.

Discrete-time models, in contrast, represent time in fixed steps, the length of which is determined by the structure of the model itself. The simplest such models represent time in generations, and were first developed for univoltine insects: insects that have one generation per year, and in which offspring are not in the adult stage at the same time as their parents. Most genetic models also represent time in discrete generations (see, for example, Hartl & Clark, 1997).

The second major class of models that represent time in discrete steps are those developed for long-lived vertebrate populations (Caughley, 1977a). In these, the time scale is conventionally one year, and whilst generations will clearly overlap in a long-lived animal, the assumption is made that events occur sufficiently slowly that a good representation of population dynamics can be obtained by taking a 'snapshot' once a year. Such models will usually also include age structure.

Age and stage structure

Unstructured models represent the population size (or possibly biomass) of a given species with a single variable, aggregating all age classes and developmental stages together. This is quite appropriate for univoltine species if a snapshot of population size once per generation is adequate, and is sufficient for many abstract strategic models. The obvious advantage is that the model can be kept relatively simple, and the number of parameters required can be kept down.

In many cases, however, age or stage structure cannot be ignored. Different ages or stages may have very different reproductive rates, survival rates etc. Omission of such detail will make quantitative predictions quite inaccurate. Furthermore, there are considerable problems in attempting to derive appropriate aggregated parameters to use in unstructured models, if the input data are in either age- or stage-structured form. These problems and possible work-arounds are discussed in Chapters 4 and 5. Finally, the qualitative behaviour of structured models may be quite different from their unstructured equivalents. (For example, an invulnerable age class may change the behaviour of host–parasitoid or predator–prey models (Murdoch et al., 1987), and age-structured disease models may have much simpler behaviour than their unstructured equivalents (Bolker & Grenfell, 1993).) Even for very general strategic models, it is therefore worth exploring structured models, usually in comparison with, or in addition to, their unstructured counterparts.

There are numerous ways in which age or stage structure can be included in models, and readers should refer to some of the excellent books or reviews on the subject for details (e.g. Gurney et al., 1983; Nisbet & Gurney, 1983; Caswell, 1989). This section will outline briefly the types of such models from

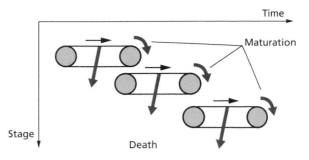

Fig. 1.1 The conveyor belt analogy for a structured population model. Each life-history stage can be imagined as a separate conveyor belt, which moves forward with time at a constant speed. The conveyor moves individual organisms through the life-history stage. They leave the stage either by falling off the belt (through death), or by reaching the end of the belt, at which point they mature into the next life-history stage. Organisms enter the first stage by birth, and the number of births is a function of the number of individuals on one or more of the conveyors. All stages, other than the first, can be entered only by completing the previous stage. Death rates in a stage may be a function of population density of that stage, or other stages, and may vary with time. The length of each conveyor belt (the maturation time) may also vary dynamically according to environmental conditions or population density. This basic pattern underlies many structured population models discussed in the text, including the Leslie model (all conveyors are one time unit long), McKendrick–von Forster model (conveyors are infinitesimally short) and Gurney–Nisbett models. However, it cannot be used for some plant species (and only for a few animal species), in which differing modes of reproduction may cause individuals to enter different life-history stages, or in which regression from later to earlier stages is possible.

the point of view of parameter requirements. A good way to understand all such models is via a conveyor belt analogy, shown schematically in Fig. 1.1.

The Leslie matrix model (Leslie, 1945) is probably the best-known and most straightforward age-structured model. The population is divided into age classes of the same length as the time step (usually both are one year long). A proportion p_x individuals in age class x are then assumed to survive into the next age class, and the age-specific fertility (the number of female offspring in the first age class per female in age class x) is given by F_x. Methods to estimate these parameters, and some technical details of definition, are discussed in Chapter 4. Given these definitions, the vector of individuals in age classes 1 to n at time $t + 1$ can be derived from the numbers present at time t from the following matrix equation:

$$\begin{pmatrix} n_1 \\ n_2 \\ n_x \\ n_k \end{pmatrix}(t+1) = \begin{pmatrix} F_1 & F_2 & F_x & F_k \\ p_1 & 0 & 0 & 0 \\ 0 & p_2 & 0 & 0 \\ 0 & 0 & p_{k-1} & 0 \end{pmatrix} \begin{pmatrix} n_1 \\ n_2 \\ n_x \\ n_k \end{pmatrix}(t). \tag{1.3}$$

This equation simply states that the only way to enter the first age class is by being born of a member of one of the age classes present at the previous time step, and that all subsequent age classes can only be entered by surviving through the previous age class in the previous time step. This simple model can generate only very simple behaviour: depending on the age-specific survivals and fertilities, the population will either grow or decline exponentially. Determining the rate at which it will do so is discussed in Chapter 5. The model is widely used in practice for short-term population projection, a task that can be accomplished very easily on a standard spreadsheet. A very full discussion of the Leslie model and others based on it is given by Caswell (1989).

The basic ideas of the Leslie matrix can be generalized fairly easily to cases in which stage classes are of differing lengths, so that individuals do not necessarily mature out of their current step in a given time step. Individuals can thus jump more than one step ahead, and even regress to a previous stage. Such models are often called Lefkowitz matrices. Again, Caswell (1989) has a full discussion of these models.

Simple matrix-based models can be very useful for practical purposes, particularly for short-term population extrapolation (see, for example, McDonald & Caswell, 1993; Heppell *et al.*, 1994; Smith & Trout, 1994). They also form the basis around which more complex stochastic (random) models for population viability analysis are based (Burgman *et al.*, 1993).

A further way of including age or stage structure in models has been developed by Nisbett and Gurney (Gurney *et al.*, 1983; Nisbet & Gurney, 1983; Nisbet *et al.*, 1989). This approach combines some of the analytic tractability of simple differential equation models with the realism of matrix-based models. In very general terms, time is treated as a continuous variable, so that there is no possibility that the model behaviour may be determined by the step length (as is the case with a matrix-based model). However, individuals cannot leave a life-history stage until a certain time period after entering it. Potentially, the maturation period may be a dynamic variable, determined by density-dependent processes or external variables. The principal disadvantage of the approach is that the model is formulated as a series of delay-differential equations. These are a little difficult to solve using either standard algorithms for the solution of differential equations or by elementary programming techniques.

McKendrick–von Forster equation

The most abstract age-structured model is the McKendrick–von Forster equation, which is a partial differential equation in which both time and age are continuous. Its derivation is not easy to explain. This version closely follows Caswell (1989). Define $N(a,t)$ as the density of individuals of age a at time t. In other words, the number of individuals between the ages a_1 and a_2 at time t is given by

$$\int\limits_{a=a_1}^{a=a_2} N(a,t)\mathrm{d}a. \qquad (1.4)$$

This means that the number of individuals in the age interval $(a, a + \mathrm{d}a)$ at time t is $N(a,t)\mathrm{d}a$. In the absence of mortality, after a short period $\mathrm{d}t$ all these individuals would be still alive, but $\mathrm{d}t$ older, so

$$N(a,t) = N(a + \mathrm{d}t, t + \mathrm{d}t).$$

Now, allow for mortality occurring at a per capita rate $m(a,t)$, which may depend both on age and time. Then, the total number of animals of age a that die in the time period $\mathrm{d}t$ will be given by $m(a,t)N(a,t)\mathrm{d}t$. Hence:

$$N(a,t) - N(a + \mathrm{d}t, t + \mathrm{d}t) = m(a,t)N(a,t). \qquad (1.5)$$

This says simply that the difference between the number of animals that would be present if there was no mortality and the number that are actually present is the number that have died. All that it is necessary to do now is to note that

$$N(a+\mathrm{d}t,t+\mathrm{d}t) \approx N(a,t) + \frac{\partial N(a,t)}{\partial t}\mathrm{d}t + \frac{\partial N(a,t)}{\partial a}\mathrm{d}t. \qquad (1.6)$$

Substituting eqn (1.6) into eqn (1.5),

$$\frac{\partial N}{\partial t} + \frac{\partial N}{\partial a} = -m(a,t)N(a,t). \qquad (1.7)$$

Animals also are born, at which time their age is 0. The total rate of offspring production by individuals of age a at time t is given by the age-specific fecundity $f(a,t)$ multiplied by $N(a,t)$. Thus, the total number of animals of age 0 at time t is given by the boundary condition

$$N(0,t) = \int\limits_{a=0}^{\infty} f(a,t)N(a,t)\mathrm{d}a. \qquad (1.8)$$

This model appears hard to understand. However, it is simply the result of taking the conveyor belt analogy in Fig. 1.1 to a limit in which each belt becomes infinitesimally short, but there is an infinite number of belts. The model is also difficult to parameterize. The task is one of choosing appropriate functional forms for both age-specific fecundity and mortality, both possibly as functions of time and other variables. Then all the parameters involved in these functional forms need to be estimated. Where the model has been used in practice, it has usually been approximated by one of the previous forms before parameter estimation was attempted. Nevertheless, the equation is worth presenting as the general form of each of the more specific versions described above.

Deterministic and stochastic models

Deterministic models are those without a random component. This means that the population size in the next time period (or after an arbitrarily short time step in a continuous-time model) is entirely determined by the population size and structure at the current time, or, in the case of time-delayed models, by the previous history of population size and structure. In contrast, stochastic models include one or more random components, so that the population size in the future follows a probability distribution. It is a truism that the real world is stochastic, but nevertheless deterministic models are very valuable for some purposes.

The main advantage of deterministic models is their simplicity. As a rule, far more progress can be made in examining a deterministic model analytically than is possible with a stochastic model. Analytic results are always helpful, as it is much easier to deduce the influence of a particular parameter on a model's behaviour if an algebraic representation of a result is available than if only a numerical result can be obtained. It is not necessary to solve a model fully, but algebraic representations of equilibria and their stability will provide very useful information. Even if only numerical solutions can be obtained, a single solution of a deterministic model represents its behaviour for a given set of parameters. In contrast, a single solution of a stochastic model may be quite unrepresentative of the 'usual' behaviour of the system for a particular parameter set. It is therefore essential to represent the probability distribution of outcomes at any particular time. Some approximations to the form of this distribution can sometimes be calculated analytically in simple cases (see Tuljapurkar, 1989), but usually it is necessary to calculate a large number of repeated solutions of a stochastic model for any set of parameters, and then to examine the frequency distribution of possible outcomes.

For some purposes, stochastic models are essential. Frequently, a problem may need to be approached with a combination of stochastic and deterministic models.

Deterministic models tend to be more suitable for strategic modelling, whereas tactical modelling is likely to require stochastic approaches. However, there are numerous examples of stochastic models being used for answering very general questions, and in many cases a simple deterministic projection model is sufficient to answer a specific problem about the likely future behaviour of a particular population. An excellent, but fairly technical discussion of stochastic modelling can be found in Tuljapurkar (1989). A much more accessible introduction to practical stochastic modelling is provided by Burgman *et al.* (1993).

Using stochastic models introduces a series of problems additional to those of deterministic models. First, it is necessary to decide on what sort of stochasticity should be included. Second, it is necessary to determine appropriate forms for the probability distributions of the stochastic elements. Third, it is necessary to decide how the results should be presented. Stochasticity is often divided into three types, although these intergrade. Demographic stochasticity is randomness that occurs because the size of real populations can take only integer values. If an age class of 20 individuals has a probability of survival over a year of 0.5, this does not mean that exactly 10 individuals will survive, only that a random number sampled from a probability distribution with a mean or expected value of 10 will survive. If only demographic stochasticity is operating, the probability distribution of this random variable will be binomial. In addition to demographic stochasticity, there is randomness introduced because the environment is not constant. Environmental variation will cause the survival of 0.5 in the example above to change from year to year, although it may have a long-run mean of 0.5. There is no reason why environmental stochasticity should follow any particular probability distribution: it will depend on the nature of the environment, the parameter concerned, and the ecology of the organism in question. It is therefore much harder to make generalizations about the influence of environmental stochasticity on populations than it is for demographic stochasticity (see, for example, Goodman, 1987; or Lande, 1993). Ways in which appropriate probability distributions for environmental stochasticity in various parameters may be added to models are discussed in several chapters. Finally, many authors (e.g. Ewens *et al.*, 1987; Lande, 1993) consider catastrophic stochasticity separately from environmental stochasticity. Catastrophic stochasticity is usually defined as the effect of random factors that may kill a large proportion of a population, irrespective of its size. It is really just an extreme form of environmental stochasticity, but appropriate ways to model it and its effects are often qualitatively different from most other forms of environmental variation.

All the above can be grouped together as *process error*. They are sources of random variation that cause the actual population to vary. In addition to process error, almost all ecological data will be subject to *observation error*. Observations of any quantity at all are unlikely to be the actual values, but instead are random numbers hopefully clustered around the actual value, with a sampling error. The distinction between process and observation error is important, as methods designed to operate in the presence of only one of these the sorts of error may be unreliable if the other source of error is dominant, or if both are important. This problem will be dealt within specific circumstances as it occurs.

For many ecological models, it is necessary not only to estimate the mean value of a parameter, but also to estimate some of the properties of its frequency distribution. This is most obvious for stochastic models: if you wish to know the influence of a varying death rate on population size, it is self-evident that you need to know something of how that death rate varies through time. However, if any process is nonlinear (and most ecological processes are), then the behaviour of a deterministic model will be influenced by the frequency distribution of the parameters in the population and not just their mean values. For example, the most crucial parameter in most epidemiological models is R_0, the reproductive number of the disease (Anderson & May, 1991). If $R_0 > 1$, the disease will increase, but if $R_0 < 1$, the disease will not persist. R_0 cannot simply be estimated from the mean number of contacts per infected individual. It is necessary also to estimate the variance in the number of contacts, and the larger the variance for a given mean contact rate, the greater is R_0 (Woolhouse *et al.*, 1997). R_0 and other parameters associated with host pathogen interactions are discussed in Chapter 10.

Together with the estimate of a given parameter, most standard statistical methods will return a standard error. This is an estimate of the precision with which the mean value of the parameter in the sampled population has been estimated, and it is thus a property of the sampling distribution. The sampling distribution may be related to, but is not the same as, the distribution of the parameter itself in the sampled population. In the simplest case, if a parameter has a normal distribution in a population, with a variance σ^2, the standard error of an estimate of the population mean from a sample of size n will be $\sqrt{(\sigma^2/n)}$. However, if the parameter itself does not have a normal distribution, the problem is not as straightforward.

Individual- and event-based models

Most stochastic models operate using a relatively small number of classes of individuals, and then use standard probability distributions (binomial, Poisson etc.) to generate the number of individuals in each class, at each successive time step. An alternative approach is to attempt to follow each individual in the population from its birth, through growth, dispersal and reproduction, to death (Rose *et al.*, 1993; Judson 1994; Lomnicki, 1988). Such a model obviously requires more computer storage space and processing than one which deals with a small number of aggregated categories, but this is no longer the serious constraint it once was. Conceptually, such models are often quite simple, as only very elementary probability theory is necessary to construct them. The number of parameters and difficulties in estimation are frequently no greater than for an equivalent structured model, as individuals will usually be grouped into categories for parameterization purposes. Individual-based models hold particular promise for combining genetic and ecological processes, as even the most elementary genetic structure rapidly complicates the structure of any ecological model (see, for example, Beck, 1984; Andreasen & Christiansen, 1995).

Event-based models are a second form of calculation-intensive stochastic model. Rather than imposing a fixed time step on the model, they model each event (birth, death etc.) as it occurs. The approach was first suggested nearly 40 years ago (Bartlett, 1961), but has not often been used (but see Bolker & Grenfell, 1993). The basic idea is very straightforward. Suppose that there are three events, A, B and C, that could occur to a population at a given time t, and that they occur at rates a, b and c per unit of time. These rates can change through time, possibly depending on population size or any other internal or external variable. Then the waiting time to the next event has a negative exponential distribution with mean $1/(a + b + c)$. Which of the events occurs after this waiting time simply depends on the relative sizes of the rates: it will be A with a probability $a/(a + b + c)$, B with a probability $b/(a + b + c)$, etc. The negative exponential is a particularly easy distribution to generate on a computer (Press $et\ al.$, 1994), and thus iteration is both straightforward and fairly quick. The event-based approach is probably the best way of numerically solving a continuous-time stochastic model, although if populations are large, individual demographic events may happen so rapidly that solution is very slow.

CHAPTER 2

Parameter estimation toolbox

Introduction

This chapter outlines some general methods and approaches that are common to many parameter estimation problems. It is mainly concerned with some general statistical models that can be used to estimate ecological parameters. The variation inherent in all biological data means that no parameter in an ecological model can simply be directly measured. Parameters must be estimated, and obtaining an estimate from data always involves using a statistical model. Sometimes, the statistical model is very simple. For example, if the mean weight of animals in a population is the parameter of interest, a sample would be selected, the animals would be weighed, and the mean calculated. The model used to assess the precision of the estimate would be that the mean weight of a sample of animals follows an approximately normal distribution. In a slightly less straightforward case, a parameter might be estimated by fitting a linear regression model to observed pairs of response and predictor variables. As you will see throughout this book, the statistical models necessary to extract biologically meaningful parameters from data that can be observed or obtained experimentally are often quite complex.

How many parameters?

Give me four parameters and I'll draw you an elephant: give me five and I'll waggle its trunk. (Attributed to Linus Pauling; Crawley, 1992)

Entities ought not to be multiplied except from necessity. (Attributed, probably incorrectly, to William of Occam (d. 1349); *Brewer's Dictionary of Phrase and Fable*)

It is tempting to believe that the inclusion of more detail and more parameters will improve a model. The art of modelling, however, is to leave out as much detail as possible and still have a model that performs its task. This generalization is equally true of strategic and tactical models, although, as noted in the previous chapter, tactical models usually require more detail.

Statistical models should seek to represent data as parsimoniously as possible. Occam's razor is generally recognized as a guide to the appropriate level of model complexity; objective methods, discussed later in the book, are available to assist with choosing the appropriate level, although none of these should be applied uncritically.

Selecting the appropriate level of complexity for mechanistic models is a much less straightforward task. Occam's razor still should be a guiding principle, but the objective is to devise a model with as few superfluous mechanisms as is possible, rather than aiming strictly to minimize the number of parameters involved. Objective statistical methods may provide some assistance with the number of parameters required to describe certain mechanisms adequately, but will rarely be able to assist with the number of mechanisms that should be included. The Pauling quote cautions that, even with relatively few parameters, almost any pattern can plausibly be represented, with little necessary connection to reality. Even relatively simple and abstract ecological models often have upwards of 10 parameters, sufficient to make the elephant ride a bicycle whilst playing a trumpet. Many models have over 100 parameters, and could draw the whole circus!

Apart from the possibility of including spurious processes with too many parameters, a very large number of parameters will make a model unworkable, as they will be impossible to estimate and it will be equally impossible to investigate the sensitivity of model solutions to these estimates.

There is an important distinction between parameters estimated extrinsically and those estimated from the same time series with which the model is being tested. If parameters are determined entirely extrinsically, there is no necessary limitation on the number of parameters, provided not too much *post hoc* reasoning has gone into the choice of mechanisms included in the model. However, estimating any more than a single parameter from the time series on which a mechanistic model is tested severely limits the ability to draw any useful conclusions about the fit of the model to those same data.

Selecting an appropriate statistical model

Most ecologists are familiar with the idea of using a hypothesis-testing approach to decide whether additional components should be included in a statistical model. For example, in a two-way analysis of variance you might examine the ratio of the interaction mean square (for two factors A and B) to the error mean square, and compare it with a tabulated F ratio. On this basis, you might decide whether it was necessary to include parameters for the interaction between A and B. The idea can easily be extended to generalized linear models (McCullagh & Nelder, 1989) by using a likelihood ratio test. These approaches are only valid, however, when comparing nested models. That is, a model is compared with another, which is the original model with added components. Likelihood ratios or F ratios cannot be used to compare models that are not nested. For example, in a mark–recapture study estimating survival, it is possible to use a likelihood ratio to compare a model in which survival is constant through time and with age with another in which survival is constant through time, but varies with age. The former model is nested within the latter. It is not possible, however, to use a likelihood ratio to compare the model in which survival is constant through time, but varies with age, with one in which survival varies through time, but not with age. The two models are not nested.

Recently the Akaike information criterion (AIC) (Akaike, 1985; Burnham & Anderson, 1992) has been suggested as a tool for comparing non-nested models. It is defined as

$$AIC = -2l + 2p, \qquad (2.1)$$

where p is the number of parameters fitted, and the log-likelihood (l) is the value calculated using maximum likelihood estimates for each of the parameters fitted in that particular model. The model with the lowest AIC is selected as optimal for the given data. The AIC is derived from information theory, but heuristically can be understood as the likelihood component, representing the discrepancy between the model and the data, which will never increase as more parameters are added, plus a penalty for adding additional parameters.

Rescaling and redimensioning models

One of the most useful general tools available to minimize problems with parameter estimation is that of redimensioning or rescaling the model. This involves re-expressing the model so that variables are defined either relative to parameters, or possibly relative to other variables. The approach is most suitable for fairly abstract models developed for the purpose of understanding the qualitative behaviour of ecological systems. It is less likely to be useful for highly detailed tactical models, as they will usually be expressed in particular units that are relevant for the specific problem at hand.

There are several potential advantages of rescaling a model. First, rescaling may reduce the number of different parameters that need to be estimated. In some cases, it may be possible to define away a problem with a particularly intractable parameter by defining the entire system in units of that parameter. Second, the process of rescaling will often show that generic features of the model dynamics are determined by particular parameter combinations, especially ratios, rather than by particular parameters alone. The rescaling exercise in itself, then, produces useful information, and essentially performs an algebraic sensitivity analysis. The very simple example in Box 2.1 illustrates some of these points.

Changing the variables in a model is a similar, but slightly more complex process. If a model includes more than one variable, then it may be worthwhile to replace one or more original variables with ratios of one variable to

Box 2.1 Rescaling the Ricker equation

The Ricker equation is one of the simplest ways of representing intraspecific competition in a difference equation. The main purpose of the model is to help in understanding the effect of overcompensation on population dynamics. Overcompensation occurs when, beyond a certain point, the more individuals there are in one generation, the fewer there are in the next. Consider a population of a univoltine species, with the population in generation t represented by N_t. The carrying capacity is K, and a parameter R describes both the rate of reproduction and the extent of overcompensation. The following equation then describes the model:

$$N_{t+1} = N_t \exp R\left(1 - \frac{N_t}{K}\right). \tag{1}$$

It is very easy to confirm that the model has an equilibrium at $N_t = K$, because the term in the brackets is then 0, and $\exp(0)$ is 1. The greater the value of R, the more humped is the relationship described by the equation.

Equation (1) has two parameters, but K can easily be removed by rescaling in the following way. Define a new variable $X_t = N_t/K$. This means that we are measuring population size in units of the carrying capacity or equilibrium population size. The original $N_t = KX_t$. Substituting this into eqn (1) gives

$$KX_{t+1} = KX_t \exp R\left(1 - \frac{KX_t}{K}\right). \tag{2}$$

On cancelling K,

$$X_{t+1} = X_t \exp R(1 - X_t). \tag{3}$$

This means that the dynamic behaviour of the original model depends only on the parameter R, and that any conclusions drawn about the behaviour of eqn (3) will apply directly to eqn (1). The general behaviour of the system can easily be explored numerically on a spreadsheet. As R, the extent of the overcompensation, increases, the model moves from a smooth approach to equilibrium through an oscillatory approach to sustained cycles and finally to chaos (see May, 1974a).

If we were interested in fitting this model to real data to see what sort of dynamic behaviour would be predicted, the rescaling tells us that only R needs to be estimated. Note, however, that slightly more complex functional forms do a much better job at representing real insect populations (see Bellows, 1981).

another, or with sums or differences involving other variables. This may simplify the model structure or make its behaviour easier to interpret. It is essential that the new variables should have direct biological meaning. For example, in a model that originally has numbers of individuals of each of three genotypes as its variables, it may well be worth considering changing variables to a model with total population size as one variable, and frequency of two of the three genotypes as the other two. (Note that changing variables will not change the number of equations necessary to specify the model fully.) In a differential equation model, the standard product and ratio rules of calculus should be used to form the new equations.

It is difficult to make general rules to determine whether a change of variable is worthwhile. It is largely a matter of trying the exercise to see if the new model is simpler or more tractable than the original. An example where a change of variable is very useful is given in Box 2.2.

Parameter uncertainty and sensitivity analysis

It is extremely rare that any ecological parameter can be measured exactly. Using the methods discussed throughout this book, the best that can be done is to estimate the most likely value that a parameter may have, given a set of observed data. However, the observed data will also be consistent with the parameter taking a range of other values, with somewhat lower likelihood. It is necessary to see how the outcome of the model may vary over the range of plausible parameter values. This is sensitivity analysis.

Understanding such sensitivity is also critical to designing any program of parameter estimation. The outcome of any model is more sensitive to variations in some parameters than it is to variation in others. There is little value in expending large amounts of effort to obtain a precise estimate of a parameter if the behaviour of the model is relatively invariant to changes in the parameter. Often, the major goal of a modelling exercise may be to perform a sensitivity analysis to guide further research. With abstract models capable of being represented algebraically, sensitivity can often be determined by careful examination of the equations themselves (see Box 2.2).

With more complex models, an extensive numerical investigation is required. A good general overview of the possible approaches is provided by Swartzman and Kaluzny (1987, pp. 217–25). A common approach is to perturb each of the uncertain parameters in turn, recording the response of the model, whilst holding all other parameters constant at their most likely point estimates. Such one-at-a-time sensitivity analysis may produce very misleading results. In many cases, the model output may be almost invariant to perturbation of either of two parameters singly, but joint perturbation of both in a particular direction may produce a major change in the response.

Box 2.2 Rescaling and algebraic sensitivity analysis in a host–parasite model

The model which follows was developed by Anderson (1980) to explore the role that macroparasites (parasitic worms, etc. that cannot multiply rapidly within their hosts) may have on the population size of their hosts. The primary objective of the model was to investigate how the extent to which parasites can depress the population size of their host depends on the sizes of the various parameters that describe the host–parasite interaction.

This example is selected to illustrate two important points. First, it shows how changing the variables in a model may lead to a form that is simpler and easier to understand. Second, it shows how examination of the algebraic forms of modelling results can be used to perform a 'sensitivity analysis', to determine which parameters or parameter combinations are most important in the outcome of the model.

The basic model is as follows:

$$\frac{dH}{dt} = (a - b - \beta H)H - \alpha P, \tag{4}$$

$$\frac{dP}{dt} = P\left[\frac{\lambda H}{H + H_0} - (\mu + b + \alpha + \beta H) - \alpha \frac{(k+1)P}{kH} \right], \tag{5}$$

where, H is the size of the host population, and P the size of the parasite population. The parameters in the model are summarized in the table below. For a more detailed explanation of the structure of models of this type, see Chapter 10.

Parameter	Meaning
a	Host birth rate
b	Host death rate
β	Density dependence in host death rate that is unrelated to parasite infection
α	Increment in death rate of host per parasite carried
λ	Rate of production of infective stages per parasite
H_0	Inverse measure of transmission efficiency of infective stages (the death rate of infective stages divided by their infectivity per unit of time)
μ	Death rate of parasites in hosts
k	Parameter of negative binomial distribution of parasites in hosts, inversely related to degree of aggregation

This model is much simpler if the parasite burden is expressed in terms of the mean parasite burden $M = P/H$. To do this, it is necessary to use the quotient rule of calculus:

continued on p. 20

Box 2.2 *contd*

$$\frac{d}{dt}(P/H) = \frac{\frac{dP}{dt}H - P\frac{dH}{dt}}{H^2}, \tag{6}$$

from which

$$\frac{dM}{dt} = \frac{dP}{dt} \times \frac{1}{H} - \frac{M}{H} \times \frac{dH}{dt}. \tag{7}$$

Thus,

$$\frac{dH}{dt} = H\{(a - b - \beta H) - \alpha M\}, \tag{8}$$

$$\frac{dM}{dt} = M\left\{\frac{\lambda H}{H + H_0} - (a + \mu + \alpha) - \frac{\alpha M}{k}\right\}. \tag{9}$$

These equations are simpler in structure than the initial forms in P and H. More importantly, they make some features of the model behaviour much more obvious. For example, it is clear from the last term in eqn (9) that parasite-induced mortality only has a density-dependent impact on mean parasite burden if the parasites are aggregated within their hosts ($k < \infty$). Furthermore, it is immediately obvious that the parasite cannot invade a host population unless dM/dt is positive for small M. There is thus a minimum host density for disease persistence H_T, which can be found by rearranging eqn (9):

$$H_T = \frac{H_0(a + \mu + \alpha)}{\lambda - (a + \mu + \alpha)}. \tag{10}$$

Determining whether a parasite is able to persist in a given host population of known density might well be a major objective of a modelling exercise. Inspection of eqn (10) shows clearly that the threshold density for parasite persistence does not depend strongly on the increment in death rate per parasite α, if that increment is substantially lower than the host birth rate a or death rate of parasites on hosts μ. However, the threshold is directly proportional to H_0, and thus in turn to the infective stage death rate.

If parasite persistence is the issue at hand, this 'algebraic sensitivity analysis' makes it clear, without simulation or solution of the equations, that experimental attention should be concentrated on measuring the infective stage death rate, amongst other parameters. Measuring the parasite-induced death rate is only important to this question if a single parasite causes an increase in host death rate of the same order of magnitude as the host birth rate.

To take a particularly simple example, consider a simple spreadsheet model that was constructed to examine the economic viability of moving use of the Australian pastoral zone away from concentration on sheep grazing towards a high-value harvest of kangaroos for human consumption. At present, the majority of the kangaroo harvest is of low value, being used for skins and pet food. Using a scenario based on current practices, a relatively small proportion of the harvest is used for game meat (25% of the harvest is for meat, most being for skins, and only 5% of the meat is for human consumption), and the value of the game meat is low ($A0.50 per kilogram for game versus $A0.45 per kilogram for pet food). Increasing the proportion of the harvest used for human consumption has little effect on the total value of the harvest, because the price of the meat is low. Holding the proportion of the harvest taken for human consumption constant but increasing the price of the game meat also has little effect, because so little of the harvest is used for human consumption. However, increasing both together produces a substantially increased total value. This is a particularly obvious example, but is based on a real simulation (Hardman, 1996). In many other circumstances, similar parameter interactions will occur, but will be hidden by the complexity of the model. Parameter interactions may also be compensatory: changes in the model's response due to alteration of one parameter may approximately cancel changes in response to alteration of another parameter.

A factorial model, in which combinations of the selected factors are varied systematically (Henderson-Sellers & Henderson-Sellers, 1993) is a more rigorous approach to sensitivity analysis than one-at-a-time parameter variations. The difficulty is that the number of parameter combinations rapidly becomes unworkable. If only two perturbations of each parameter are considered ('high' and 'low'), the number of combinations of n parameters is 2^n. One way to reduce this number is to use fractional factorial models, in which some multifactor interactions are ignored. For example, the full factorial design with six factors and two levels requires 64 runs, but a design with only 16 runs could potentially identify all the main effects and nine of the 15 two-factor interactions. For specific guidelines on constructing such designs, see Henderson-Sellers and Henderson-Sellers (1993).

An alternative approach is to select combinations of values of the parameters for use in sensitivity analysis at random (Swartzman & Kaluzny, 1987, p. 223), sampling each parameter from an assumed probability distribution. The approach will generate the sensitivity of the output to variation in each parameter, averaged across the sensitivity in the response to all the other variable parameters. This is in contrast to looking at each parameter in turn, holding all others at their average value. A difficulty of entirely random sampling is that, by chance, some areas of parameter space may be missed entirely. Random sampling of parameter space may also be misleading if some parameters are correlated.

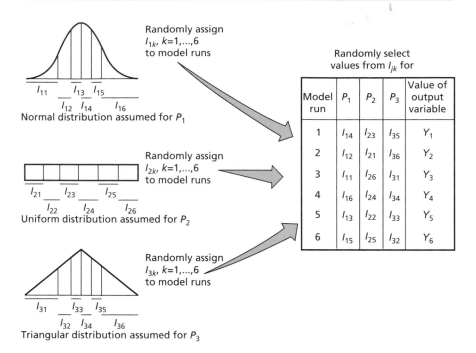

Fig. 2.1 Latin Hypercube sampling. The diagram illustrates a sensitivity analysis for three parameters. Parameter P_1 has a normal distribution, P_2 has a uniform distribution and P_3 has a triangular distribution. For six model runs, each distribution is divided into six segments of equal area (probability). For each model run, a value for each parameter is selected at random from within one of these segments. Segments are sampled for successive runs without replacement, leading to the table of parameter values shown to the right of the figure. Note that not all possible parameter combinations are used (this would require 6^3 rather than 6 model runs). From Swartzman and Kaluzny (1987).

Latin hypercube sampling (Iman & Conover, 1980) is one solution to the problem that random sampling may miss areas of parameter space. Suppose that there are k parameters for which sensitivity is required to be analysed, and n simulation runs available for the sensitivity analysis. For each parameter, the range over which the sensitivity is to be examined is then divided into n intervals into which the parameter might fall with equal probability. If the hypothesized distribution of the parameter were uniform, these would be of equal width. More usually, a distribution such as a normal distribution or perhaps triangular distribution would be used, so that the intervals are wider at the boundaries of the distribution (Fig. 2.1). For each parameter, one of the n intervals is selected at random without replacement, and within that interval a value is selected at random. When this process is repeated for each parameter, the first parameter combination is formed. The whole process is repeated n times. As intervals are sampled without replacement, a complete coverage for each individual parameter is obtained, and the combinations of parameter values that are investigated are selected at random. This process seems almost baroque in its complexity, but can be programmed without too much difficulty, and is the default sampling procedure in the risk analysis package @Risk (Palisade Corporation, Newfield, New York).

Sensitivity analysis is particularly important for stochastic models, such as population viability analysis models, because little guidance is likely to be provided by algebraic analysis. However, the computational requirements may be very large indeed, given that 500–1000 replications may be necessary to gain a good description of the probability distribution of outcomes for a single parameter combination. Statistical approaches such as logistic regression are helpful for interpreting such results, provided it is recognized that the significance levels are a function of the number of replicates used in the simulation. For an example of this approach, see McCarthy *et al.* (1995).

Statistical tools

Accuracy and precision

The ideal for any estimator (estimation procedure) is that it is unbiased and of minimum variance. 'Minimum variance' means that repeated estimates are as close as possible to each other. This is often called 'precision'. An estimator is unbiased if its expected value (loosely, the average of many repeated estimates) is equal to the true value of the entity being estimated. Lack of bias, together with precision, is often termed 'accuracy' (Fig. 2.2). Ideally, an estimate should satisfy both criteria simultaneously, but on occasions, it is necessary to compromise one to achieve major gains in the other. All

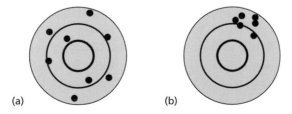

(a)   (b)

Fig. 2.2 The target analogy for bias and precision. Estimating a parameter can be imagined as aiming for a target, in which the bull's-eye is the value of the parameter itself. Each estimation is represented by one shot at the target. (a) An estimator (estimation procedure) that is unbiased, but relatively imprecise compared with (b). The estimates are clustered around the correct value, but there is considerable scatter. (b) An estimator that is more precise than (a), but is biased. The estimates are tightly clustered together, with little scatter, but they are not clustered around the correct parameter value.

estimation problems are subject to a constraint of limited available effort, whether effort is measured in time or money. The design problem is therefore to allocate limited effort in such a way that the estimate is as accurate as possible.

Most biologists are familiar with the standard statistical methods of analysis of variance and regression. There is a plethora of books on these approaches for biologists, and the subject is covered well in almost all elementary statistics texts. I will not elaborate on the details of the methods here. There are, however, a couple of general points that are worth making.

First, statistical methods are usually taught to biologists with a heavy emphasis on hypothesis testing. Hypothesis testing using analysis of variance, regression, etc. uses a so-called parametric model, meaning that it proceeds by a process of estimating parameters in a statistical model. Hypothesis tests are essentially a process of determining whether particular parameters or groups of parameters could possibly be zero (or less commonly, some other specified value), given the observed data. This means that it is straightforward to turn any parametric statistical test around so that it estimates one or more parameters, rather than simply tests hypotheses. Usually, a parameter estimate is more useful than a straightforward hypothesis test. For example, Box 2.3 shows how a t test can be used to obtain an estimate for the size of the treatment effect, and shows how this approach provides substantially more information than does the hypothesis test alone.

Second, analysis of variance and regression are versions of an identical statistical model: the general linear model. All this means is that the parameters of the model are separated from each other by plus signs, and that the random variation associated with each observation (or possibly group of observations) is added on to the systematic (nonrandom) part. Each source of random variation or error is also assumed to have a normal distribution, although the method performs fairly well (it is 'robust') with many nonnormal error distributions. An important consequence of the assumption of additive normal error is that the amount of variability in 'treatments' should not be related to their predicted mean response. Linear models can be built with complex nested or crossed error structures (see Cochran & Cox, 1957; Sokal & Rohlf, 1995; Zar, 1999). The principal advantages of linear models should be familiar to most ecologists. They are very well understood and they can be solved algebraically, using almost all standard statistical packages. In many cases, conventional linear models will be perfectly adequate to estimate parameters for ecological models. Even if the ecological model appears nonlinear, or the errors are non-normal, a transformation of the original data may be able to be represented linearly. By a substantial margin, logarithmic transformations are the most common, but a variety of other functions can be used.

Box 2.3 Using a *t* test for estimation

Student's *t* test is usually the first hypothesis test taught to students of statistics. Its purpose is to test whether two treatments (or a treatment and control) differ in their effect on the mean of some response. Alternatively, it can be used to test if two samples could have been drawn from populations with the same mean. More formally, it tests

H_0: $\mu_1 = \mu_2$ versus H_A: $\mu_1 \neq \mu_2$.

The standard test statistic is

$$t = \frac{\bar{y}_1 - \bar{y}_2}{se_{\bar{y}_1 - \bar{y}_2}},$$

where $\bar{y}_1$ and $\bar{y}_2$ are the means of the response for treatments 1 and 2, and $se_{\bar{y}_1 - \bar{y}_2}$ is the standard error of the difference between the treatment means. Conventionally, the null hypothesis is rejected if the test statistic is greater in magnitude than $t_{0.05}$, the critical two-tailed value of the tabulated t distribution with $n_1 + n_2 - 2$ degrees of freedom and a significance level of 0.05.

It is essential to recognize, however, that statistical significance is not the same as biological importance. It is possible for a statistically significant result to be of no biological importance. The extent of the difference, whilst real, may be too small to matter. It is equally possible for a statistically insignificant result to occur when the experimental results are consistent with a difference of biological importance. Converting the hypothesis test into a parameter estimate with confidence limits makes this clear, and supplies far more useful information than does the hypothesis test alone.

It is very easy to turn this around so that a confidence interval for the difference between the two treatment effects is generated:

$$\mu_1 - \mu_2 = (\bar{y}_1 - \bar{y}_2) \pm t_{0.025} se_{\bar{y}_1 - \bar{y}_2}.$$

The following diagram shows a range of possible outcomes from calculating such a confidence interval. The horizontal axis shows the difference between the two treatments, with the hatched region around zero indicating the range of differences that are of no biological importance. This region would be determined before the experiment is undertaken, using biological, not statistical, criteria. Any difference between −0.7 units and +1 unit would not matter in terms of altering a biological conclusion, or changing a management action. Note that this interval need not be symmetrical about 0.

Case **a** is an unequivocal result. It is statistically insignificant, as the confidence interval includes 0 and, at the specified confidence level, could not be biologically important, as the interval does not extend beyond the hatched region.

continued on p. 26

Box 2.3 *contd*

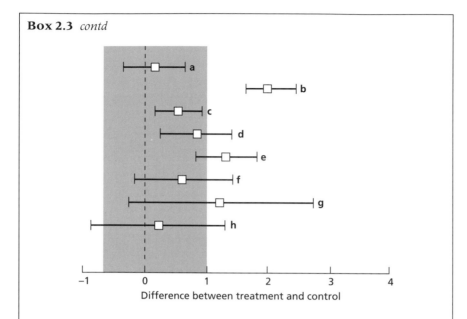

Case **b** is similarly unequivocal. It is statistically significant, as the interval does not include 0, and is biologically important at this confidence level, as the error bars do not leave the region of biological importance.

Case **c** is statistically significant, but is biologically unimportant, as the error bars do not leave the hatched region.

The remaining cases are equivocal. In **d** and **e**, the result is statistically significant, but may or may not be biologically important. Cases **f** and **g** are not statistically significant, but do not preclude the possibility of a biologically important difference in one direction. Case **h** is similarly not statistically significant, but biologically important differences in either direction are not inconsistent with the result.

Logistic regression

Many parameter estimation problems in ecology involve estimating proportions. For example, estimating a survival rate is a problem of estimating the proportion of individuals in a particular class that survive from one time period to the next. Elementary statistical theory states that, if the probability of an event happening to any one individual is p, and there are n independent trials, the number of individuals x to which the event happens follows a binomial distribution. As this distribution is not a normal distribution, and its variance is related to the mean, standard linear methods may be inappropriate. A further problem is that a linear model can easily generate estimated probabilities that are either less than zero or greater than one.

Logistic regression is a statistical method that handles these problems. It gets its name because it attempts to estimate a logistic transformation of the proportion in question as a linear function of one or more predictor variables. The basic equation is of the following form:

$$\ln\left(\frac{p}{1-p}\right) = \mu + \beta x + \alpha_i. \tag{2.2}$$

Here, p is the proportion being predicted (for example the probability of survival), x is a continuous predictor variable (for example, body weight), and μ, β and the α_i are parameters to be estimated. In this case, the α_i could correspond to two different sexes. Assuming that the observed data follow a binomial distribution with p given by eqn (2.2), the parameters can be estimated by the method of maximum likelihood. Most statistical packages will have procedures to do this. An excellent technical explanation of the general approach can be found in McCullagh and Nelder (1989), and Crawley (1993) provides a much more accessible explanation for ecologists.

Despite their power, standard logistic modelling approaches have some drawbacks in many ecological contexts. A crucial assumption of the binomial distribution is that each 'trial' should be independent. In many ecological situations they will not be. For example, suppose that eqn (2.2) was being fitted to survival data that were obtained by following the fate of fledgling birds, which had been marked and weighed at the nest. It is quite likely that survival might vary between nests, as well as between individual fledglings. Observations of survival from two or more fledglings from the same nest would then not be independent. In standard linear models, such mixed or nested designs are relatively easy to deal with, by having two error terms, one for 'nests' and the other for 'individuals within nests'. The corresponding theory for logistic models is much more poorly developed. The result of such 'extra-binomial variation' is that the data may be overdispersed, meaning that the variance is greater than that predicted by a standard binomial distribution. This problem, and some solutions, is discussed briefly in Box 2.4.

Box 2.4 Overdispersion and binary data

In any problem with a binary response (success or failure), the data will come in i groups, within which the values of each of the predictor variables are the same. This means that the predicted probability of success p_i will be the same for all observations in that group. For example, in the black mollie data of Table 2.1, there are eight such groups. In some cases, particularly if there are continuous predictors, each group may contain only one observation. Such data are called 'sparse' and the following discussion does not apply.

If a model with N_p parameters fits the data well, the residual deviance should have approximately a χ^2 distribution, with $i - N_p$ degrees of freedom,

continued on p. 28

Box 2.4 *contd*

provided the groups each contain a reasonable number of observations. The number of observations in each group that is 'reasonable' can be determined by the same rule of thumb used in conventional contingency table analysis: the expected number of both success and failures in at least 80% of groups should be 5 or greater.

If the residual deviance is greater than the upper 5% tail of a tabulated χ^2 distribution with $i - N_p$ degrees of freedom, then there is evidence of overdispersion, also known as extra-binomial variation or heterogeneity. It may sometimes be reduced or removed by fitting an additional covariate, factor or interaction. Frequently, this will not be possible, but you may still wish to use the model, both to make inferences about which explanatory variables are important influences on the response, and to estimate parameters.

There are several ways in which overdispersion can be handled (see McCullagh & Nelder, 1989; Crawley, 1993, Collett, 1991, Wilson & Grenfell, 1997). The simplest solution is to assume that the binomial variance has been inflated by some constant scale factor ϕ. This can be estimated by:

$$\phi = D_0/(i - N_p),$$

where D_0 is the residual deviance. It is probably slightly better to use Pearson's X^2 for the full model, in place of the residual deviance, but it is less convenient, and the improvement will be slight (Collett, 1991). ϕ can then be used to adjust changes in deviance to see if particular terms can be omitted from the model. For example, suppose you wished to determine whether a factor with k levels could be omitted from a full model with N_p parameters, and a residual deviance D_0, and that the residual deviance with the factor omitted was D_1. Instead of comparing $D_0 - D_1$ with a χ^2 distribution with $k - 1$ degrees of freedom, you would form the ratio

$$F = \frac{(D_0 - D_1)/(k-1)}{\phi},$$

and compare it with an F distribution with $k - 1$ and $i - N_p$ degrees of freedom. The parallels with a standard analysis of variance should be obvious.

Overdispersion has a very small effect on the parameter estimates themselves, but will cause their standard errors to be underestimated. An approximate solution is to multiply the standard errors produced in the standard analysis by $\sqrt{\phi}$ (Collett, 1991).

Some packages (e.g. GLIM and PROC GENMOD in SAS) allow a scale parameter ϕ to be specified in a logistic regression. Running a standard analysis, estimating ϕ from the equation above, and then rerunning the analysis specifying that value of ϕ as the scale parameter will perform very similar adjustments to those just described (Crawley, 1993). More sophisticated ways of handling overdispersion are discussed by Collett (1991).

Maximum likelihood estimation

Linear models are not adequate for all estimation problems in ecology. There are two problems, frequently linked. First, many ecological processes and therefore models are intrinsically nonlinear. This means that there is no transformation capable of producing a model with all parameters separated by plus signs. Second, data may have a non-normal error structure, which means that the data cannot satisfactorily be modelled as a systematic part plus a normally distributed random part. The result of these two problems is that the statistical theory upon which linear model theory is based is invalid, possibly causing biased parameter estimators, and more probably, incorrect measures of the precision of the resulting estimates. Logistic regression, described above, is a common example of a statistical model for which conventional least squares estimation cannot be used.

In such cases, a much more general approach, that of maximum likelihood, is often helpful. It is important to note that least squares estimation, applied to a linear model with normally distributed errors, generates maximum likelihood estimates for the model parameters. Maximum likelihood estimation thus encompasses linear least squares estimation. It is not an alternative method. There is an outstanding review of the general approach and philosophy in Edwards (1972), and a recent discussion from an ecological perspective is provided by Hilborn and Mangel (1997).

It is worth digressing a little to explain the basic approach, because it underlies most estimation problems discussed in this book. Most ecologists will probably treat the process as a 'black box', using one of a number of computer packages that can handle maximum likelihood estimation for common models. The basic idea is, however, relatively straightforward, and there will be occasions when it is necessary to compute a maximum likelihood estimate from first principles, rather than relying on a computer package.

The principle is best explained first with a specific example, rather than being tackled abstractly. The data in Table 2.1 come from an experiment investigating the mortality induced in the fish *Poecilia latipinna* by the protozoan parasite *Ichthyophthirius multifiliis*. The table shows, for a range of intensities of infection x_i, the number y_i of fish surviving five days after infection out of a total number m_i that were in the infection class originally. The objective is to fit, using maximum likelihood, the logistic regression model

$$\ln\left(\frac{p_i}{1-p_i}\right) = \alpha + \beta x_i, \tag{2.3}$$

where p_i is the probability of a fish surviving five days, given it is in infection class x_i.

Table 2.1 Survival of black mollies (*Poecilia latipinna*) infected with *Ichthyophthirius multifiliis*. The following data are from an experiment conducted by McCallum (1982b) (also summarized in McCallum, 1985). Black mollies were subjected to *Ichthyophthirius* infective stages at several densities. Two days after infection, the parasite burden established on each fish was assessed by counting the number of the ectoparasitic ciliates visible on the fish's flank within a counting frame with an area of 0.16 cm^2. The number of days the fish survived after infection, up to a maximum of 10 days, was then recorded. The data are abstracted from this information, and were obtained by grouping the fish into ranges of infection intensity, and then recording the proportion of the fish in each of these infection classes surviving to day 6 after infection. (A better, but more complex, way to analyse these data is to model the survival time of each fish as a function of parasite burden. This approach is taken in Chapter 4)

Midpoint of infection class (parasites per 0.16 cm^2)	Survivors to day 6	Number of fish in class
10	11	11
30	9	10
50	15	18
70	12	14
90	6	7
125	4	9
175	4	13
225	1	9

Assuming that the survival of one fish does not affect the survival of another, the probability of a given number of fish in each infection class surviving should follow a binomial distribution:

$$P(y_i) = \binom{m_i}{y_i} p_i^{y_i} (1 - p_i)^{(m_i - y_i)}. \tag{2.4}$$

The joint probability of observing each number of surviving fish, given known p values, is simply a series of equations of the form above multiplied together:

$$P(y_1, y_2, \ldots, y_n) = \prod_{i=1}^{n} \binom{m_i}{y_i} p_i^{y_i} (1 - p_i)^{(m_i - y_i)}. \tag{2.5}$$

This equation could calculate the probability of any given outcome, provided the probabilities p_i were known. What we actually have, however, is one set of observed outcomes y_i, and our task is to estimate the probabilities p_i.

The basic idea of maximum likelihood estimation is very simple and logical. The objective is to find the set of parameters that are the most likely of all possible sets to have generated the observed result. This will be the set for which the probability of detecting the observed outcome is as large as possible. In other words, we need to find values of α and β that maximize the value of eqn (2.5) when the solution of eqn (2.3) is substituted into it. When a joint probability distribution or density function is considered as a function of a set of parameters to be estimated, it is called the *likelihood function* of those parameters. In this case, then, the likelihood of the parameter set (α, β) is

$$L(\alpha,\beta; y_1, y_2, \ldots, y_n) = \prod_{i=1}^{n} \binom{m_i}{y_i} \left[\frac{e^{\alpha+\beta x_i}}{1+e^{\alpha+\beta x_i}} \right]^{y_i} \left[\frac{1}{1+e^{\alpha+\beta x_i}} \right]^{(m_i-y_i)} \tag{2.6}$$

It would be perfectly possible to maximize eqn (2.6), and this would generate maximum likelihood estimates for α and β. It is more convenient, however, to maximize the log of the likelihood (which will also maximize the likelihood itself).

Noting that the $\binom{m_i}{y_i}$ terms do not depend on the parameters, and can be pulled out as constants, the problem is now to maximize the log-likelihood l:

$$l(\alpha,\beta; y_1, y_2, \ldots, y_n) = \sum_{i=1}^{n} \left\{ y_i \ln \left[\frac{e^{\alpha+\beta x_i}}{1+e^{\alpha+\beta x_i}} \right] + (m_i - y_i)\ln \left[\frac{1}{1+e^{\alpha+\beta x_i}} \right] \right\} \tag{2.7}$$

With a little simplification and rearrangement, eqn (2.7) becomes:

$$l(\alpha,\beta; y_1, \ldots, y_n) = \sum_{i=1}^{n} y_i(\alpha + \beta x_i) - \sum_{i=1}^{n} m_i \ln(1 + e^{\alpha+\beta x_i}). \tag{2.8}$$

Figure 2.3 shows eqn (2.8) plotted as a likelihood surface for the data in Table 2.1. It can be seen that there is a maximum at about $\alpha = 3.2$ and $\beta - -0.02$. However, the maximum lies on the top of a rather flat ridge. This means that a variety of other combinations of the parameters will fit the data almost as well. It is a relatively straightforward task to maximize eqn (2.8) numerically. Box 2.5 shows how to do it using a standard spreadsheet such as Excel.

In general, the process of maximum likelihood estimation of a parameter set β, given a set of observed data y can be summarized as follows:

1   Obtain a function that gives the joint probability of any particular set of outcomes y, given known values of the parameters β.

2   Consider this as a likelihood function $L(\beta;y)$ for the unknown parameters β, given the particular observed outcomes y.

3   Take the natural logarithm of L, and discard any terms in the resulting sum that do not involve the parameters β. Call the result $l(\beta;y)$.

4   Treating the data values y as known constants, use an algorithm to find values of the parameters β that maximize $l(\beta;y)$.

The fit of a model obtained by maximum likelihood is often expressed as the *residual deviance*. This is twice the difference between the maximum

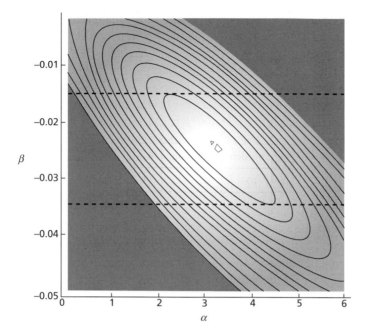

Fig. 2.3 Contour plot of the likelihood surface for the black mollie survival data in Table 2.1. β is plotted on the vertical axis, and α on the horizontal axis. The contour interval is 2, and hence the first contour encloses an approximate 95% confidence region for the joint distribution of the parameters. The profile likelihood confidence limits for β are shown as dashed lines.

achievable log-likelihood (obtained when the predicted and observed values are identical) and that attained by the model under consideration (McCullagh & Nelder, 1989, p. 118). The adequacy of the fitted model is then often assessed by comparing the residual deviance with a χ^2 distribution with $i - N_p$ degrees of freedom, where i is the number of classes in the data, and N_p is the number of fitted parameters. This is only appropriate, however, under restrictive conditions (see McCullagh & Nelder, 1989, p. 118; and Box 2.4). In the particular example considered above, the maximum achievable log-likelihood is that obtained by substituting the observed frequencies themselves into eqn (2.5) (Box 2.5).

A more reliable procedure to assess whether a model is adequate is to examine the change in deviance when additional parameters are added to the model. If the true value of the additional parameters is zero, the change in deviance should have a χ^2 distribution with degrees of freedom equal to the number of additional parameters. This will be the case asymptotically (that is, when the sample size is sufficiently large). In practice, it means that you can determine whether additional parameters can be justified by comparing the change in deviance with the upper tail of a tabulated χ^2 distribution (see Box 2.4).

Box 2.5 Maximum likelihood estimates using a spreadsheet

This example uses Excel, and requires the use of the add-in 'Solver', which allows simple optimizations to be calculated. The data are obtained from Table 2.1.

Step 1. Input basic data:

	A	B	C
1	burden	survivors	n
2	10	11	11
3	30	9	10
4	50	15	18
5	70	12	14
6	90	6	7
7	125	4	9
8	175	4	13
9	225	1	9

Step 2. Guess values for parameters (cells K2 and L2), use these to calculate the linear predictor (cell D2) and (using eqn (2.8)) individual terms in the likelihood equation (cell F2).

	A	B	C	D	E	F	G	H	I	J	K	L
1	burden	survivors	n	lin pred		log likeli					intercept	slope
2	10	11	11	= K2 + L2*A2		= B2*D2 – C2*LN(1 + EXP(D2))					4	–0.1
3	30	9	10									
4	50	15	18									
5	70	12	14									
6	90	6	7									
7	125	4	9									
8	175	4	13									
9	225	1	9									
10												

continued on p. 34

Box 2.5 *contd*

Step 3. Copy the linear predictor and log-likelihood terms down columns, and sum the log-likelihood terms in cell F10:

	A	B	C	D	E	F	G	H	I	J	K	L
1	burden	survivors	n	lin pred		log likeli					intercept	slope
2	10	11	11	3		−0.534 46					4	−0.1
3	30	9	10	1		−4.132 62						
4	50	15	18	−1		−20.638 7						
5	70	12	14	−3		−36.680 2						
6	90	6	7	−5		−30.047						
7	125	4	9	−8.5		−34.001 8						
8	175	4	13	−13.5		−54						
9	225	1	9	−18.5		−18.5						
10						= SUM(F2:F9)						

Step 4. Use Solver (Tools Solver) to maximize cell F10 by changing cells K2 and L2:

continued on p. 35

Box 2.5 *contd*

	A	B	C	D	E	F	G	H	I	J	K	L
1	burden	survivors	n	lin pred		log likeli					intercept	slope
2	10	11	11	2.975 699		−0.547 29					3.213 425	−0.023 77
3	30	9	10	2.500 246		−3.288 96						
4	50	15	18	2.024 793		−8.306 46						
5	70	12	14	1.549 34		−5.794 97						
6	90	6	7	1.073 887		−3.131 33						
7	125	4	9	0.241 845		−6.424 89						
8	175	4	13	−0.946 79		−8.049 24						
9	225	1	9	−2.135 42		−3.140 86						
10						−38.684						

The maximum likelihood estimators are now in cells K2 and L2, with the maximized log-likelihood in cell F10.

Step 5. Calculate the fitted number of survivors (cell E2) and observed (cell G2) and fitted (cell H2) proportions of survivors. Calculate components of the maximum likelihood achievable by substituting observed frequencies into the likelihood equation (column I). Note that the entries in column I are just logs of the terms in eqn (2.4). Cell I2 is explicitly set to 0 to avoid the problem of calculating log(0).

	A	B	C	D	E	F	G	H	I	J	K	L
1	burden	survivors	n	lin pred	fitted	log likeli	p	fitted p	max likeli		intercept	slope
2	10	11	11	2.975 699	= C2*EXP(D2)(1 + EXP(D2))		= B2/C2	= E2/C2	0		3.213 425	−0.023 77
3	30	9	10	2.500 246					= B3*LN(G3) + (C3 – B3)*LN(1 – G			
4	50	15	18	2.024 793								
5	70	12	14	1.549 34								
6	90	6	7	1.073 887								
7	125	4	9	0.241 845								
8	175	4	13	−0.946 79								
9	225	1	9	−2.135 42								
10												

continued on p. 36

Box 2.5 *contd*

Step 6. Copy these formulae down the columns, sum column I in cell I10, and calculate the residual deviance in cell G12 as = (I10-F10)*2.

	A	B	C	D	E	F	G	H	I	J	K	L
	burden	survivors	n	lin pred	fitted	log likeli	p	fitted p	max likeli		intercept	slope
1												
2	10	11	11	2.975 699	10.466 11	−0.547 29	1	0.951 464	0		3.213 425	−0.023 77
3	30	9	10	2.500 246	9.241 591	−3.288 96	0.9	0.924 159	−3.250 83			
4	50	15	18	2.024 793	15.900 76	−8.306 46	0.833 333	0.883 376	−8.110 1			
5	70	12	14	1.549 34	11.547 46	−5.794 97	0.857 143	0.824 818	−5.741 63			
6	90	6	7	1.073 887	5.217 348	−3.131 33	0.857 143	0.745 335	−2.870 81			
7	125	4	9	0.241 845	5.041 514	−6.424 89	0.444 444	0.560 168	−6.182 65			
8	175	4	13	−0.946 79	3.633 908	−8.049 24	0.307 692	0.279 531	−8.024 14			
9	225	1	9	−2.135 42	0.951 315	−3.140 86	0.111 111	0.105 702	−3.139 49			
10						−38.684			−37.319 7			
11							Residual deviance					
12							2.728 681					

Step 7. Plot observed and predicted proportions:

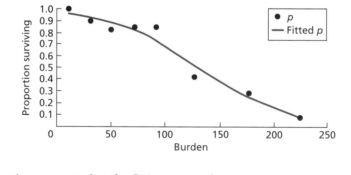

Visual inspection suggests that the fit is very good.

Maximum likelihood and nonlinear least squares

The parameter estimates produced by ordinary least squares analysis of linear models are maximum likelihood estimates, provided that the error can be assumed to be additive, normally distributed and with a constant variance. In fact, it is possible to go considerably further than this. If you want to fit a non-linear relationship $f(x, \beta)$ to a data set with a response y, predictor variables x and parameters β, then, provided the relationship can be assumed to be of the form

$$y = f(x, \beta) + \varepsilon, \tag{2.9}$$

where ε are normally distributed errors with constant variance, a standard nonlinear least squares fitting procedure will produce maximum likelihood estimates for the parameters β (see, for example, Seber & Wild, 1989).

In many ecological problems, the assumption of additive normal error with constant variance in eqn (2.9) will not be reasonable, particularly if y is a measure of abundance. Taylor (1961) showed that the variance in animal abundance is usually related to mean abundance via a power relationship with an exponent close to 2. If Taylor's power law applies, and the exponent is close to 2, the error may be approximately lognormal. This means that $\log(y)$ will be the appropriate dependent variable to use in the nonlinear least squares procedure, and the resulting parameters should be maximum likelihood estimators. Log transformations should be familiar to most readers from linear regression, where they are employed both to linearize relationships and to normalize errors. In linear regression, there may be a conflict between the most appropriate transformation to normalize errors and the most appropriate transformation to linearize the relationship. If nonlinear regression is used, this conflict does not arise, and you are free to select the transformation that does the best job of producing normally distributed residuals with approximately equal variance. Standard residual analysis should always follow a non-linear regression, as it should follow any regression.

Selecting the most appropriate model from several alternatives is rather harder with nonlinear regression than it is for linear regression. This is partially because statistical theory is better developed for linear models, but also because the potential range of possible models is far greater, once nonlinear relationships are allowed. Some software may produce an R^2 analogous to that from linear regressions. Although this statistic has some use, its interpretation is not straightforward in nonlinear regression (Ryan, 1997). Comparing nested models can done with some semblance of statistical rigour. (Recall that two models are nested if the simpler one can be obtained from the more complex one by fixing one or more parameters at particular values.)

Suppose we are comparing a full model, with f parameters estimated from

Table 2.2 Extra sum of squares analysis for nonlinear models with normal errors

Source	Sum of squares	Degrees of freedom	Mean square	F ratio
Extra parameters	$SS_e = SSE_p - SSE_f$	$f - p$	$s_e^2 = \dfrac{SS_e}{f - p}$	$\dfrac{s_e^2}{s_f^2}$
Full model	SSE_f	$n - f$	$s_f^2 = \dfrac{SSE_f}{n - f}$	
Partial model	SSE_p	$n - p$		

n data points, with a simpler, partial model, with only p parameters estimated from the data. Let SSE_f be the residual sum of squares from the full model, and SSE_p be the residual sum of squares from the partial model. The extra sum of squares analysis in Table 2.2 can be used to determine whether the partial model is adequate, or whether the full model should be used (Bates & Watts, 1988). This analysis is only approximate, but will usually be a reasonable guide. It is almost always a better approach to deciding if additional factors should be included than is examining whether the added parameter is 'significantly different from zero'. If two models are not nested, comparing the fit with a single statistic is not particularly useful, although R^2 may give a rough idea. Examination of the residuals, together with thinking about the biology of the relationship, will usually be a more reliable guide to model selection than R^2 alone.

Confidence intervals for maximum likelihood estimates

There are three methods that can be used to calculate confidence intervals for maximum likelihood estimates. These are listed in order of increasing computational effort, but also increasing generality and accuracy. First, an estimated standard error can be used to form a confidence interval in the usual way, assuming that the parameter estimate has an approximately normal distribution. As almost all computer packages that perform maximum likelihood estimation will generate an estimated standard error as part of their output, this is the most straightforward approach. However, it may fail if estimates are close to a boundary of biological plausibility (for example, if an estimated survival rate is close to 1), or if the parameter estimate has a very asymmetrical distribution. It is also quite difficult to calculate standard errors if you are not using a standard package. The process involves calculating the Fisher information matrix for the parameter set $\boldsymbol{\beta}$:

$$-E\left(\frac{\partial^2 l}{\partial \beta_r \partial \beta_s}\right), \; r = 1,\ldots,N_p, \; s = 1,N_p, \tag{2.10}$$

where l is the log of the likelihood function, evaluated at its maximum, and there are N_p parameters. In some cases, calculation of the matrix might be possible algebraically. Otherwise, a routine for calculating partial derivatives numerically could be used. The approximate variance–covariance matrix of the parameter set is then the inverse of the matrix (McCullagh & Nelder, 1989, p. 119). This is not a procedure that many ecologists would want to follow from first principles.

A second approach is to use profile likelihood (Venzon & Moolgavkar, 1988; McCullagh & Nelder, 1989; Lebreton *et al.*, 1992). The idea here is that, as any particular parameter is altered, the log-likelihood will decrease on either side of the maximum likelihood estimate. Twice the change in likelihood approximately has a χ^2 distribution, with one parameter of interest, and thus one degree of freedom. The 95% confidence bounds for the parameter are thus the values of the parameter estimate in each direction at which twice the change in log-likelihood equals the upper 5% tail of a χ^2 distribution with one degree of freedom (that is, 3.84).

Graphically, the objective is to find the maximum and minimum values of each parameter that form the extreme bounds of the 5% likelihood contour (see Fig. 2.3). In algebra, using the black mollie survival logistic regression example, the upper 95% profile likelihood bound $\hat{\beta}_u$ will satisfy:

$$\hat{\beta}_u = \max \beta \colon \{l(\hat{\beta},\hat{\alpha}) - l(\beta,\alpha) = 1.92\}. \tag{2.11}$$

Here, $l(\hat{\beta},\hat{\alpha})$ is the maximum log-likelihood achievable using the model in question, and $\hat{\beta}$ is the value of β at which this log-likelihood occurs. Note that the maximization of β subject to the constraint $l(\hat{\beta},\hat{\alpha}) - l(\beta,\alpha) = 1.92$ must be done varying both parameters.

If you are investigating a few parameters only, this is not a difficult procedure on a standard spreadsheet (see Box 2.6). An algorithm to speed the process up if there are many parameters is given by Venzon and Moolgavkar (1988).

The third approach is to use resampling procedures. These are described in the following section.

Computer-intensive methods: jackknives and bootstraps

Conventional statistical analysis assumes that data follow a relatively small number of standard probability distributions and then uses the theoretical properties of those distributions for inference. Computer-intensive methods (Diaconis & Efron, 1983) use the power of modern computers to derive the probability distribution of parameter estimates from data by 'brute force'. Most involve resampling, in which the observed data are repeatedly subsampled, and the variability of the resulting parameter estimates is then examined.

Jackknife methods repeatedly recalculate estimates of parameters omitting one observation in turn (Bissell & Ferguson, 1974; Crowley, 1992; Lipsitz et al., 1994). The principle is simply that, if the parameter can be estimated precisely, its value should not vary too much depending on which particular observation is omitted, whereas if its value varies markedly, this indicates that the parameter cannot be estimated precisely.

More technically, suppose the parameter of interest is Π, and an estimate p has been generated based on n observations. The jackknife proceeds by calculating estimates of the parameter Π omitting each observation in turn. Let p_{-i} be the estimate formed by omitting the ith observation from the sample. These are then used to calculate pseudo-values π_i:

$$\pi_i = p - (n - 1)(p_{-i} - p). \tag{2.12}$$

The basic idea is that the pseudo-value for observation i quantifies the change occurring in the estimate when observation i is omitted. Obviously, if omitting any observation changes the estimate only slightly, all pseudo-values will be similar, and close to p. The only slightly non-intuitive part is the multiplication by $n - 1$. This is needed to make the variance work. The mean of the pseudo-estimates $\bar{\pi}$ is the jackknifed estimate of Π, and the variance of the estimate is:

$$s_{\bar{\pi}}^2 = \frac{\sum_{i=1}^{n}(\pi_i - \bar{\pi})^2}{n(n - 1)}. \tag{2.13}$$

This is just the usual variance of the estimated mean that would be obtained if the pseudo-values were treated as ordinary observations. A confidence interval for the parameter can then be constructed using the standard approach for an approximately normally distributed estimate. Box 2.6 shows a worked example of a simple jackknifed estimate.

The principal advantage of the jackknife is that it requires only modest amounts of calculation (you need to calculate as many separate estimates as there are data points). For small data sets, it is therefore feasible with either a calculator, or a standard statistics package. It can also reduce bias. The main disadvantage is that it relies on the central limit theorem to obtain confidence limits, and its behaviour in some cases is not well understood. In particular, the jackknife will not work if small changes in the data do not cause correspondingly small changes in the estimate. For this reason, the median is not an appropriate statistic to jackknife (Efron & Tibshirani, 1993, p. 148). For more detail, see Crowley (1992).

Bootstrap methods, in contrast, require considerably more calculation, but do not rely on the data or estimate following any particular probability distribution. As above, suppose the parameter to be estimated is Π, and the sample contains n observations. Bootstrap samples also of size n are taken by sampling

Box 2.6 Confidence intervals

Profile likelihood
If a spreadsheet has been used to calculate maximum likelihood estimates, it is straightforward to calculate upper and lower confidence bounds using profile likelihood.

As an example, consider finding bounds for β in the black mollie survival data analysed in Box 2.5. Following eqn (2.11), the upper 95% limit for β can be found by finding values of α and β that maximize β subject to the constraint that:

$$l(\hat{\beta},\hat{\alpha}) - l(\beta,\alpha) - 1.92 = 0;$$

i.e. enter
$= -38.684 - \text{SUM(F2:F9)} - 1.92$ in cell F10, and ask Solver to maximize β in cell L2, by modifying the estimates of α and β in cells K2 and L2, subject to the constraint that cell F10 = 0. To ensure that Solver can find the appropriate solution, starting values on the correct side of the maximum likelihood estimates should be used. The correct solutions to the confidence bounds for β correspond to the vertical bounds drawn on Fig. 2.3. The results are given in the table at the end of this box, together with the limits obtained by using the standard error generated by GLIM, and jackknife and bootstrap estimates.

Jackknife estimate
The jackknife is calculated by leaving one observation out in turn, recalculating the maximum likelihood estimates, and then converting them into pseudo-values using eqn (2.12). In this case, the individual observations are the 91 fish, not the eight entries in Table 2.1. There will, however, be only 15 distinct values the 91 pseudo-values can take, as each row in Table 2.1 corresponds to two combinations of the predictor variable and binary response variable, and all fish in row 1 survived. The estimates of β obtained by omitting each distinct combination of burden and survival are given in the table below, together with the corresponding pseudo-values generated by eqn (2.12). The mean and standard error of these pseudo-values are the bias-corrected estimate of β and its estimated standard error. These are given in the table at the end of the box.

continued on p. 42

Box 2.6 *contd*

Burden	Survived*	Freq.	p_{-i}	π_i
10	1	11	−0.023 66	−0.033 89
30	1	9	−0.023 63	−0.036 35
30	0	1	−0.025 61	0.141 596
50	1	15	−0.023 61	−0.038 14
50	0	3	−0.025 07	0.092 579
70	1	12	−0.023 62	−0.037 86
70	0	2	−0.024 56	0.046 979
90	1	6	−0.023 67	−0.033 36
90	0	1	−0.024 11	0.006 889
125	1	4	−0.023 96	−0.007 03
125	0	5	−0.023 56	−0.043 26
175	1	4	−0.025 01	0.087 282
175	0	9	−0.023 33	−0.063 99
225	1	1	−0.026 55	0.226 243
225	0	8	−0.023 47	−0.050 58
Sum		91		

*0 = No, 1 = Yes.

Bootstrap estimates

The following figure shows the distribution of estimates of β obtained from 2000 bootstrap samples from the 91 observations of fish survival. The bootstrapping was performed by the SAS macro jackboot, which can be downloaded from the SAS Institute web site.

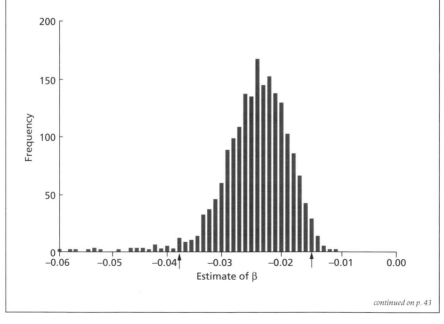

continued on p. 43

Box 2.6 *contd*

These estimates can be used either to calculate a standard error, and hence a 95% confidence interval based on a normal approximation, or to calculate a confidence interval based directly on the 2.5th and 97.5th percentiles of this distribution. The percentiles are marked with arrows on the figure.

Comparison of estimates
The following table shows the estimates and 95% confidence intervals obtained for β using the black mollie survival data. Which estimates and limits are most 'correct' cannot be determined for real, as distinct from simulated, data. The most obvious point is that the intervals are not the same. The two intervals that permit asymmetry (profile and percentage bootstrap) are substantially so (see also the distribution of the bootstrapped estimates).

	Normal (GLIM)	Profile likelihood	Jackknife	Bootstrap (normal)	Bootstrap (percent)
$\hat{\beta}$	−0.023 77	−0.023 77	−0.022 59	−0.022 54	−0.023 77
$se(\hat{\beta})$	0.004 833	–	0.005 22	0.005 47	–
Lower 95%	−0.029 07	−0.034 22	−0.032 81	−0.033 89	−0.038 12
Upper 95%	−0.009 74	−0.015 07	−0.012 36	−0.011 20	−0.015 86

with replacement from the original sample. From each of these, the parameter in question is estimated, using a plug-in estimate. (A plug-in estimate is simply one that would generate the parameter itself, if the sample were the whole population. For example, $\Sigma(Y - \bar{Y})^2/n$ is a plug-in estimate of a population variance, because $\Sigma(Y - \mu)^2/n$ is the definition of a population variance, whereas the usual $\Sigma(Y - \bar{Y})^2/(n - 1)$ is not a plug-in estimate.) As the samples are taken with replacement, each bootstrap sample will contain some observations more than once, and omit others entirely. It is intuitively obvious that, if the estimate can be made precisely, the value of the estimate based on each bootstrap sample will not greatly depend on which observations are included or omitted. The overall bootstrap estimate is simply the mean of all the individual estimates from each sample. The standard error of the bootstrap estimate is just the standard deviation of the individual estimates. To obtain a good bootstrap estimate, together with its standard error, the number of bootstrap samples needs to be in the region of 50–200 (Efron & Tibshirani, 1993, p. 52).

There are two ways that a confidence interval can be obtained from a bootstrap estimate (Efron & Tibshirani, 1993, pp. 168–201). One approach is simply to use the estimate and its standard error, together with a tabulated t statistic, in the standard formula for a confidence interval. This approach requires only modest computation, but relies on the estimate having a roughly normal distribution.

A better approach is to obtain the confidence interval directly from the frequency distribution of the bootstrap estimates themselves: a 95% confidence interval for the parameter is simply the range between the 2.5th and 97.5th percentiles of the distribution. This approach is much more demanding computationally than the previous one, requiring at least 1000 replications to provide reliable results, but can handle non-normal distributions without trouble. This approach does not, however, necessarily produce the best confidence intervals possible, particularly if the plug-in estimate is biased.

The most appropriate way of bootstrapping confidence limits for biased estimates is an area of some controversy. Readers should consult Efron and Tibshirani (1993) or Shao and Tu (1995) for further details. A further caution is that the bootstrap can be badly affected by small sample sizes. It is a way of estimating the variability of estimates that have complex derivations or non-normal sampling distributions, not a panacea that can be applied uncritically to any parameter estimation problem.

Another complication is that the level at which resampling should be done is not always obvious. Many sampling or experimental designs have a nested structure, in which several individual observations or replicates are taken from each of a number of subgroups. Resampling at the level of the individual replicate will result in an underestimate of the true error variance: the subgroup is the appropriate level for resampling.

Box 2.6 shows a worked example of a bootstrap estimate using the same data as were used for the jackknife example.

Bootstrap estimates are probably preferable to jackknife estimates in most circumstances (Crowley, 1992), as they make fewer assumptions about the source data. They do require considerably more computational power. However, this is unlikely to be a significant problem, given the power of modern personal computers. They also require special programs, although risk analysis add-ins to spreadsheets can perform the analysis very easily. For further details, see Efron and Tibshirani (1993).

Bayesian methods

Bayes's theorem is one of the basic tenets of probability theory, and entirely uncontroversial. It states that the probability that event A occurs, given that event B has occurred, is equal to the probability that both A and B occur, divided by the probability of the occurrence of B. Symbolically,

$$P(A|B) = \frac{P(A \cap B)}{P(B)}.$$ (2.14)

That said, the extent to which Bayesian methods have their place in statistical inference, either in addition to, or as a replacement for, conventional ('frequentist') statistical methods, is an area of continued, and sometimes quite heated controversy – compare, for example, Ellison (1996) with Dennis (1996). Bayesian methods are not new, and have had a limited impact on mainstream applied statistics to date. However, in the last few years they have been advocated strongly for parameter estimation and decision-making in ecology (Hilborn & Walters, 1992; Ellison, 1996; Link & Hahn, 1996; Ludwig, 1996; Hilborn & Mangel, 1997).

The basic idea of Bayesian parameter estimation is to reframe eqn (2.14) so that event A is obtaining a parameter estimate or set of estimates θ, and event B is obtaining the data at hand x. Thus

$$P(\theta|x) = \frac{P(x|\theta) \times P(\theta)}{P(x)}.$$ (2.15)

$P(\theta|x)$ is the *posterior probability* of obtaining a parameter estimate θ, given the data obtained in the experiment or survey. $P(x|\theta)$ is the probability of obtaining the observed data x, given a particular parameter estimate θ, and is thus the likelihood function for the parameter. $P(x)$ is the probability of obtaining the data under all admissible parameter estimates. It is essentially a scaling constant to ensure that the posterior probabilities sum or integrate to 1, and will usually not need to be calculated explicitly.

The quantity $P(\theta)$ is the *prior probability* of θ, and it is this component which is the source of much of the controversy about Bayesian methods. The prior probability quantifies the knowledge or belief about the probability distribution of θ before the data were collected. In some cases, this may be a subjective belief of the investigator, and can be thought to undermine the objectivity of the parameter estimation process (Dennis, 1996). However, many Bayesians argue that the objectivity of conventional statistics is illusory (Berger & Berry, 1988). It is merely that Bayesians admit it. Furthermore, if a reasonable amount of data are available, the posterior distribution depends only weakly on the prior (Ellison, 1996). The inclusion of prior information in the Bayesian approach also accommodates the incremental nature of scientific progress.

A Bayesian parameter estimation process will generate, as output, a posterior probability density for the unknown parameter or parameters. This can be used to generate a point estimate of the parameter, at the maximum of the probability density. A 95% confidence interval – more correctly termed a *credibility interval* (Ellison, 1996) – for the parameter is simply the region containing 95% of the probability density. The parallels with maximum likelihood and profile likelihood should be obvious. Technically, a conventional 95% confidence interval is a region which will include the true parameter value in 95% of all possible samples, whereas a Bayesian credibility interval is a region within which, given the data in the particular sample, 95% of possible values for the parameter will fall (Ellison, 1996). These are very different concepts, although most ecologists in fact use something close to the Bayesian idea to understand a conventional confidence interval.

One view is that Bayesian methods are a revolution that threatens to overturn conventional inference, either for the better (Ellison, 1996) or worse (Dennis, 1996). Alternatively, they may be merely a different perspective from which to view current methods, with the additional ability to include prior information if it is helpful to do so (Hilborn & Walters, 1992). At the moment, there is no consensus as to which view is correct. Certainly, the Bayesian perspective can be a useful one to apply to many areas of parameter estimation. For example, sensitivity analysis can be viewed as taking prior distributions for the model parameters, based on the best estimates of uncertainty and their distribution in nature, and using them to generate posterior distributions of the model's output. In the majority of cases, despite the very different philosophies of Bayesian and conventional approaches, the practical differences between them are small. There are, however, exceptions to this generalization (e.g. Ludwig, 1996).

Computer packages

The most basic computer tool for ecological modelling is a standard spreadsheet. Spreadsheets were originally developed for accountancy, but as demography is essentially accountancy with individual animals rather than money, it is not surprising that a spreadsheet is a valuable ecological tool. In addition to the ability to perform simple arithmetic operations and to display simple graphics, all modern spreadsheets have at least some basic statistical operations built in, and usually they can also perform straightforward minimization or maximization iteratively. The randomization routines that are included in most spreadsheets are fairly rudimentary, limiting their value for stochastic modelling. However, a variety of 'risk analysis' add-ins are available (e.g. @Risk) which provide a wide range of random number distributions, built-in sensitivity analysis and automatic model iteration.

Spreadsheets are not really suited to complex modelling or statistical operations on large data sets. The statistical methods frequently are not good at handling missing values or unequally replicated treatments (both inevitable features of real biological data). Convoluted spreadsheet models are also quite difficult to check for errors. Nevertheless, spreadsheets are a good way to explore the behaviour of simple mathematical equations, and are an excellent way to 'demystify' mathematical processes and models. They are therefore used quite extensively in this book for illustrative purposes. For the final analysis of large data sets or iteration of complex models, however, most people are likely to require more specialized software.

Statistical software

A number of packages are now available for statistical analyses more complex than those possible using a standard spreadsheet. Some are large packages available for a wide variety of computers and operating systems – for example, SAS (SAS Institute, Cary, North Carolina); SPSS (SPSS Inc, Chicago); GLIM (Royal Statistical Society, London) and S-Plus (MathSoft Inc, Seattle). Each of these packages will perform the basic analysis required for general parameter estimation problems. The choice between them is largely a matter of personal taste and local availability. All of the above packages operate primarily from a command-line interface. This means that you tell the computer what to do, it does it and returns the results. The disadvantage of this sort of interface is that you need to know the commands and syntax. The advantage is that what you have done will be fully documented in a 'log file'. It also means that writing programs or macros to automate repetitive tasks or to perform nonstandard analyses is fairly straightforward.

  The alternative possibility is to use a package specially written for personal computers (whether IBM compatible or Macintosh). Such packages – for example, Statistica (StatSoft, Tulsa, OK) and Systat (SPSS Inc, Chicago) – are usually menu-driven, meaning that using the program is a matter of pulling down menus and ticking boxes. This approach is easier for the novice user, but it is easy to lose track of exactly which boxes have been ticked. There is little point in being prescriptive about which package should be used. The important consideration is that the package must be capable of the analysis requested, and sufficiently flexible that nonstandard analyses are possible and that intermediate results and parameter estimates can be obtained if necessary. It is also important that the documentation should be sufficient that it is clear which algorithms have been used for particular analyses.

Modelling software

A number of packages are now available specifically for ecological modelling. Most of these are designed for population viability analysis. Such programs are usually discrete-time structured models, with added stochastic elements. The major programs in this category include VORTEX (Lacy, 1993), RAMAS (Ferson & Akçakaya, 1988; Akçakaya & Ferson, 1990; Ferson, 1991) and ALEX (Possingham et al., 1992a; 1992b). As with statistics packages, there is little point in recommending particular packages over others. All packages make explicit assumptions about the systems that they model, and other assumptions are implicit in the structure of the models themselves. Rather than using the models as 'black boxes', it is essential to examine these assumptions, to ensure that they are appropriate for your particular system.

CHAPTER 3

Population size

Introduction

Population size is not really a parameter in many ecological models. It is a variable, and usually the variable of principal interest. Nevertheless, it is essential to commence a book such as this with a discussion of methods of estimating population size, for at least two reasons. First, because population size is the fundamental variable, unless you have estimated it, there is little you can do to see if the model approximates reality. Second, estimation of many parameters is a matter of comparing population estimates between two or more times or classes of individuals. There is a very large and continually expanding literature on methods for estimating population size. This chapter will not go into specific methods in great depth, but aims to provide a comparative analysis of the advantages and disadvantages of the various methods, so that you can be directed to appropriate software, or to the primary literature on the best method for the particular problem at hand. Nor will it attempt to deal with the actual field methods for marking, capturing or counting animals. These are specific both to the type of organism and the habitat. Sutherland (1996) provides a handbook of taxon-specific methods.

The first issue is to decide whether an estimate of actual population size is required, or whether an index (that is, a number proportional to population size) will suffice. People tend to assume that an index is more straightforward to obtain, and it is quite common to see a statement in the literature along the lines of 'although our method does not estimate actual population size, we are confident that it provides a good index of population size'. This may well be correct, but an index is only useful for comparing two or more areas, times, categories of animals, etc., and it is only valid for this purpose if the proportionality constant connecting the actual population size to the index remains the same for each of the areas, times, etc. This crucial assumption must be tested or justified if an index is to be used, but it is very hard to do so if only the index is available (Pollock, 1995).

The second problem is to define the population for which the estimate is required, and to ensure that the experimental or survey design estimates the size of this population, and not just a subset of it. Population definition involves defining both the spatial boundaries of the population, and the size, age or developmental stage classes that are to be included in the population. Note that, for population estimation problems, the statistical population is often not the same thing as the biological population. The statistical population is the total number of units (quadrats, transects, etc.) that could be sampled to estimate population size, not the number of animals in the total population.

Frequently, some classes of animals will be poorly sampled, or not sampled at all, with a particular sampling design. These may be individuals of particular size or age classes, or those living in particular places. For example, it is common that only a proportion of any given population is trappable. Often, investigators may recognize these problems, but will defend the chosen estimation method as nevertheless providing an index of population size.

Subsampling methods: quadrats, belt and line transects

The most elementary method of estimating population size in a given area is to count the individuals in a number of subareas, providing an estimate of the average number per unit area. The requirement for this approach to work is that organisms should be visible, and that they should not move before they can be counted. Within this overall approach, two sets of decisions need to be made. First, it is necessary to decide on a sampling strategy: how should the total sampling effort be distributed between subareas, and how should these be distributed across the total study area? Second, decisions also need to be made about the shape of the subareas: approximately square (quadrats); long and thin (transects); or lines.

Sampling strategies

The fundamental objective of any sampling strategy is to produce a sample that is representative of the population under investigation. There is a large body of theory and a substantial literature concerning sampling strategies. The classic reference is Cochran (1977). An excellent, more recent review can be found in Thompson (1992), which is the source for all formulae, etc. in this section that are not specifically attributed elsewhere. A simpler and more ecologically oriented review is provided by Krebs (1989). Andrew and Mapstone (1987) provide a review oriented towards marine estimation problems.

Sampling strategies were primarily developed for market research. It is fair to say that the vast majority of ecologists are quite unaware of the range of methods that exists, despite the potential value of many of the approaches in ecological sampling. Part of the problem is confusion with terminology. For example, two-stage sampling is totally different from double sampling, despite the similarity in their names. Table 3.1 provides a guide to the main approaches discussed in the following sections.

Table 3.1 Selecting a sampling strategy for estimating population size. Throughout this table, the term 'quadrat' is used for the sampling unit

Strategy	Principle	Applications	Advantages	Limitations and disadvantages
Simple random sampling without replacement (SRS)	Quadrats selected at random. Each quadrat in study area has equal probability of being in sample	Best for relatively uniformly distributed organisms	Simplicity	Relatively poor for patchily distributed organisms. High risk of unconscious bias in selecting random points, when using imprecise navigation
Systematic sampling (SystS)	Quadrats selected on a regular grid or line	As for SRS, to which it is roughly equivalent in most cases	Very simple. Sample covers whole study area, and is resistant to bias in selecting locations	May fail if there is periodicity in distribution of organism. Disliked by statisticians
Stratified random sampling (StrRS)	Study region divided into strata of known size, chosen so that variation between strata is maximized, and variation within strata is minimized	Patchily distributed organisms, where patchiness is predictable from mappable environmental characteristics	If patchiness is predictable, estimate is considerably more precise than that produced by SRS or SystS	Slightly more complex than SRS. Areas of strata must be known. Does not help with unpredictable patchiness
Cluster sampling (CS)	Primary points located at random. Cluster of quadrats located in fixed conformation around these. Cluster designed so that variability within it is maximized	Useful if cost of locating random quadrats is high compared with cost of sampling quadrats. Can help with unpredictable patchiness	Less chance of unconscious bias in quadrat location than SRS. Often produces better precision for same effort than SRS	If number of primary points is too small (< 10), confidence limits may be wide. Open to the error of treating each quadrat as an independent unit
Adaptive sampling (AS)	Initial quadrats selected at random. If population density above a threshold, neighbouring quadrats sampled too	Patchily distributed organisms, where locations of patches cannot be predicted before sampling	Major gains in precision possible relative to SRS	Highly complex, and performance in real situations not fully understood. Potential gains dependent on quadrat size and threshold rule used

Table 3.1 *contd*

Strategy	Principle	Applications	Advantages	Limitations and disadvantages
Two-stage sampling (2Stg)	Initial sample selected using SRS. From each unit selected , a sample of subunits is taken, also by SRS.	Useful if enumerating the population of an entire primary unit is not feasible. Size of primary unit usually not under investigator's control (e.g. trees, ponds)	Major gains in precision possible if primary units selected with probability depending on size	Size of all primary units in the population must then be known
Ratio estimation (REst)	Sample selected using SRS. Auxiliary variable x recorded in addition to population size. Ratio of y to x used to form estimate	Auxiliary variable usually some feature of environment that is easily measurable, and to which population size is roughly proportional	If population size y is nearly proportional to auxiliary variable, major gains in precision relative to SRS	Mean value of auxiliary variable over total study area must be known. Small problem with bias
Double sampling for stratification (DStrRS)	Proportion of study area falling into strata determined from sample selected using SRS. Population size determined from a subsample from each stratum	As for StrRS, but possible even if stratum areas cannot be determined in advance	As for StrRS	Only helpful if allocating samples into strata is easier than measuring population size itself
Double sampling for ratio estimation (DREst)	Initial sample taken using SRS. Auxiliary variable measured on this sample. Subsample taken from which population size is recorded too. Then proceed as for REst	As for REst, but can be used if auxiliary variable overall mean is unknown. Can also be used to calibrate or ground-truth survey methods	As for REst	Only helpful if it is cheaper and easier to estimate auxiliary variable than population size, and if there is a sufficiently strong relationship between auxiliary and population size

Simple random sampling

As its name suggests, this is the simplest possible sampling design, and one with which almost all biologists are familiar. It is rarely the case, however, that simple random sampling is the best design for a given problem. In a simple random sample, each element in the statistical population has an equal probability of appearing in the sample. The most straightforward way to ensure that a sample is random is to number all individuals in the population, and then to use a random number generator to select a sample of the required size from that list. This will rarely be possible in practice. However, if the sampling problem is one of selecting locations for transects or quadrats, essentially the same thing can be done by placing a coordinate system over a map of the study area, and then selecting coordinates from a random number table. If the study area has an irregular outline, coordinate pairs outside the bounds are simply rejected. However, finding a particular point in the field from a map coordinate pair with sufficient accuracy to locate a quadrat is not a straightforward task. It has become easier recently with the development of Global Positioning System (GPS) satellite receivers, which can provide the coordinates of a point to an accuracy of 20–100 m. Nevertheless, the potential for unwitting bias when quadrat dimensions are considerably smaller than the accuracy with which points can be located is very high.

Technically, the study area should be gridded into sampling units before sampling commences, and a unit should be sampled if a random coordinate pair falls within that unit. It is only in this way that independence of sampling units and non-overlap can be guaranteed (Caughley & Sinclair, 1994, p. 198). However, if the sampling intensity is low (that is, if only a small proportion of the total area is sampled), the practical consequences of taking random points and constructing quadrats around them will be small. It is worth mentioning in this context that the precision of an estimate depends primarily on the sample size, and provided less than 10% of the population is sampled, precision depends only very weakly on the proportion of the total population sampled.

In a simple random sample, a 95% confidence interval for the true population density μ is given by the familiar formula

$$\mu = \bar{y} \pm t_{0.05}\sqrt{\frac{s^2}{n}}, \qquad (3.1)$$

where $\bar{y}$ is the mean density per sampling unit, n is the number of units sampled, s^2 is the variance in density between units, and $t_{0.05}$ is the critical point of the two-tailed t distribution with a significance level of $\alpha = 0.05$ and $n - 1$ degrees of freedom. Provided n is reasonably large (greater than 10), $t_{0.05}$ will not change much with sample size (see Fig. 3.1). Maximizing the overall precision is thus a matter of either increasing sample size or decreasing s^2.

If a population is distributed randomly throughout its environment, quadrat size or shape is immaterial. The larger quadrats are, the more they 'average' density, and the smaller will be the variance in mean density between quadrats. If the same total area is included in the quadrats, this decline in variance will exactly match the effect of reduction in n on the standard error. Provided

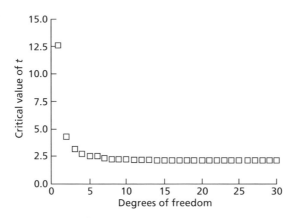

Fig. 3.1 The critical value of t, for a two-tailed test with $\alpha = 0.05$, as a function of degrees of freedom.

n is large enough that the critical value of t does not change much, it is the total area sampled that will determine the precision of the estimate, and the division of this into quadrats can be made following purely logistic considerations. However, very few organisms are distributed randomly. Almost all populations are distributed patchily, and it is designing appropriate sampling methods in the face of this patchiness that is the main problem in sampling design.

Some patchiness is predictable. Characteristics of the habitat can be used to give at least some indication of the likely population density. Provided these characteristics are mappable, so that the relative areas of each habitat type can be determined, stratified sampling will almost always give large gains in precision. In addition to this predictable patchiness, there will be unpredictable patchiness within strata. This is harder to handle. Adaptive sampling designs may be valuable, and the size and shape of quadrats or transects should be selected according to the 'grain' or scale of such patchiness.

A major problem in ecological sampling is that simple random sampling is often very difficult in the field. Even if random points can be located, the cost per sampling unit (whether in effort, time or money) of finding a position and moving to it is likely to be very high compared with the cost of actually taking the sample. For this reason, ecologists often use sampling strategies in which each individual sampling point does not need to be located at random.

Stratified random sampling

If at least some of the patchiness in population density is predictable from environmental properties, then major gains in precision can be made by dividing the total area for which the population estimate is required into strata. Strata are simply subregions of known area into which samples are allocated separately. Within strata, simple random sampling or most of the methods described later in this chapter can be used. The basic idea of stratification is that as much of the variation in the data set as is possible should be *between* strata, thus minimizing the amount of variation *within* strata. As both the estimated mean and its variance are weighted averages of the within-stratum mean and variance, minimizing variation within strata maximizes the precision of the overall estimate.

Stratification is almost always worthwhile. It is worth doing if there is even a suspicion that population density varies between subareas. If a randomly distributed population is incorrectly apportioned into strata, the resulting estimate will be only slightly worse than one based on simple random sampling over the entire area. Provided the relative areas of strata are known, the correct formulae will ensure that the resulting estimates are unbiased. They will not, of course, be of minimum variance unless the optimum stratification is used. Stratification is frequently valuable even if it is carried out after the sample is collected. The critical issues that need to be discussed are thus: (i) the formulae necessary to calculate overall means and standard errors from stratified data; (ii) deciding on the appropriate number of strata and allocation of effort between them; and (iii) post-sampling stratification.

Estimators of mean and variance in stratified random samples The estimated mean density and its estimated variance in a stratified design are simply the average of the within-strata values, weighted by the proportion of the total area within each stratum.

Suppose the total study area is A units, and that the area of the hth stratum is a_h, and that there are L strata. The sample size within the hth stratum is n_h, and the estimated mean and variance within the hth stratum, calculated by standard formulae, are $\bar{y}_h$ and s_h^2. The stratum weight for the hth stratum is then

$$w_h = \frac{a_h}{A},$$
(3.2)

and an estimate of the overall population mean density is

$$\bar{y}_{st} = \sum_{h=1}^{L} w_h \bar{y}_h$$
(3.3)

with an estimated variance of

$$\text{var}(\bar{y}_{st}) = \sum_{h=1}^{L} w_h^2 \frac{s_h^2}{n_h}.$$
(3.4)

Equations (3.3) and (3.4) neglect the finite-population correction, a modification which should be used when sampling without replacement and if the samples are an appreciable proportion of the total number of units in each stratum. With this correction, the formulae become:

$$\bar{y}_{st} = \frac{1}{N}\sum_{h=1}^{L}N_h\bar{y}_h \tag{3.5}$$

and

$$\text{var}(\bar{y}_{st}) = \sum_{h=1}^{L}\left(\frac{N_h}{N}\right)^2\left(\frac{N_h - n_h}{N_h}\right)\frac{s_h^2}{n_h}. \tag{3.6}$$

Here, N is the total number of units ('quadrats') in the population ('area') being sampled, and N_h is the total number of units in the hth stratum (in contrast to n_h, the number of units in the *sample* taken from the hth stratum).

These can be used in the usual way to calculate a confidence interval, provided the sample size in each stratum is reasonably large. If not, a rather messy formula for the approximate degrees of freedom d needs to be used:

$$d = \left(\sum_{h=1}^{L}q_h s_h^2\right)^2 \bigg/ \left[\sum_{h=1}^{L}(q_h s_h^2)^2 \bigg/ (n_h - 1)\right], \tag{3.7}$$

where $q_h = N_h(N_h - n_h)/n_h$. If the finite-population correction can be neglected, a marginally simpler form for d is:

$$d = \left(\sum_{h=1}^{L}\frac{w_h^2}{n_h}s_h^2\right)^2 \bigg/ \left[\sum_{h=1}^{L}\left(\frac{w_h^2}{n_h}s_h^2\right)^2 \bigg/ (n_h - 1)\right]. \tag{3.8}$$

Box 3.1 shows an example using these formulae.

Allocation of sampling effort For standard stratified sampling to work, it is necessary to know the proportion of the total population of units to be sampled that falls within each stratum. If the problem is to estimate total population size or density in a specified area, this means that the strata must be able to be mapped. A method known as double sampling can be used if the proportions are not known – see Thompson (1992) and later in this chapter.

There is little formal statistical guidance available concerning the number of strata to include in a stratified sampling design. General advice seems to be that there should not be too many, with a maximum of perhaps six strata (Krebs, 1989, p. 223). Iachan (1985) provides a way of calculating stratum boundaries that may be helpful in some cases.

Given that the number of strata has been decided on, there are well-established rules for deciding how sampling effort should be allocated. The optimum allocation is quite different from the principle of simple random sampling that each individual sampling unit in the total population should

Box 3.1 Stratified random sampling

This example is based on simulated data. A study area 20 units long by 20 units wide is divided into 400 cells or quadrats: 20% of the area (80 cells) forms a high-density stratum, 30% of the area (120 cells) is a medium-density stratum, and the remaining 50% of the area (200 cells) is a low-density stratum. The objective of the exercise is to estimate the mean density of animals per cell. The following table shows the actual mean densities and standard deviations per stratum, for the full population from which the data were sampled. The actual mean density is thus 11.0195, and the coefficient of variation (the standard deviation divided by the mean) is approximately 50% in each stratum. The data were generated using a normal distribution, rounded to the nearest integer, or to 0 for negative values.

Stratum	1	2	3
Mean	29.76	9.325	4.54
Standard deviation	16.693	5.011	2.399
Number of cells	80	120	200

Initially, suppose that a sample of 50 units is taken at random from the entire study area, and that which stratum these belonged to was determined only after the unit had been selected. Here are the means, variances and units per stratum for one such sample. Also shown are, the stratum weights N_h/N; the finite-population correction, which is the proportion of each stratum not sampled $(N_h - n_h)/N_h$; and q_h (see eqn (3.7)).

Stratum h	1	2	3
Mean $\bar{y}_h$	25.266 67	9.764 706	4.722 222
Variance s_h^2	259.209 5	25.566 18	5.624 183
Sample size n_h	15	17	18
Stratum size N_h	80	120	200
Stratum weight w_h	0.2	0.3	0.5
Finite-population correction $\left(\dfrac{N_h - n_h}{N_h}\right)$	0.812 5	0.858 333	0.91
$q_h = N_h(N_h - n_h)/n_h$	346.666 7	727.058 8	2022.222

Using these, it is straightforward to calculate an estimate of the overall mean density with eqn (3.5),

$$\bar{y}_{st} = \frac{1}{400}\{80 \times 25.667 + 120 \times 9.764 + 200 \times 4.722\} = 10.424;$$

continued on p. 58

Box 3.1 *contd*

and its estimated variance, using eqn (3.6),

$$\text{var}(\bar{y}_{st}) = \left\{0.2^2 \times 0.8125 \times \frac{259.21}{15} + 0.3^2 \times 0.8583 \times \frac{25.566}{17} + 0.5^2 \right.$$
$$\left. \times 0.910 \times \frac{5.624}{18}\right\} = 0.74888.$$

The degrees of freedom are given by eqn (3.7):

$$d = \frac{(346.67 \times 259.21 + 727.06 \times 25.566 + 2022.2 \times 5.6241)^2}{\left[\frac{(346.67 \times 259.21)^2}{15-1} + \frac{(727.06 \times 25.566)^2}{17-1} + \frac{(2022.2 \times 5.6241)^2}{18-1}\right]} = 23.69.$$

The critical value of t with 23 degrees of freedom and $\alpha = 0.05$ (two tailed) is 2.069. Hence, the 95% confidence interval for the mean density is $10.424 \pm 2.069\sqrt{0.74888}$. These calculations can most easily be done using a spreadsheet.

The following table compares the 95% confidence interval thus obtained with the interval obtained from exactly the same sample using simple random sampling formulae, and also with another sample of 50 units taken using sampling effort proportional to the total population in each stratum. In this case, post-sampling stratification has produced a major gain in precision. This will often, but not always, be the case. The particular optimally allocated sample used here is, however, only marginally better than the particular randomly allocated sample. In general, the benefits of approximately optimal allocation are substantial:

	$\bar{y}$	$t \times se$	Lower 95%	Upper 95%
Simple random sample	10.488 89	2.910 65	7.578 239	13.399 54
Post-sample stratification	10.423 6	1.790 468	8.633 132	12.214 068
Optimal stratification	11.449 8	1.626 964	9.822 833	13.076 76

The mean, variance and sample size for the optimally allocated sample used in the above table are:

Stratum	1	2	3
Mean	30.210 53	9.692 308	5.000
Variance	283.731	30.397 44	7.777 778
Sample size	27	13	10

have an equal probability of turning up in the sample. The general rule is that the variance of the overall estimate is minimized if sampling effort is concentrated in larger and more variable strata. Given that variance in population density is usually roughly proportional to the mean density squared – Taylor's power law, Taylor (1961) – a good rule of thumb in estimating population size is to allocate sampling effort proportionally to the estimated fraction of the total population within each stratum. Clearly, this requires some preliminary idea of the answer to the question you are seeking, but the reasoning is not circular because a mistake will not bias the final result.

More formally, if the total sample size available to be distributed is n, and the population standard deviation in the hth stratum is σ_h, then the optimum allocation is

$$n_h = \frac{n w_h \sigma_h}{\sum\limits_{k=1}^{L} w_k \sigma_k}. \tag{3.9}$$

Here, w_h is as defined in eqn (3.2). To use eqn (3.9) in practice, some guess would need to be made about the value of the σ_h.

It is also possible to optimize allocation subject to a fixed cost, in cases where the sampling cost may differ between strata (Thompson, 1992).

Post-sampling stratification. In some cases, the stratum to which an observation belongs may not be known until the sample is taken. This is particularly likely in marine sampling, where the characteristics of the bottom at a sampling location may not be known until the bottom is reached. Nevertheless, stratification may still be worthwhile, provided the relative areas of each bottom type are known. The sample sizes n_h in each stratum are random variables in post-sampling stratification, as they are a result of the sampling itself. The estimated mean density $\bar{y}_{st}$ is still given by eqn (3.3) and the variance of $\bar{y}_{st}$ is still given by eqn (3.4) or eqn (3.6). The sample allocation will not, of course, be optimal, except by chance.

Systematic sampling

Many ecological surveys use systematic samples, in which transects or quadrats are taken a fixed distance apart, often on a grid. There are good practical reasons for using a strategy of this type. Samples are certain to be spread across the entire survey area, and moving from one sampling location to another is easy. Location of quadrats or transects is straightforward and much less likely to be affected by unintentional bias than is location of 'random' sampling positions with an imprecise navigation system.

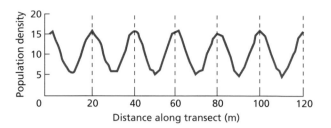

Fig. 3.2 The possible effect of periodicity in density on a systematic sample. In this example, samples are always taken 20 metres apart along a transect, in the positions shown with dashed lines. Sampling locations are always close to the maximum level of the periodically varying density, leading to a severely biased sample.

Statisticians tend to be wary of systematic sampling (Cochran, 1977; Thompson, 1992). There is certainly a substantial problem if there is any form of periodicity in the data, particularly if the periodicity is close to a multiple or submultiple of the sampling interval (Fig. 3.2). If there is a trend running across the survey area, the estimate will be biased, but this effect will probably be small. In most cases, a systematic sample can be treated as a random sample without too many problems (Krebs, 1989; Caughley & Sinclair, 1994). For a more technical treatment of systematic sampling, see Bellhouse (1988). Many of the practical advantages of systematic sampling, without the disadvantages, can be gained by using cluster sampling (see below).

Adaptive sampling

Almost all biological populations are aggregated, but a substantial proportion of that aggregation is difficult to predict from properties of the habitat. For example, the crown of thorns starfish, *Acanthaster planci*, occurs in dense aggregations on coral reefs, causing massive damage to corals (Moran, 1986), but at any given time will be absent or rare between aggregations. Any conventional sampling technique will be extremely inefficient at estimating population density. Ecologists have often addressed patchiness by locating aggregations in an initial survey, and then estimating the population size or density within aggregations – see, for example, Endean and Stablum (1975) for *Acanthaster*. Such approaches are understandable, but do not yield an unbiased estimator of population size.

Recently, an adaptive sampling approach has been developed, which appears to allow ecologists to do what they have always wanted: to sample more intensively in areas where aggregations have been located (Thompson, 1990; 1991a; 1991b; 1992). The basic approach is as follows:

1 Divide the area to be sampled into units (usually on a grid pattern).
2 Take a random sample of n primary units.

3 If the population density in a primary unit is above some threshold (often zero), then sample all adjacent units.

4 If any one of these exceeds the threshold, then in turn sample all units adjacent to it.

5 Continue the process until all adjacent units are below the threshold (these units are called edge units).

The final sample thus consists of a series of networks or clusters of varying sizes, which will each contain at least one primary unit.

The problem is to derive an unbiased estimator of the total population size and an estimate of its variance from these data. There are several ways in which this might be done (Thompson, 1990). The one that appears usually to be the best relies on a modification of the principle that an unbiased estimator for a population total can be obtained by dividing the y value of each unit by the probability of its inclusion in a sample (the Horwitz–Thompson estimator).

The following recipe for calculating the estimated population size and its variance is for a primary sample taken with simple random sampling without replacement, and is based on Thompson (1990; 1992), which should be consulted for the theory underlying the method. Thompson (1992) also includes estimators for more complex cases, such as systematic samples or stratified designs.

A network is a cluster of sampling units, such that, if any one of them is included in the original sample, the inclusion rule will ensure that all the rest are as well. Note that this means that a primary unit with a population size below the threshold is a network of size 1. Define K to be the total number of networks in the population (this will not be known in a real sampling problem). Let Ψ_k be the set of units that make up the kth network, and x_k be the number of units in this network. Then the probability α_k that any one of the units in the kth network will be used in the estimator is

$$\alpha_k = 1 - \binom{N - x_k}{n} \Big/ \binom{N}{n}. \tag{3.10}$$

Let the indicator variable z_k equal one if any unit of the kth network is included in the initial sample, and let z_k equal zero otherwise. Then the estimated average value of y per unit is given by

$$\hat{\mu}_2 = \frac{1}{N} \sum_{k=1}^{k} \frac{y_k^* z_k}{\alpha_k}, \tag{3.11}$$

in which y_k^* is the total of the y values in the kth network. Note that the indicator variable means that the y_k^*/α_k values should simply be summed for those networks which have been sampled, and that edge units around networks (those that do not reach the threshold) should not be included.

The estimated variance for $\hat{\mu}_2$ is rather messy. Defining α_{kh} as the probability that the initial sample contains at least one unit in each of the networks k and h,

$$\alpha_{kh} = 1 - \left\{ \left[\binom{N-x_k}{n} + \binom{N-x_h}{n} - \binom{N-x_k-x_h}{n} \right] \right\} \bigg/ \binom{N}{n},$$ (3.12)

the estimated variance is

$$\text{var}(\hat{\mu}_2) = \frac{1}{N^2} \left[\sum_{k=1}^{K} \left(\frac{1}{\alpha_k^2} - \frac{1}{\alpha_k} \right) y_k^{*2} z_k + \sum_{k=1}^{K} \sum_{h \neq k} \left(\frac{1}{\alpha_k \alpha_h} - \frac{1}{\alpha_{kh}} \right) y_k^* y_h^* z_k z_h \right]$$ (3.13)

Equation (3.13) is quite daunting, but the indicator variables mean that it has only to be calculated for those networks h and k that do occur in the sample. A worked example is shown in Box 3.2.

Box 3.2 Adaptive sampling

The following example uses data collected by Smith *et al.* (1995) and used in a simulation study of the efficiency of adaptive sampling. The data are a full enumeration of ring-necked ducks in a 5000 km² area of central Florida. The area was surveyed by air, and the data shown below are essentially complete counts of the ducks in two hundred 25 km² quadrats, arranged in a 10 × 20 grid.

	1	2	3	4	5	6	7	8	9	10	11	12	13	14	15	16	17	18	19	20
1	0	0	20	0	0	0	0	0	0	0	0	0	0	0	0	0	675	0	0	0
2	0	0	0	0	100	100	75	0	0	0	0	0	0	0	0	0	0	0	0	0
3	0	4000	13 500	0	0	154	120	200	0	0	0	0	0	0	0	0	0	0	55	0
4	0	0	0	0	0	80	585	430	0	4	0	0	0	0	0	35	0	0	0	0
5	0	0	0	0	0	0	0	0	40	0	0	0	0	2	0	0	0	1615	0	0
6	0	0	0	0	0	0	0	0	57	0	0	0	0	2	0	0	0	200	0	0
7	0	0	0	0	0	0	0	0	0	0	0	0	0	0	0	0	0	0	0	0
8	0	0	0	0	0	0	0	0	0	0	0	0	0	0	0	0	0	0	0	0
9	0	0	0	0	0	0	0	0	0	0	0	0	0	0	0	0	0	0	1141	13
10	0	0	0	0	0	0	0	0	0	0	0	0	0	0	0	0	0	0	107	22

continued on p. 63

Box 3.2 *contd*

It is immediately obvious that the ducks are very strongly clustered. The overall mean abundance is 116.67 per quadrat, with a standard deviation of 1001. The following diagram shows an initial random sample of 20 quadrats (darker shading), and the adaptive sample (lighter shading) that results if all quadrats are counted that share an edge with any quadrat already found to have a nonzero count.

New example

	1	2	3	4	5	6	7	8	9	10	11	12	13	14	15	16	17	18	19	20
1	0	0	20	0	0	0	0	0	0	0	0	0	0	0	0	0	675	0	0	0
2	0	0	0	0	100	100	75	0	0	0	0	0	0	0	0	0	0	0	0	0
3	0	4000	13500	0	0	154	120	200	0	0	0	0	0	0	0	0	0	0	55	0
4	0	0	0	0	0	80	585	430	0	4	0	0	0	0	0	35	0	0	0	0
5	0	0	0	0	0	0	0	0	40	0	0	0	0	2	0	0	0	1615	0	0
6	0	0	0	0	0	0	0	0	57	0	0	0	0	2	0	0	0	200	0	0
7	0	0	0	0	0	0	0	0	0	0	0	0	0	0	0	0	0	0	0	0
8	0	0	0	0	0	0	0	0	0	0	0	0	0	0	0	0	0	0	0	0
9	0	0	0	0	1	0	0	0	0	0	0	0	0	0	0	0	0	0	1141	13
10	0	0	0	0	0	0	0	0	0	0	0	0	0	0	0	0	0	0	107	22

Four networks, shown outlined with thick lines, are included in the final adaptive sample. The number of cells in each, x_k, together with the total count in each network, y_k, are shown below. Note that edge cells, that is, cells that were counted because they were adjacent to cells in the network, but which were found to be empty, are not included. (Zero cells would not always be omitted: if network cells surrounded them, they would form part of the network. This particular population contains no such cells.)

Cluster	1	2	3	4
y_k	1844	4	1815	1
x_k	9	2	2	1

continued on p. 64

Box 3.2 *contd*

To calculate the estimated mean density, using eqn (3.11), it is first necessary to use eqn (3.10) to calculate α_k, the probability that the kth sampled network would be used in any single sample of this size:

$$\alpha_k = 1 - \binom{N - x_k}{n} \bigg/ \binom{N}{n}.$$

Here, $N = 200$ and $n = 20$, and thus $\binom{N}{n} = 1.613\ 59 \times 10^{27}$. All these calculations can be done fairly easily using a spreadsheet. The intermediate calculations are shown in the following table. Note that, as one would expect, the chance of network 4, which is a single cell, being included in the sample is simply the initial sampling fraction, 0.1.

Network	1	2	3	4
$\binom{N - x_k}{n}$	6.123 8E + 26	1.306 3E + 27	1.306 3E + 27	1.452 2E + 27
α_k	0.620 485 93	0.190 452 26	0.190 452 26	0.1
y_j^* / α_k	2971.864 32	21.002 638 5	9529.947 23	10

The estimated density is simply the sum of the final row divided by N. Thus,

$\hat{\mu}_2 = 62.614$.

To calculate the estimated variance, it is necessary to calculate α_{kh}, the probability of any pair of these networks being included in a given sample, using eqn (3.12). These probabilities are shown in the following triangular matrix:

Network	2	3	4
α_{1k}	0.114 946 84	0.114 946 84	0.060 260 31
α_{2k}		0.034 793 95	0.018 227 5
α_{3k}			0.018 227 5

Next, a weighted sum of the y_k^{*2} and their cross-products needs to be calculated following eqn (3.13). The components of this equation are below:

continued on p. 65

Box 3.2 *contd*

Network	1	2	3	4
$\left(\dfrac{1}{\alpha_k^2}-\dfrac{1}{\alpha_k}\right)y_k^{*2}$	3351859.74	357.100 271	73523040	90
Cross terms				
$\left(\dfrac{1}{\alpha_1\alpha_k}-\dfrac{1}{\alpha_{1k}}\right)y_1^*y_k^*$		−1751.79861	−794878.62	−881.931 109
$\left(\dfrac{1}{\alpha_2\alpha_k}-\dfrac{1}{\alpha_{2k}}\right)y_2^*y_k^*$			−8502.920 2	−9.422 236 33
$\left(\dfrac{1}{\alpha_3\alpha_k}-\dfrac{1}{\alpha_{3k}}\right)y_3^*y_k^*$				−4275.339 73

The variance is then the sum of all elements in this table divided by N^2, $var(\hat{\mu}_2) = 1901.626$, the square root of which is the standard error, $se(\hat{\mu}_2) = 43.608$.

These results should be compared with the actual population mean of 116.67, and the standard error that would be expected from a simple random sample with the same number of cells counted (48), which is 144.55. This particular sample appears to have considerably underestimated the true mean, although the true mean lies well within two estimated standard errors of the estimated mean. Note, however, that the single highest-density cell (13 500) contributes 67.5 to the overall mean density, and that it is not included in this particular adaptive sample. As that cell is in a network of 2, its chance of being sampled in an adaptive sample of 20 initial random cells is 0.19.

Extensive simulations by Smith *et al.* (1995) suggest that adaptive sampling is actually little better than simple random sampling for this particular population. It is so highly aggregated that without using some information on the cause of this aggregation to guide sampling, both adaptive sampling and simple random sampling perform very poorly. However, if the proportion of suitable habitat in each cell is used to direct sampling intensity, the adaptive method provides significant further gains.

The efficiency of adaptive cluster sampling relative to more conventional sampling regimes was examined in a simulation study by Smith *et al.* (1995). They enumerated three species of ducks in an area of 5000 km² in Florida, and then simulated sampling using adaptive and simple random designs. Two quadrat sizes, two thresholds and a range of sampling intensities were investigated. In some cases, adaptive sampling produced major gains in efficiency when compared with simple random samples with the same number of units sampled, but this was not invariably the case. Quadrat size and the threshold had major impacts on efficiency, but no single combination was best for each of the three species investigated. In some cases, adaptive sampling was actually worse than simple random sampling, although this comparison was in terms of quadrats sampled, and in most cases it would be cheaper to use adaptive sampling than to enumerate the same number of quadrats using random sampling.

In conclusion, the approach has significant potential, but appropriate guidelines for threshold rules, quadrat size and initial sample size have yet to be developed fully. It is applicable only to cases in which the entire area to be sampled can be divided into grid cells before sampling commences, and where locating a particular grid on the ground is not a problem. The method was originally described for square sampling units, but should work equally well with hexagonal units, or any other equilateral unit that tiles a plane. It would not, however, work easily for strip transects, unless it happened that the species being sampled could be expected to exist in approximately linear patches with the same orientation as the transects. For example, it could be adapted to sample a species that lived patchily along a river.

The nature of the method is such that one would expect it to be most efficient relative to simple random sampling if most patches were several sampling units in size. If the species is invariably found in groups, an alternative approach of broad-scale sampling to locate patches, followed by estimation of patch size (a two-stage sample, see below) should also be considered.

Cluster sampling

This involves choosing primary units, usually at random, and then measuring all secondary units that are part of that primary sample. Secondary units might, for example, be a series of quadrats in some fixed conformation around a randomly selected point. Formulae for means and variances are simply obtained by averaging over all secondary units for each primary unit, and treating those means as data for a simple random sample. Cluster sampling is worth considering if the cost of locating random points is high (as is often the case in ecological sampling). A cluster sample with M secondary units in each of n primary units will actually return a more precise estimate than a simple random sample with nM units if there is a negative correlation between population densities within clusters (Thompson, 1992). In ecological terms, this means that, if clusters can be designed so that they sample over the grain of the distribution of the organism, rather than within patches, substantial gains in precision may be possible. One proviso with using cluster sampling is that the number of primary units should not be so small that the critical value of t used for constructing confidence intervals becomes too large. However, as can be seen from Fig. 3.1, it is only if the number of primary units is very small indeed that this is a problem.

Cluster designs are quite commonly used in ecological sampling. For example, a series of quadrats may be taken at fixed intervals on a number of transect lines. However, people are often quite unaware that the design is a cluster design, and may often analyse it incorrectly, using the quadrats as the sampling unit. For this reason, the 'quadrats along transects' design is listed as an example of 'how not to sample' by Caughley and Sinclair (1994). As they state themselves, if the analysis is done using transects as units, there is no problem with lack of independence, and little information is lost.

Given the costs and possibilities for unconscious bias in locating large numbers of random quadrats in most landscapes, cluster samples using a modest number (perhaps 20) of randomly selected starting points, with quadrats in a fixed conformation around each, should be considered more often in ecological studies.

Two-stage designs

These are similar in some ways to cluster designs, in that there are primary units, and secondary units within them. However, subsamples within primary units are selected randomly. Designs of this type are very common in ecology. For example, crustacea may be counted in subsamples that are taken from a series of plankton tows, or leaves may be subsampled from trees, and the number of leaf miners per leaf might be counted.

In a two-stage design with simple random sampling at each stage, deriving the estimated total population is fairly straightforward, provided that the total number of primary units is known and that the total number of secondary units in each of the *sampled* primary units is known (Thompson, 1992). Satisfying these two conditions, however, will often not be trivial. Suppose the total number of primary units is N, and the number of secondary units in the ith primary unit is M_i. Of these, n primary units are sampled, m_i secondary units are sampled in the ith selected primary unit, and the population size y_{ij} in each of these is determined.

The total estimated population size in the ith primary unit, $\hat{y}_i$, is simply the observed sum of all counts of secondary units in the ith unit divided by the fraction of all subunits that was sampled,

$$\hat{y}_i = \frac{M_i}{m_i} \sum_{j=1}^{m_i} y_{ij}; \tag{3.14}$$

and the estimated total population $\hat{\tau}$ is simply these quantities, summed over all primary samples taken, divided by the fraction of all primary units that actually was sampled,

$$\hat{\tau} = \frac{N}{n} \sum_{i=1}^{n} \hat{y}_i. \tag{3.15}$$

The estimated variance of this quantity contains a component due to variation between subunits, and a component due to variation within subunits:

$$\text{var}(\hat{\tau}) = N(N-n)\frac{s_u^2}{n} + \frac{N}{n} \sum_{i=1}^{n} M_i(M_i - m_i)\frac{s_i^2}{m_i}, \tag{3.16}$$

where the estimated between-primary-unit variance is

$$s_u^2 = \frac{1}{n-1} \sum_{i=1}^{n} (\hat{y}_i - \hat{\mu}_1)^2 \tag{3.17}$$

and the estimated within-primary-unit variance for unit i is

$$s_i^2 = \left(\frac{1}{m_i - 1}\right) \sum_{j=1}^{m_i} (y_{ij} - \bar{y}_i)^2. \tag{3.18}$$

In these,

$$\hat{\mu}_1 = (1/n) \sum_{i=1}^{n} \hat{y}_i \text{ and } \bar{y}_i = (1/m_i) \sum_{j=1}^{m_i} y_{ij}.$$

In many applications of two-stage sampling, the number of secondary units in each primary unit will vary considerably. In such cases, the precision of the estimates can be improved by sampling primary units with probability proportional to the number of secondary units they contain (probability proportional to size or PPS sampling). This approach is straightforward if the size of all units in the population is known. However, this may not often be the case in ecological sampling. If sizes are not known, the estimates may be biased. For further details, see Cochran (1977) or Thompson (1992).

Ratio estimation

The idea behind ratio estimation is that the variable of interest in a survey (for example, population size) may be approximately proportional to some other variable, the value of which is known for each sampling unit in the population. This second variable is called an auxiliary variable. Ratio estimation does not appear to have been used often in population estimation problems – with the exception of its use as a means for handling aerial transect lines of differing lengths (Caughley, 1977b) – but it has considerable potential. For example, population size or density will often be approximately proportional to the percentage cover of whatever vegetation an animal uses for shelter or food. The limitation of ratio estimation is that the auxiliary variable needs to be known, not just estimated, for every unit in the sample. For broad-scale surveys of terrestrial animal populations, remote sensing technology may well be able to provide a suitable auxiliary variable without too much difficulty. If an auxiliary variable can only be estimated, but can be done so with more precision or less cost than population size itself, it still may be helpful, but the appropriate method is double sampling (see below).

In ratio estimation, rather than estimating the mean population density μ with the sample mean $\bar{y}$, the ratio of y to the auxiliary variable x is estimated, and this is used to estimate μ, given that μ_x, the overall mean of the auxiliary variable, is known. The sample ratio of y to x is simply

$$r = \frac{\sum_{i=1}^{n} y_i}{\sum_{i=1}^{n} x_i} = \frac{\bar{y}}{\bar{x}} \qquad (3.19)$$

and the ratio estimate of the population mean of y is

$$\hat{\mu}_r = r\mu_x \qquad (3.20)$$

If the relationship between x and y is perfectly linear and passes through the origin, the ratio estimate will be perfect: x is assumed to be known, and a perfect correlation would mean that y was known too. In general, the tighter the correlation, the more precise the estimate.

Unfortunately, the ratio estimator is biased, although this bias is typically negligible if the sample size is large. As a rule of thumb, Cochran (1977) suggests that $n > 30$ and coefficients of variation for $\bar{x}$ and $\bar{y}$ less than 10% should be adequate.

There are several possible expressions for the variance of this estimate, the differences occurring because of this problem of bias. The following (Thompson, 1992, p. 61) is probably the best simple estimate available:

$$\text{var}(\hat{\mu}_r) = \left(\frac{\mu_x}{\bar{x}}\right)^2 \left(\frac{N-n}{n}\right) \left(\frac{\sum_{i=1}^{n}(y_i - rx_i)^2}{n(n-1)}\right). \qquad (3.21)$$

This can be used in the normal way to obtain a confidence interval. (Here N is the total number of sampling units in the population.)

Ratio estimation is closely related to stratification. If remote sensing was used to determine vegetation cover, the options would be either to use the cover per sampling unit as a basis for stratification, or to use it in a ratio estimate. If the relationship between the variables really was linear and passed through the origin, a ratio estimate would do a better job of reducing the error in the estimate than a stratification, which uses a relatively small number of discrete levels. However, the stratification could deal with any systematic relationship between cover and population density. The situation is analogous to comparing regression and analysis of variance. In fact, a method known as regression estimation is an elaboration of ratio estimation, and can in turn be elaborated to allow for any general linear relationship between population density and continuous or categorical auxiliary variables. This is beyond the scope of this book. For details, see Thompson (1992).

Double sampling

Ratio estimation and stratification may give major gains in precision, but the standard versions of both rely on perfect knowledge about some aspect of the population being sampled. In stratified sampling you need to know the areas of the strata, and in ratio estimation the value of the auxiliary variable is assumed to be known, not estimated, for every unit. Perfect knowledge of any aspect of the area being studied is a tall order in much ecological research. Double sampling is a technique that uses an initial survey to collect information about the stratification or auxiliary variable, and then uses a subsample to collect information on the variable of primary interest. It is obviously only of value if the auxiliary variable or stratification variable is easier to estimate than the primary variable, and if there is a reasonably strong relationship between the primary and other variables. Cochran (1977) provides a detailed discussion of optimum allocations, costs and benefits of double sampling. For most ecological applications, subjective judgement should be a good guide as to whether the benefits are worth the trouble.

In double sampling for ratio estimation, an initial sample of size n' is drawn at random from the total population of N sampling units. For each of these units, the auxiliary variable x is recorded. From this initial sample, a subsample of size n is drawn, also at random, and y is observed in addition to x. From the subsample, the sample ratio r can be estimated, using the standard formula (eqn (3.19)).

The population mean of the auxiliary variable can be estimated using the following formula, which is just the usual estimate for a mean:

$$\hat{\mu}_x = \frac{\sum_{i=1}^{n'} x_i}{n'}.$$

(3.22)

From this, the ratio estimate of the mean value of y can be obtained using an equation analogous to eqn (3.20):

$$\hat{\mu}_r = r\hat{\mu}_x. \tag{3.23}$$

The estimated variance of this is

$$\text{var}(\hat{\mu}_r) = \left(\frac{N-n'}{N}\right)\frac{s^2}{n'} + \left[\frac{n'-n}{n'n(n-1)}\right]\sum_{i=1}^{n}(y_i - rx_i)^2, \tag{3.24}$$

where s^2 is the sample variance of the y values in the subsample.

Thompson (1992) describes a hypothetical application of this method to ground-truthing aerial surveys for wildlife. An initial sample of survey blocks is surveyed from the air, the counts forming the auxiliary variable. A sub-sample of these is then taken and counted from the ground (a process he assumes occurs with negligible error). The ground counts are then the y variables, and the ratio of ground to aerial counts produces a correction factor for obtaining a total estimate.

Double sampling for stratification follows a similar procedure. It is only applicable to post-sampling stratification: if strata are already known, the process is redundant. A simple random sample of size n' is selected from the total population of N sampling units. These are then classified into L strata, using an appropriate objective and predetermined criterion. If n'_h samples fall into the hth stratum, the sample stratum proportion is

$$w_h = \frac{n'_h}{n'}, \tag{3.25}$$

and this estimates the population stratum weight W_h.

A subsample of n_h units is then taken from the n'_h units in the hth stratum of the initial sample. This will usually, but not necessarily, be an equal proportion from each stratum. The variable of interest y is then measured for each unit in the subsample. Sample means and variances $\bar{y}_h$ and s_h^2 can be calculated using standard formulae. The estimated population mean is exactly the same as in standard stratified sampling (eqn (3.3)):

$$\bar{y}_d = \sum_{h=1}^{L} w_h \bar{y}_h. \tag{3.26}$$

The estimated variance of this estimator is considerably more complex than the previous form, because additional error associated with estimating the stratum weights needs to be incorporated:

$$\text{var}(\bar{y}_d) = \left(\frac{N-1}{N}\right)\sum_{h=1}^{L}\left(\frac{n'_h-1}{n'-1} - \frac{n_h-1}{N-1}\right)\frac{w_h s_h^2}{n_h} + \left(\frac{N-n'}{N(n'-1)}\right)\sum_{h=1}^{L} w_h(\bar{y}_h - \bar{y}_d)^2. \tag{3.27}$$

Number, shape and orientation of quadrats or transects

The accuracy of a population estimate based on subsampling does not depend strongly on the overall proportion of the total population that is sampled, provided the survey is properly designed. Inspection of any of the formulae for standard errors earlier in the chapter shows that, for a given number of quadrats, quadrat size, mean density and variance per quadrat, the precision of an overall estimate is essentially the same whether 5% or 0.001% of the total population has been counted. It is sample size, rather than sampling fraction, which is important. Conversely, even if 50% of the total population has been counted, an unrepresentative sample will provide a poor estimate.

Any sampling design for estimating population density must seek to optimize accuracy, subject to a constraint of cost (whether in money or time). This truism is worth reiterating, because many ecologists tend to maximize degrees of freedom, with a implicit assumption that accuracy will thereby be optimized. However, maximizing the number of independent quadrats used in the estimate will not necessarily maximise the accuracy or precision of an estimate of population size within a constraint of fixed cost. As previously discussed, if a population is distributed randomly and if the total area that can be sampled is the constraint, precision gained by taking more, small random quadrats will be exactly balanced by increased variation in density between these smaller quadrats. Given that the cost of quadrat location is rarely negligible, overall precision would tend to be optimized by taking few large quadrats, rather than a large number of smaller ones. This conclusion needs to be qualified by three caveats. First, if the number of quadrats is too small (less than 20 or so), the increased value of t will make confidence limits wider. Second, if quadrats or strip transects are too large or wide, it may not be possible to search them properly. Missed individuals will obviously bias results downwards. Third, this simple conclusion assumes that the organisms are distributed randomly. As they usually are clumped, it is essential to consider the nature of their distribution.

The general principle concerning quadrat shape with patchily distributed populations is identical to that of cluster sampling. Quadrat shape and size should be such that variation *within* quadrats is maximized, and variation *between* quadrats is minimized. This means that quadrats should ideally be larger than the grain of the distribution, so that quadrats will usually combine some high-density areas, and some low-density matrix. It also means that long thin transects are usually preferable to a square quadrat of the same area (see Fig. 3.3). Thin quadrats are also usually easier to search without missing individuals or double counting. A disadvantage, however, of thin transects compared with square quadrats is that they have a higher ratio of edge to area. This means that there may be more problems with deciding whether objects

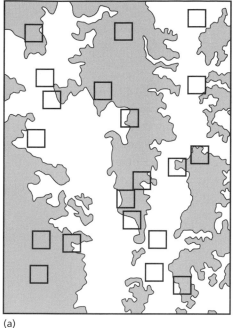 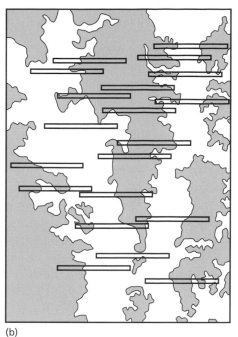

(a)   (b)

Fig. 3.3 The influence of quadrat shape on the variance of an estimated population density. The shaded areas represent areas of low population density, and the unshaded areas represent high population density. (a) Twenty square quadrats at randomly chosen locations. These quadrats tend to be entirely in areas of high density or entirely in areas of low density. The result is substantial variation between quadrats, and hence a high error variance. (b) Twenty strip transects of the same area as the square quadrats, also placed at randomly chosen locations, and aligned so that they run across the spatial pattern in the area being sampled. These are likely to include areas of both low and high density, and hence the mean population density will vary less between sampling units, increasing the precision of the estimated mean density.

are inside or outside the sampling unit. It is human nature to want to count objects that are on the edge of the sampling units, or to count mobile animals that look as if they might have been inside the transect before they moved. This may be an important source of positive bias. Variation between transects will be minimized if they run across the grain of dominant habitat features rather than parallel to them. For a more detailed discussion, see Krebs (1989, pp. 64–72).

Line transects

Rationale

The term 'line transect' can mean at least three quite different things:

1 A transect in which all individuals within a certain distance on either side of a line of known length are counted. This is more correctly called a strip transect, and does not differ fundamentally from any other quadrat.

2 A line-intercept transect, often used to determine percentage cover, for example of vegetation or coral. A line is laid and the percentage of the length of the line which crosses coral or vegetation is determined. This is a good way to determine percentage cover, because estimating proportions of area in quadrats is much harder to do objectively. Line-intercept transects are less suitable for estimating population size, although they have sometimes been used for this purpose (Kimura & Lemberg, 1981; Thompson, 1992).

3 A transect in which all individuals *visible* on either side of a line are counted, and their distance from the line is also recorded.

This section is concerned with this last type of line transect.

Strip transects have been the traditional means of estimating population size in large vertebrates. Observers move along straight lines of length L, counting all animals present, out to a distance w from either side of the line. This distance is the half strip width. If n animals are present in the strip, the density of the population D on the transect is simply the number seen divided by the area searched,

$$D = \frac{n}{2Lw}. \tag{3.28}$$

Selecting an appropriate strip width is not easy. If the width is too large, individuals towards the edge of the strip, furthest from the observer, will tend to be missed more often than individuals closer to the observer, biasing the estimate. On the other hand, if the strip is too narrow, many individuals that are seen will be beyond the strip boundary and unable to be counted. Information is thus wasted. The basic idea of a line transect is that the width of the strip is self-adjusting. The 'effective strip width' is determined by how rapidly the distribution of sightings drops off with distance from the line.

Line transect methods have gained increasing acceptance over the last few years, following the publication of an influential paper (Burnham *et al.*, 1980) and book (Buckland *et al.*, 1993), and the availability of associated software. Nevertheless, they have been dismissed as a fad by Caughley and Sinclair (1994) who cite the difficulty in estimating sighting distances and technical problems in fitting appropriate curves to data as fatal flaws. These certainly are not trivial difficulties, but their proposed solution to sighting difficulties in strip transects of using two observers on either side of a strip would at least halve efficiency (each observer would only count on one side) and introduce problems of double counting.

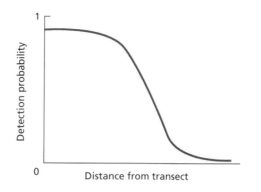

Fig. 3.4 The general shape of a sightability curve.

Method

Only the basic principles are outlined here. Anyone seeking to use this method should read Buckland *et al.* (1993) for further details, and will presumably use a computer package such as DISTANCE to perform the calculations.

Each transect is sampled by moving along a straight line of set length and recording, for each individual seen, the perpendicular distance it is away from the line (this is often determined from the sighting angle and sighting distance using elementary trigonometry). There is no necessary limit on how far individuals may be from the line, although it is usually convenient to set a maximum strip width. This information is sufficient to estimate abundance, provided certain assumptions are satisfied:

(i)   Animals on the transect line must always be seen.
(ii)   Animals do not move before being counted, and none is counted twice.
(iii) Distances and angles are measured without error.
(iv)   Sightings of individuals are independent events.

If you use this approach, then obviously some individuals, particularly those a long way from the line, will be missed. The probability of an individual being sighted, given that its distance away from the transect line is x, can be represented as $g(x)$, which will have a general form similar to that in Fig. 3.4. By assumption (i) above, $g(x)$ is assumed to be 1 when $x = 0$. It will then usually have a 'shoulder', such that all individuals within a certain distance can be seen, and then will drop off towards 0 the further the individuals are from the line. The curve $g(x)$ itself cannot be observed. What can be recorded is a frequency distribution of sightings as a function of distance away from the line (Fig. 3.4). This should have the same shape as $g(x)$, and can be used to derive the population estimate as follows.

The probability P_a of detecting an individual in a strip of width w, given that it is present (at an unknown distance from the line), is

$$P_a = \frac{\int_{x=0}^{w} g(x) \, dx}{w}. \tag{3.29}$$

Note that eqn (3.29) will be 1 if the detection probability is uniformly 1 over the whole strip, and in general eqn (3.29) can be understood as the mean detection probability, across the whole strip width.

By dividing eqn (3.29) into eqn (3.28), we can adjust eqn (3.28) for the proportion of individuals that will, on average be missed:

$$\hat{D} = \frac{n}{2L \int_{x=0}^{w} g(x) \, dx} \tag{3.30}$$

(Note that the w has cancelled out.) The problem is thus to estimate the integral $\int_{0}^{w} g(x) \, dx$. This is often called the *effective half strip width* because it takes the place of the half strip width w in eqn (3.28). What we have available to do this is observed data on the frequency distribution of sighting densities, which can be used to estimate $f(x)$, the probability density function of sightings as a function of distance, given that the individuals have been seen:

$$f(x) = \frac{g(x)}{\int_{x=0}^{w} g(x) \, dx}. \tag{3.31}$$

This has the same shape as $g(x)$. All that it is necessary to do is to ensure that $f(x)$ integrates to 1 over the total strip width.

The denominator of eqn (3.31) is the effective half strip we need, and by assumption (i), we know that $g(0) = 1$. Hence,

$$\int_{x=0}^{w} g(x) \, dx = 1/\hat{f}(0), \tag{3.32}$$

where $\hat{f}(0)$ is the estimated value of the probability density of sightings, evaluated at 0. The problem is then to fit an appropriate form of $f(x)$ to the sighting distances, and extrapolate it back to zero.

There are a number of ways that $f(x)$ might be fitted to observed data. One possibility is to use a parametric curve, such as a negative exponential distribution or half normal distribution. In general, such curves are too restrictive and are rarely used. A more satisfactory procedure is to use a curve-fitting process that allows more flexibility in the possible shape of the sighting curve.

For example, Buckland *et al.* (1993) suggest using series expansions such as Fourier series or polynomials to fit sighting curves. Thompson (1992) recommends the use of a kernel estimate, which essentially involves using a function called a kernel to perform a weighted moving average across the observed data. Buckland *et al.* recommend against using kernels. This is no doubt an issue over which statistical debate will continue. For the time being, ecologists will be constrained by available software.

Clustered distributions

One very obvious limitation with the basic model just described is that sightings are assumed to be independent. This does not mean that the organisms themselves must necessarily be distributed independently or randomly. Almost all animals will be clustered, at least to some extent, in areas of preferred habitat. If sighting one member of a group increases the likelihood of detecting another member of the same group, there is a problem with the standard model. There are two consequences. First, the variance of the resulting population estimate will not be estimated correctly. Second, the model will be overspecified (too many parameters are likely to be included), with an overelaborate form of the detection curve being fitted (Buckland *et al.*, 1993).

An obvious solution to the problem is to record sightings of groups rather than individual organisms, and then to convert group density to population density with an estimate of average group size. Provided sightability of groups does not depend on their size, this approach is straightforward, although an additional source of variation is added to the estimated density. However, it is often the case that large groups are easier to see some distance from the transect line than are small groups. As small groups far from the line are missed, average group size will be overestimated, with a resulting positive bias to the final population estimate. There are several ways in which this problem can be addressed (Buckland *et al.*, 1993, pp. 125–30).

Confidence intervals from line transects

The error in estimated density from a line transect has two major components, one due to sampling error in the number n of organisms seen, and the second due to error in the probability density of sightings at 0, $\hat{f}(0)$. In addition, there may be error associated with estimation of group size and a correction if sightability on the line is less than one. The one component in common with a strip transect estimate is variation in the actual number of organisms seen. As is the case with a strip transect, it is not possible to obtain a sensible estimate of the variation in n from a single strip or line. Buckland *et al.* (1993, p. 88) suggest that it may be possible to assume that n follows a Poisson distribution (in which case, var(n) = n), and further recommend an *ad hoc* doubling of this variance. As they themselves state, this is not really an adequate solution. Few ecologists would defend estimating sampling error from a single quadrat by assuming a Poisson distribution, and the line transect is simply a quadrat with complications. Error in $\hat{f}(0)$ can be estimated from a single line, but this is one component only of the sampling error, and usually not the dominant component.

Replicate lines are essential to obtain an adequate estimate of sampling error. As with any other sampling method, if there are fewer than 10 replicates there may problems with lack of degrees of freedom (see Fig. 3.1). The simplest and most straightforward approach to analysing replicates is to run the entire estimation process on each replicate, and then to generate a standard error in estimated density from these using standard formulae. The catch is that Buckland *et al.* (1993) recommend that at least 60–80 sightings are necessary to gain an adequate estimate of $\hat{f}(0)$. Estimation of $\hat{f}(0)$ separately for each line would then require a minimum of 600–800 sightings overall, restricting the method to very common species or very expensive surveys.

More usually, $\hat{f}(0)$ would be estimated from pooled replicates within a stratum, with variation in n estimated from differences between the replicates. Obtaining an analytic form for the variance from a compound process like this is not straightforward, although Buckland *et al.* (1993) provide some suggestions on how to approximate it. A better alternative is to use a bootstrap or jackknife (see Chapter 2), with the resampling being conducted at the level of replicate lines, not individual sightings. Resampling based on individual observations would supply no information whatsoever about variation in n, as bootstrap samples would all have a fixed sample size of n.

Box 3.3 shows an application of line transect sampling to the estimation of the population density of wallaroos (*Macropus robustus*).

Box 3.3 Line transect sampling of a wallaroo (*Macropus robustus*) population

This example is based on data provided by Tony Pople (University of Queensland). In an area of western Queensland, eight replicate transect lines, each of 5 km, were surveyed on foot. The raw data consist of distances from the transect line to each wallaroo sighted. The following figures summarize the distribution of sightings per transect.

Each transect could be treated independently, with a sighting distribution generated from the observations from that transect only, and the estimates could then be treated as simple replicates, with a mean and variance calculated using standard formulae. However, it is clear that some of the transects have insufficient observations to derive a reasonable estimate of the sighting distribution. An alternative approach is

continued on p. 79

Box 3.3 *contd*

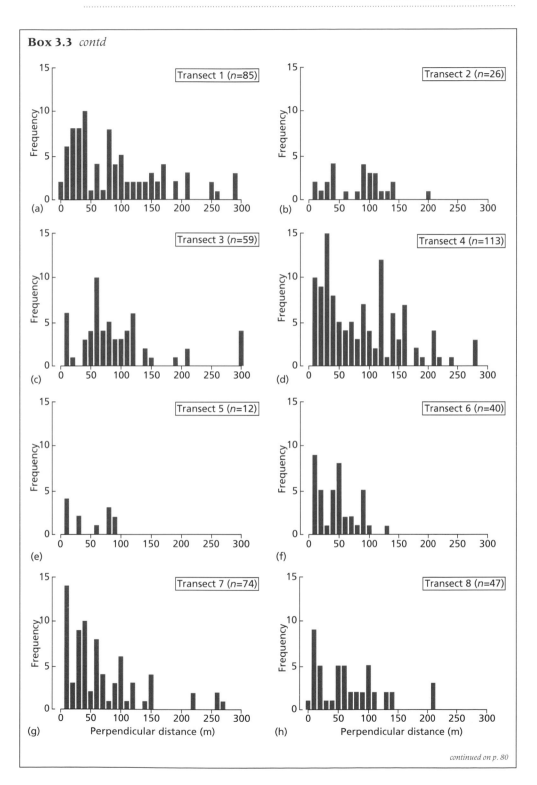

continued on p. 80

Box 3.3 *contd*

to base the sighting distribution on all the data pooled, and then to use the numbers of animals seen per transect to derive some idea of the standard error. If this latter approach is used, a simple mean and variance of the estimated densities per transect may not be a reliable indication of the sampling error, as the sighting distribution is not known, but will be estimated with error. The following analysis was carried out using the program DISTANCE (Buckland *et al.*, 1993).

The first problem is to fit a sighting curve to the observed data. DISTANCE allows for several possible functional forms, each of which consists of a relatively simple parametric function of the perpendicular distance y, termed a key function, multiplied by a series expansion with an adjustable number of terms. The table below summarizes some of the forms available. The fit of the different curves to observed data can be compared using the Akaike information criterion (see Chapter 2), with the most suitable model being that for which the AIC is minimized. The final column of the table shows the AIC obtained by fitting each model to the wallaroo data.

In the table, w is the maximum distance from the line at which observations are included (the truncation line), y is the sighting distance, σ is a dispersion parameter that is estimated from the data, b is a shape parameter, m is the (adjustable) number of terms in a series expansion, and the a_j are the constants in the series. Hermite polynomials are special polynomials that are generated from the equation

$$H_n(x) = (-1)^n e^{x^2/2} \frac{d^n e^{-x^2/2}}{dx^n}.$$

Key		Series expansion		
Name	Function	Name	Function	AIC
Uniform	$1/w$	Cosine	$\sum_{j=1}^{m} a_j \cos\left(\dfrac{j\pi y}{w}\right)$	4464
Uniform	$1/w$	Polynomial	$\sum_{j=1}^{m} a_j \left(\dfrac{y}{w}\right)^{2j}$	4467
Half normal	$\exp(-y^2/2\sigma^2)$	Hermite polynomial	$\sum_{j=2}^{m} a_j H_{2j}(y/\sigma)$	4469
Hazard rate	$1 - \exp(-[y/\sigma]^{-b})$	Cosine	$\sum_{j=1}^{m} a_j \cos\left(\dfrac{j\pi y}{w}\right)$	4470

The first two combinations are entirely empirical. They make no assumptions whatsoever about the form of the data, but simply use as many terms as are necessary to obtain a good fit to the data. It is sensible to impose the additional restriction (as

continued on p. 81

Box 3.3 *contd*

DISTANCE does) that the sightability should decrease monotonically. The half normal–
Hermite combination has its shape based on theoretical considerations. Finally, the
hazard rate key function may perform well for data that are spiked at zero distance.

In this particular case, it is clear that the uniform key–cosine model has the lowest
AIC. It is shown below fitted to the observed pooled sighting distances. (This is $g(x)$.) It
appears to be a reasonably good fit.

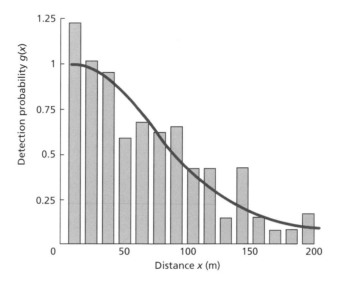

The estimated value of $f(0)$ is 0.0096, leading to an effective strip width (ESW) of
103.76 m.

Following eqn (3.30), and recalling that the total number of wallaroos sighted was
435, over a total of 40 km of transect, the overall population density is:

$$D = \frac{n}{2L \times ESW} = \frac{435}{2 \times 40 \times 0.10376} = 52.40 \text{ wallaroos km}^{-2}$$

All estimators produced densities varying from this by no more than 5 km^{-2}.

There are several ways that confidence limits could be placed on this overall
estimate. Buckland *et al.* (1993, p. 87) provide an approximate standard error, built out
of the estimated variances of each of the components of the final density estimate.
Alternatively, the entire estimation procedure could be run independently on the eight
replicate lines, and the mean and variance of these individual estimates calculated.
Finally, the data can be bootstrapped by resampling, leaving out entire replicate lines.
The table below compares confidence intervals for each of these approaches in this
particular case.

continued on p. 82

Box 3.3 *contd*

Method	Estimated density	Standard error	Lower 95% bound	Upper 95% bound
Variance components	52.403	10.743	32.825	83.657
Replicate lines	49.393	9.014	32.191	75.787
Bootstrap normal	52.403	12.729	32.776	83.784
Bootstrap quantiles	52.403		33.065	82.828

The bootstrap was based on 2000 replicates, and took over 30 minutes on a 200 MHz Pentium computer. The 'bootstrap normal' uses the bootstrap to calculate a standard error, and then uses a normal approximation, whereas the bootstrap quantiles are limits based on the lowest 2.5% and highest 97.5% estimates obtained in the 2000 bootstrap runs.

In this case, there is clearly little difference between the various procedures.

Mark–recapture and mark–resight

Almost all ecologists are familiar with the Lincoln–Petersen model for estimating population size. A sample of n_1 individuals is captured, marked and released. A short time later, a second sample of n_2 individuals is captured and examined for marks. The basic idea is that the proportion of individuals in the second sample with marks should be the same as the proportion of all individuals in the population with marks. Thus, if r individuals are recaptured or resighted,

$$\frac{r}{n_2} = \frac{n_1}{N}, \tag{3.33}$$

where N is the unknown total population size. The assumptions on which this simple idea rests are discussed in most basic ecology texts (e.g. Krebs, 1989) and are implicit in the argument just presented. The population must be closed, with N constant between first and second captures, marked and unmarked individuals must mix, there must be equal probabilities of capture for marked and unmarked animals, and marks must not be lost or missed. Box 3.4 shows how to estimate population size and confidence limits using the Lincoln–Petersen method.

Beyond this basic and intuitive method, there is an enormous and sometimes bewildering array of methods and models (Cormack, 1989; Pollock *et al.*, 1989; Pollock, 1991; Lebreton *et al.*, 1992). This section will not attempt a full discussion of each method, but will attempt to direct readers towards the most appropriate methods.

It is important to be clear on the purpose of a mark–recapture study. If repeated capture times are available, the method can be used to estimate

Box 3.4 The Lincoln–Petersen method

The simple estimator derived from eqn (3.33),

$$\hat{N} = \frac{n_1 n_2}{r}, \tag{1}$$

is biased for small values of r. A better estimator with less bias is

$$\hat{N}_C = \frac{(n_1 + 1)(n_2 + 1)}{r + 1} - 1 \tag{2}$$

(Seber, 1982, p. 60). Despite the simplicity of the method, generating confidence intervals for a Lincoln–Petersen estimate is not entirely straightforward.

An approximate estimator of the coefficient of variation $(s/\hat{N})$ of $\hat{N}$ is

$$CV(\hat{N}) \approx 1/\sqrt{\mu}, \tag{3}$$

where μ is the expected number of recaptures. This is very helpful in planning mark–recapture experiments, making it clear that a reasonable estimate cannot be obtained unless 20 or so recaptures are expected. This approximation is not sufficient, however, to place a 95% confidence interval on an estimated population size.

An approximate variance of $\hat{N}_C$ is given by

$$\text{var}(\hat{N}_C) = \frac{(n_1 + 1)(n_2 + 1)(n_1 - r)(n_2 - r)}{(r + 1)^2 (r + 2)}. \tag{4}$$

However, the sampling distribution of $\hat{N}$ is asymmetrical (it is bounded below by $n_1 + n_2 - r$, the number of different animals caught, but has no upper bound). Basing a confidence interval on the variance of $\hat{N}_C$ will frequently produce a lower limit that is less than the number of animals that have been caught. A better approach in almost all cases is to obtain a confidence limit for the expected number of animals recaught, or the proportion of animals recaught, and then to use that limit to back-calculate limits for N itself.

For example, Seber (1982, p. 63) suggests that, if r and r/n_2 are both fairly small (say, $r/n_2 < 0.1$ and $r < 50$), 95% confidence limits for μ, the expected number recaught, can be generated by looking up r in a table giving confidence limits for a Poisson mean, given an observed count. These limits can then be used in association with the observed n_1 and n_2 to generate limits for N itself using eqn (2).

Alternatively, if $r > 50$, a 95% confidence interval for p, the proportion of marked animals in the population at recapture, is given by

continued on p. 84

Box 3.4 *contd*

$$\hat{p} \pm \left\{ 1.96 \sqrt{\frac{\hat{p}(1-\hat{p})}{n_2 - 1}} + \frac{1}{2n_2} \right\}. \tag{5}$$

These limits can be substituted into eqn (1) in place of r/n_2 to obtain confidence limits for N. For example, the upper 95% bound for N would be

$$N_\mathrm{u} = \frac{n_1}{p_1},$$

where p_1 is the lower bound on p from eqn (5).

The best approach, however, is to use profile likelihood confidence intervals (see Chapter 2). These can be generated from a package such as CAPTURE (Rexstad & Burnham, 1992).

survival and recruitment rates, in addition to, or instead of, population size. In fact, estimating survival rates requires less restrictive assumptions than does estimating population size. Using mark–recapture methods to estimate survival is reviewed by Lebreton *et al.* (1992), and is further discussed in Chapter 4.

Models can be divided into those for closed populations, with no immigration, emigration, death or recruitment, and models for open populations, in which any of these may occur. The Lincoln–Petersen method is the most basic closed-population model. The most elementary open population method is the Jolly–Seber method, which requires at least three sampling occasions. In essence, it is a series of Lincoln–Petersen estimates, in which the number of marked individuals currently in the population must be estimated at each sampling occasion.

Closed populations

Closed-population approaches are appropriate for a series of captures and recaptures conducted over a short period of time, so that it can be assumed that there is no birth, death, immigration or emigration over that period. A corollary of this is that the number of marked individuals in the population is known exactly: it is the number that have been released. Even in studies conducted over a short period of time there is usually a problem of geographic closure, unless the population is contained in a discrete area, like a pond, from which it cannot move. The study area will have edges which animals move over, meaning that the population size of some larger, unspecified area is being estimated. There is a discussion of this problem in Otis *et al.* (1978). The most straightforward solution is to assume that the effective area being sampled is the study area itself, plus a boundary strip of about half a home range. A more sophisticated method is to analyse the trapping data in a series of nested subgrids.

If animals are individually marked, the raw data in many mark–recapture studies can be summarized as a series of *capture histories*. A capture history is a series of 0s and 1s, recording whether an animal was captured (1) or missed (0) on each trapping occasion. For example, in a study with five capture occasions, a capture history

0 1 0 0 1

would mean that the animal was first captured on the second trapping occasion, missed for the next two, and then recaptured on the final occasion. In a study with t sampling occasions, there will be 2^t distinct capture histories possible, although one of these,

0 0 0 . . . 0,

is unobservable: the animals with this history have never been caught. There are various ways to analyse capture histories. One of the most elegant statistically is to analyse the number of individuals with each capture history as counts in a log-linear model. For details on this method, and in particular on relating the parameters of the model back to parameters of ecological interest (such as population size), see Cormack (1989).

Mark–recapture methods can be classified according to the assumptions they make about capture probabilities, which are the probability of any given individual turning up in the sample on a given sampling occasion. Probably the most critical assumption with either the Lincoln–Petersen or Jolly–Seber method is that capture probabilities of all animals on each sampling occasion are assumed to be equal (Pollock, 1991). That is, every individual in the population, whether marked or unmarked, has an equal probability of being captured at any sampling occasion. This is rarely the case in any real animal population. In addition to inherent heterogeneity in capture probability between individuals, there may be a problem with trap response, in which individuals that have been caught once become either trap-happy or trap-shy for some period after they are caught. Little can be done statistically to handle capture heterogeneity in a single-recapture experiment, although some experimental design modifications (such as moving the trap positions between initial and recaptures) may help. If there are multiple recapture occasions, however, it is possible to test for and adjust for some forms of capture heterogeneity.

Otis *et al.* (1978) suggest a hierarchy of models with increasingly complex assumptions about capture probabilities. A more expansive treatment of the same suite of models can be found in White *et al.* (1982), and these can be implemented via the program CAPTURE (Rexstad & Burnham, 1991). The most elementary model is M_0 in which the probability of capture is constant both between individuals and for each sampling occasion. This will rarely be adequate in practice, as it requires not only all individuals to be identical, but also sampling effort and catchability to be identical between sampling times. It may be possible to control effort, but controlling the weather, which will affect activity and thus catchability, is a rather ambitious objective.

The next level in complexity is a series of single-factor models. Model M_t allows catchability to vary between sampling occasions, but holds capture probability constant for each individual. The Lincoln–Petersen index is an M_t model with two sampling occasions, as is the Schnabel method, another classical method which is based on more than two sampling occasions. Model M_b allows for behavioural responses to trapping (trap-happiness or trap-shyness) and model M_h allows for capture probabilities to be variable between individuals. The latter two models are discussed in detail in Otis *et al.* (1978), but their practical applications are rather limited because they do not allow for any time dependence in catchability.

Two-factor models that allow any two of the above factors to be included have also been developed. Model M_{ht}, for example, allows both for heterogeneity between individuals in capture probability, and also for capture probabilities to vary through time. Unfortunately, many of these complex models are fairly intractable statistically, particularly those that allow capture probabilities to vary through time (Pollock *et al.*, 1990).

Faced with this rather bewildering array of models, with the most useful ones being the most intractable, it may be hard to know how to proceed. First, it is necessary to be realistic: if there is extreme heterogeneity in catchability, no method will work well. A method based on capturing animals cannot estimate how many animals there are that cannot be caught at all. Second, it is important not to rely too heavily on 'objective' statistical tests to pick the best model. Model selection is inherently a tricky business (Cormack, 1994), and frequently very different models giving very different results may fit almost equally well (Agresti, 1994). Simulation studies show that automatic model selection procedures such as those in program CAPTURE (Rexstad & Burnham, 1991) may usually select the wrong model (Menkens & Anderson, 1988). A good model is parsimonious, but also biologically plausible. Very basic models such as the Lincoln–Petersen model or Schnabel methods are at least easy to understand, so that any biological implausibility is obvious.

Time-dependent methods

In most cases, it will be necessary to assume that capture probabilities vary between sampling occasions. I will therefore only discuss methods with time dependence in detail. The Lincoln–Petersen model has already been discussed.

A Schnabel estimate is essentially a weighted average of a series of Lincoln–Petersen estimates (Seber, 1982; Krebs, 1989). Individuals need not be marked with individually recognizable marks, but it is necessary that the number of individuals with each capture history should be known. At any time, assuming the population is closed, the number of marked individuals in the population is simply the number of distinct individuals that have been marked and released up to that time. Thus, a simple Lincoln–Petersen estimate could be calculated at any stage after the first recapturing session.

There are various ways of deriving a single population estimate from this information, each of which will return a slightly different estimate. The 'best' approach is to use maximum likelihood, which will return an estimate and model that can usefully be compared with estimates based on other models. The maximum likelihood estimate does not have a simple closed form, and will usually be calculated by a computer program such as CAPTURE (Rexstad & Burnham, 1991). Algorithms are also provided by Seber (1982, p. 139) and Otis et al. (1978, p. 106).

An intuitive understanding of how the method works can be gained by treating it as a regression problem. Letting C_t be the number of animals caught on the tth trapping occasion, M_t be the number of animals in the population known to have marks at that time and R_t be the number of already marked individuals captured,

$$\frac{R_t}{C_t} \approx \frac{M_t}{N},$$

(3.34)

following the idea of a Lincoln–Petersen estimate. Thus, a plot of the ratio R_t/C_t versus M_t should be a straight line passing through the origin with a slope of $1/N$. Standard regression methods can then be used to estimate $1/N$ and its standard error. Two complications are that the regression must be constrained to pass through the origin, and that each point should be weighted proportional to C_t. Box 3.5 illustrates these calculations, which should be possible with most standard statistics packages. If the plot is markedly curved, it suggests violation of one or more of the basic assumptions of the Lincoln–Petersen method. Unfortunately, which assumption is violated cannot be determined from the shape of the curve itself. For Tanaka's model, a modified approach that may be helpful for curved relationships, see Seber (1982, p. 145).

There have been several recent attempts (Chao et al., 1992; Agresti, 1994) to derive useful estimators for situations where capture probabilities vary both through time and between individuals (model M_{th}). These are important, as this model will often be the minimum plausible model. In general, the relative catchability of animals might vary through time, but a model in which capture probabilities vary both between individuals and through time, but without interaction, is worth pursuing. In such a model p_{ij}, the probability of individual i being captured on any particular occasion j, would be represented by

Box 3.5 The Schnabel estimator

Tony Pople obtained the following data in a mark–recapture study of *Mus musculus* in an area of regenerating heath in Myall Lakes National Park, Australia. Details of the study can be found in Fox and Pople (1984). The raw data are shown below, where t1, . . . , t9 represent nine capture occasions, and the numbers in the body of the table are the individual tag numbers of the mice caught on each trapping occasion.

t1	t2	t3	t4	t5	t6	t7	t8	t9
163	164	80	80	20	80	80	100	81
164	71	164	88	81	164	106	81	100
165	72	163	164	100	81	107	163	101
51	74	81	163	52	105	72	52	164
52	76	51	81	103	103	115	14	102
	75	52	165	165	83	82	164	73
	78	71	15	71	71	170	111	165
	77	75	76	102	165	165	115	76
		76	72	191	16	16	82	170
		72	71	72	76	75	112	77
		82	170	170	170	78	191	78
		170	16	76	78	74	73	103
		83	74	16	51	110	71	51
		74	78	75	194	194	76	
		78	77	74			170	
				77			77	
				194			113	
							51	

From these raw data, the basic variables C_t (the number of animals caught at time t), M_t (the number of marked animals in the population), and R_t (the number of recaptures) can be calculated. They are as follows:

t	t1	t2	t3	t4	t5	t6	t7	t8	t9
C_t	5	8	15	15	17	14	14	18	13
M_t	0	5	12	17	20	26	27	31	36
R_t	0	1	10	12	11	13	10	13	12

Here the ratio R_t/C_t is plotted versus M_t, together with a regression without intercept, weighted proportional to C_t. There is an indication of curvature in

continued on p. 89

Box 3.5 *contd*

this plot, suggesting that one or more of the assumptions of the Schnabel method is being violated. In this particular case, the most likely problem is that the population is not closed. Unmarked animals moving on to the trapping grid, or marked animals leaving it, will both cause R_t/C_t to increase more slowly with M_t than would be expected given a closed population. The gradient of regression line and its standard error are 0.029 594 and 0.002 805 46. Inverting this, the resulting estimate of N is 33.79, with a 95% confidence interval (27.73, 43.24). Box 3.6 applies the Jolly–Seber method to these data.

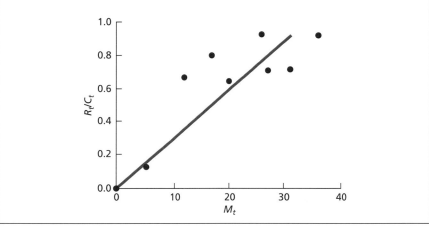

$$\ln\left(\frac{p_{ij}}{1-p_{ij}}\right) = \mu + \alpha_i + \beta_j, \tag{3.35}$$

where μ is an overall mean, α_i is an effect due to the individual animal, and β_j is the effect of the sampling occasion. Estimating population size in a closed-population model is essentially the problem of trying to estimate the number of animals in the population that have never been caught (those with the capture history 0 0 0 . . . 0), by using the numbers with each other capture history. Equation (3.35) can be fitted to all other capture histories, but cannot in general be used to estimate the number of individuals that have not been caught (Agresti, 1994). However, if certain mathematical restrictions are placed on the parameters, a population estimate can be obtained, and if the fit of the restricted model is similar to that of the general model to capture histories other than 0 0 0 . . . 0, use of the restricted model to estimate population size can be justified. Agresti (1994) supplies GLIM code necessary for fitting such models.

Open-population models

Two important complications are introduced if the assumption of population closure is relaxed. First, the population size being estimated at each recapture will vary from one occasion to the next. Second, the number of marked individuals will no longer be known, but must be estimated. It is the second complication that causes the real problems. In the standard Lincoln–Petersen method and others derived from it, the number of marked individuals in the population is the number that have ever been marked and released, and if this is known the population size at any sampling occasion can be estimated. However, if the population is open, some of these marked individuals may have died or left the population.

The Jolly–Seber method is a fairly straightforward generalization of the Lincoln–Petersen method. A good intuitive explanation of how it works is given by Pollock *et al.* (1990).

The model aims to estimate three parameters at each sampling time i. These are N_i, the population size at time i; B_i, the number of new individuals entering the population between i and $i + 1$; and ϕ_i, the probability of an individual surviving from i to $i + 1$. As will be seen shortly, all of these cannot be estimated for all sampling times. In addition, two other parameters need to be estimated: M_i, the number of marked individuals in the population at sampling time i; and p_i, the capture probability of each individual at time i. Implicit in this parameterization are the assumptions that catchability is the same for all individuals at any particular time (i.e. this is an M_t model) and that survival may vary through time, but is the same for all individuals at any given time. Table 3.2 summarizes the parameter and variable definitions for Jolly–Seber methods.

The raw data necessary to calculate a Jolly–Seber estimate are a series of capture histories (see above), one for each individual that has ever been caught. From these, several basic statistics can be calculated: m_i, the number of marked individuals caught at time i; u_i, the number of unmarked individuals caught at i; $n_i = u_i + m_i$, the total number of individuals caught at time i; R_i, the number of marked individuals released at i ($R_i < n_i$ if there are losses or removals on capture); r_i, the number of individuals released at i that are ever recaptured; and z_i, the number of individuals captured before i, not recaptured at i, but recaptured at a later date. This last statistic is crucial in estimating the number of marked individuals missed at each sampling occasion. Table 3.2 summarizes these statistics, and Box 3.6 shows an example of a calculation.

To derive an estimate of population size, imagine, for the time being, that the problem of estimating the number of marked individuals M_i in the population at time i is solved, and that the estimate is $\hat{M}_i$. Then, the population size at time i could be estimated using the standard Lincoln–Petersen method and

Table 3.2 Variables and symbols used in Jolly–Seber estimation. Note that the formulae in this table are unbiased estimators that differ slightly from the intuitive estimators given in the text. The unbiased estimators have tildes instead of hats. From Pollock *et al.* (1990)

Symbol	Definition	Derivation	Restrictions
n_i	Number of individuals caught at time i	Basic input data	
m_i	Number of marked individuals caught at time i	Basic input data	
u_i	Number of unmarked individuals caught at time i	Basic input data	
R_i	Number of marked individuals released at time i	Basic input data. Same as n_i, unless there are losses on capture	
r_i	Number of marked individuals released at time i that are ever recaptured	Obtained from capture histories	Not defined for final capture
z_i	Number of individuals captured before i, *not* recaptured at i, but recaptured later	Obtained from capture histories	Not defined for first or last capture occasions
$\tilde{M}_i$	Estimated number of marked individuals in the population at time i	$\tilde{M}_i = m_i + \dfrac{(R_i + 1)z_i}{r_i + 1}$	Cannot be estimated for first or last capture occasions
$\tilde{\phi}_i$	Estimated survival probability of survival from time i to time $i + 1$	$\tilde{\phi}_i = \dfrac{\tilde{M}_{i+1}}{\tilde{M}_i - m_i + R_i}$	Cannot be estimated for last two capture occasions
$\tilde{N}_i$	Estimated population size at time i	$\tilde{N}_i = \dfrac{(n_i + 1)\tilde{M}_i}{m_i + 1}$	Cannot be estimated for first or last capture occasions
$\tilde{B}_i$	Number of additions to the population between times i and $i + 1$	$\tilde{B}_i = \tilde{N}_{i+1} - \tilde{\phi}_i(\tilde{N}_i - n_i + R_i)$	Cannot be estimated for first or last two capture occasions

would be

$$\hat{N}_i = \frac{n_i \hat{M}_i}{m_i}. \tag{3.36}$$

Survival can be estimated by considering only the marked individuals. The proportion of marked individuals surviving from i to $i + 1$ is simply the ratio of the marked individuals present at time $i + 1$ to those present immediately after release at time i,

$$\hat{\phi}_i = \frac{\hat{M}_{i+1}}{\hat{M}_i - m_i + R_i}. \tag{3.37}$$

Box 3.6 The Jolly–Seber method

The capture history matrix for Pople's mouse mark–recapture data from Box 3.5 is:

		Capture occasion								
	A	B	C	D	E	F	G	H	I	J
1	ID	1	2	3	4	5	6	7	8	9
2	14	0	0	0	0	0	0	0	1	0
3	15	0	0	0	1	0	0	0	0	0
4	16	0	0	0	1	1	1	1	0	0
5	20	0	0	0	0	1	0	0	0	0
6	51	1	0	1	0	0	1	0	1	1
7	52	1	0	1	0	1	0	0	1	0
8	71	0	1	1	1	1	1	0	1	0
9	72	0	1	1	1	1	0	1	0	0
10	73	0	0	0	0	0	0	0	1	1
11	74	0	1	1	1	1	0	1	0	0
12	75	0	1	1	0	1	0	1	0	0
13	76	0	1	1	1	1	1	0	1	1
14	77	0	1	0	1	1	0	0	1	1
15	78	0	1	1	1	0	1	1	0	1
16	80	0	0	1	1	0	1	1	0	0
17	81	0	0	1	1	1	1	0	1	1
18	82	0	0	1	0	0	0	1	1	0
19	83	0	0	1	0	0	1	0	0	0
20	88	0	0	0	1	0	0	0	0	0
21	100	0	0	0	0	1	0	0	1	1
22	101	0	0	0	0	0	0	0	0	1
23	102	0	0	0	0	1	0	0	0	1

continued on p. 93

Box 3.6 *contd*

24	103	0	0	0	0	1	1	0	0	1
25	105	0	0	0	0	0	1	0	0	0
26	106	0	0	0	0	0	0	1	0	0
27	107	0	0	0	0	0	0	1	0	0
28	110	0	0	0	0	0	0	1	0	0
29	111	0	0	0	0	0	0	0	1	0
30	112	0	0	0	0	0	0	0	1	0
31	113	0	0	0	0	0	0	0	1	0
32	115	0	0	0	0	0	0	1	1	0
33	163	1	0	1	1	0	0	0	1	0
34	164	1	1	1	1	0	1	0	1	1
35	165	1	0	0	1	1	1	1	0	1
36	170	0	0	1	1	1	1	1	1	1
37	191	0	0	0	0	1	0	0	1	0
38	194	0	0	0	0	1	1	1	0	0

The shaded row and column labels are used in Excel formulae below. A simple way to produce a capture history matrix in later versions of Excel (Version 6 or higher), is to use the 'Pivot Table' facility. First, copy the data so that they occupy two columns, one for animal ID and the other for capture occasion (Capt). Next, highlight both columns and request a pivot table (Data PivotTable) with ID as the row variable, Capt as the column variable, and 'Count of Capt' in the body of the table. In the final step, click 'Options' and set empty cells to 0.

A capture history matrix is the basic input format required for most mark–recapture analysis programs. It can also be used to produce the necessary variables for a Jolly–Seber estimation. For example, consider z_i, which is the number of marked animals caught before i, missed at i, but caught later. The following formula will return a 1 if the individual in row 2 falls into this category for the capture occasion in column D, and a 0 otherwise: $= IF(AND(SUM(\$B2:C2),NOT(D2),SUM(E2:\$J2)),1,0)$. If this is copied into a matrix to the right of the capture histories themselves,

continued on p. 94

Box 3.6 *contd*

each column total will be z_i. Similarly, = IF(AND(D2,SUM(E2:\$J2)),1,0) will calculate elements of r_i, and = IF(AND(SUM(\$B2:C2),D2),1,0) will calculate elements of m_i. The components required for Jolly–Seber estimation, and the resulting parameters derived from this capture history, are:

	Capture occasion								
	1	2	3	4	5	6	7	8	9
n_i	5	8	15	15	17	14	14	18	13
m_i	0	1	10	12	11	13	10	13	12
r_i	5	8	15	13	16	12	5	8	
z_i		4	2	5	7	10	12	4	
$\tilde{M}_i$		5	12	17.7	18.4	24.5	40	21.4	
$\tilde{\phi}_i$	1	1	1.04	0.89	1.01	1.57	0.49		
$\tilde{N}_i$		22.5	17.5	21.8	27.6	26.3	54.5	29.1	
$\tilde{B}_i$			3.61	8.24	−1.5	13.4	2.52		
p_i		0.2	0.83	0.68	0.6	0.53	0.25	0.16	

The recapture rate for these mice is a little too low for the Jolly–Seber method to work well (r_i and $m_i > 10$ is recommended; Pollock *et al.*, 1990), although the estimated probabilities of capture, p_i, are very high. Note that some of the survival rates exceed 1 (survival from occasion 6 to occasion 7 is estimated to be 1.57!), and some of the addition terms $\tilde{B}_i$ are negative, which also should not be possible. The tendency of the Jolly–Seber method to produce biologically impossible parameter estimates is a major disadvantage. The method also requires estimation of separate survival rates for each interval, tending to produce overparameterized models. Methods based on log-linear models or direct maximum likelihood estimation are therefore preferable.

The $-m_i + R_i$ term in the denominator of eqn (3.37) is simply an adjustment to allow for losses (or additions) of marked animals in the capture process at time i.

The number of additions to the population between i and $i + 1$, B_i, is similarly straightforward. It is simply the difference between the two successive population estimates, with an adjustment for possible mortality and removals at capture,

$$\hat{B}_i = \hat{N}_{i+1} - \hat{\phi}_i(\hat{N}_i - n_i + R_i) \tag{3.38}$$

The problem that remains is to estimate the size of the marked population at any time. This can be done on the assumption that the recovery probability of all marked individuals in the population is the same. The recovery probability is the chance of ever seeing a given marked individual again. There are two groups of marked individuals just after the release at time i: those that have just been released (there are R_i of these), and those which were not caught at time i. The size of the latter group is $M_i - m_i$. The numbers of each group that are recaptured later are known, and are r_i and z_i respectively. Thus, equating recovery probabilities for these two groups,

$$\frac{z_i}{M_i - m_i} \approx \frac{r_i}{R_i}, \tag{3.39}$$

or, by rearranging eqn (3.39),

$$\hat{M}_i = m_i + \frac{R_i z_i}{r_i}. \tag{3.40}$$

A little thought will make it clear that these cannot be defined for each sampling occasion. Equation (3.40) cannot be calculated for either the first or last samples, because z_i cannot be calculated. Hence, the population size (eqn (3.36)) cannot be estimated for either the first or last samples either, and following from this, the number of immigrants cannot be estimated from either the first or last two samples. Survival cannot be estimated from the last two samples. It can, however, be estimated for between the first and second samples. Although $\hat{M}_1$ cannot be estimated using eqn (3.40), it enters eqn (3.37) only to determine the number of marked individuals missed in the first capture. If no individuals are marked, this will obviously be 0, so eqn (3.37) simplifies to

$$\phi_1 = \frac{\hat{M}_2}{R_1}. \tag{3.41}$$

Equations (3.36) to (3.40) are derived by intuitive arguments, but produce biased estimates (Pollock *et al.*, 1990). Table 3.2 gives minor modifications of these that are approximately unbiased. Variances can be calculated for each of the parameter estimates. They are, however, rather messy algebraically, and as most ecologists will use a package such as JOLLY to perform these calculations, I have not included them here. They can be found in Pollock *et al.* (1990).

An alternative approach is the log-linear method proposed by Cormack (1989), which can be modified to apply to open as well as closed models. The method is essentially a different parameterization of the standard Jolly–Seber method, with the disadvantages that it is not as intuitively understandable as the standard model, and that the model parameters must be transformed to generate biologically meaningful results. There are, however, substantial benefits. The model fits into a relatively well-understood area of statistical theory. It is straightforward to generate modified models in which some parameters are constrained to be the same or made functions of external variables. For example, the survival rates ϕ could be set constant for all time intervals, and the fit of this model examined relative to the standard unconstrained model. For the remainder of this book, this general approach will be called the Cormack–Jolly–Seber (CJS) method.

A detailed explanation of the log-linear approach, together with code necessary to perform the analysis in the computer package GLIM, is provided in Cormack (1985). A method for generating variances of the estimates using the log-linear approach, together with GLIM code, is provided in Cormack (1993).

As is noted in the explanation of the Jolly–Seber method above, survival rate estimation depends only on the marked individuals in the population. Thus, provided it can be assumed that survival does not differ between marked and unmarked animals, it is possible to estimate survival with rather fewer assumptions about homogeneity in catchability than are needed for population estimation. The use of mark–recapture techniques to estimate demographic parameters is discussed further in Chapter 4.

Designing mark–recapture studies

As with the design of any experiment, it is essential to decide on the purpose of the study before proceeding. Mark–recapture studies can supply information on population size, survival and recruitment. However, some methods (closed-population models) provide information only on population size, and others (see the following chapter) provide information only on survival. The choice of a suitable technique thus depends on the relative priorities associated with each approach.

Sample size

The precision of mark–recapture estimates is essentially determined by the capture probability of each individual on each sampling occasion. This is in contrast to quadrat-based subsampling methods, in which the sampling fraction (the probability of any individual in the population being included in the sample) does not strongly affect precision. The capture probability of each individual will not, of course, be known in advance, although all methods will estimate it as part of the estimation process for other parameters. Nevertheless, a pilot experiment should provide at least an order-of-magnitude estimate of capture probabilities. The general conclusion is that capture probabilities must be quite high. For example, in a simple Lincoln–Petersen estimate for a population of about 100 animals, close to 50 individuals must be marked, and about 50 recaught (corresponding to a capture probability of 0.5) to achieve an estimate with a 95% confidence interval of ±25%. Graphs from which sample sizes for specified precision can be read off were calculated by Robson and Regier (1964) and have been reproduced in Krebs (1989) and Seber (1982).

If there are several recapture occasions, then capture probabilities on each occasion do not need to be quite so high. Pollock *et al.* (1990) provide a series of graphs for the Cormack–Jolly–Seber method showing the coefficient of variation of the estimated population size as a function of capture probability for a range of numbers of capture occasions, survival rates and population sizes. In general, the capture probability necessary for a given precision declines as a function of population size, number of capture occasions and survival. This is as one would expect. If survival is high, and there are several capture occasions, there are many opportunities to learn about the existence of any given animal. Nevertheless, even given 90% survival between trapping occasions and 10 trapping events, a capture probability of 0.2 on each occasion is necessary to achieve a coefficient of variation of 20% for population size in the middle of the series, given a true population of about 100 animals. Estimating survival and recruitment from a Cormack–Jolly–Seber method needs, in general, even higher capture probabilities to generate acceptable coefficients of variation.

Timing and allocation of sampling effort

In closed-population models, capture probabilities that vary through time are particularly intractable (see above). It therefore makes sense to keep sampling effort as constant as possible across a series of sampling times, to increase the likelihood that a time-invariant model may be an adequate fit to the data. Time dependence is similarly a problem for open-population models, but if population size changes, keeping capture probability constant is essentially impossible.

For purely logistic reasons, most capture–recapture studies of open populations use a series of 'bouts' or study periods, which are fairly widely separated in time, but within which there are closely spaced sample times (for example, trapping nights). This design also turns out to have a number of theoretical advantages, and has been promoted as 'the robust design' by Pollock and Otto (1983) and Pollock *et al.* (1990). The basic layout of the sampling times is as represented in Fig. 3.5. Within each sampling bout, it is reasonable to assume that the population is more or less closed. A standard closed-population estimate can therefore be calculated for each sampling bout independently. Alternatively, the data from each sampling bout can be pooled and treated as

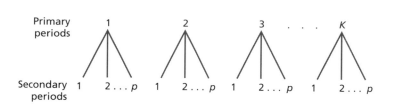

Fig. 3.5 The 'robust' design for a mark–recapture study. From Pollock *et al.* (1990).

a single capture event. Animals would be recorded as either captured at least once during the period, or not captured. These data could then be analysed through a Cormack–Jolly–Seber or similar method. Following this approach, two separate estimators will be obtained for the population size for all sampling bouts except the first and last (for which only the closed-population estimates will be generated). Pollock and Otto (1983) suggest that the closed-population estimates will normally be more reliable than the open-population estimates, as it is possible at least to examine and maybe to allow for heterogeneity in catchability. Only the Cormack–Jolly–Seber estimates, however, can estimate survival. Kendall *et al.* (1995) describe this approach as the *ad hoc* approach, and describe a more elaborate method that uses all the data together to generate the estimates. This 'complete data' approach appears to perform somewhat better than the *ad hoc* approach, but the difference in performance does not appear to be marked, and may not be justified by the increased complexity.

Removal methods

A variety of methods operate by removing individuals from the population, either permanently or by marking individuals and thereafter removing them from further consideration. The idea behind removal methods is very straightforward. If an initial removal sample leaves some proportion of the population uncaught, then the number of individuals caught in a second sample with the same effort should be smaller than the size of the initial sample, by that same proportion. This will be the case provided the probability of catching any individual remains the same at both times. Symbolically, if u_1 is the number caught on the first occasion, and u_2 is the number caught on the second occasion, then

$$\hat{N} \approx \frac{u_1}{1 - u_2/u_1},$$

(3.42)

where $\hat{N}$ is the estimated population size prior to the removal.

This idea is relatively easy to extend to more than two trapping occasions, in which case it is possible to determine how well the data fit the model (see Seber, 1982; or White *et al.*, 1982). An obvious limitation of the method is that catchability can be expected to vary between individuals, and the individuals that are harder to catch are those that will remain in the population. 'Generalized removal methods' (Otis *et al.*, 1978; White *et al.*, 1982) can be used in such cases. Because they use no information about individuals after first capture, these methods can also be used to derive estimates for population size in mark–recapture models under models with trap response alone (M_b), trap response and time dependence (M_{tb}) or trap response and inter-individual heterogeneity (M_{bh}). If capture probability varies through time, the method will fail, although it can be used in the particular case where effort varies, but capture probability can be assumed to be proportional to effort (see below).

Catch per unit effort

The standard index of population size in fisheries management is catch per unit effort. This is simply the number of individuals or amount of biomass caught per standardized unit of fishing effort. The method is based on the reasonable assumption that, all things being equal, if fish are twice as common at one time, compared with another, twice as many will be caught by the same means (trawl or whatever). The problem is that things are often not equal. In particular, learning and targeting by fishers will usually mean that catch per unit effort can be maintained or even increased as a stock declines, at least in the short term. A full discussion of the many problems associated with using catch per unit effort as a population index will be found in any recent textbook of fisheries biology (for example, see Hilborn & Walters, 1992).

Catch per unit effort is also frequently used as a quick and dirty population index in many terrestrial population studies. For example, the proportion of traps occupied has been used as an index for house mouse population size (Singleton, 1989) and the number of individuals spotlighted on a standard transect has been used as an index of rabbit and fox population size (Newsome *et al.*, 1989). Such methods are defensible if comparing populations through time at the one site, provided the method of capture or sighting remains identical and behavioural changes that affect catchability or sightability do not occur. They are likely to be quite unreliable for comparing different areas. In general, a mark–recapture or mark–resight method will provide more reliable results, but may be impractical (see Krebs *et al.*, 1994).

Catch effort data through time can be used to estimate actual population size, provided the target population is closed, or is subject to natural mortality at a rate that can be estimated independently. For details, see Seber (1982).

Indirect methods

There is an enormous variety of methods for indirect estimation of population density. These include scat counts, track counts, burrow counts, calls, etc.

Almost all of these provide an index of population size rather than an actual estimate. As is the case with any index, there is a problem in ensuring that the proportionality constant connecting the index to absolute population size does, in fact, remain a constant.

Track counts are frequently used as an index for mammals, particularly sparsely distributed, highly mobile and cryptic species such as small to medium-sized carnivores. The sampling method may involve sand plots across tracks (a passive monitoring strategy) or plots around baits (an active strategy) (Allen *et al.*, 1996). Such methods are adequate for presence/absence monitoring, but are more problematical for estimating abundance. The basic problem is that track counts are a measure of total animal activity, not total animal numbers. Activity levels may change markedly for any number of reasons, including control measures. For example, poisoning of foxes in Western Australia appeared to remove territory-holding foxes. This may have caused the foxes remaining to move, increasing overall activity levels and track counts, despite the fact that a large proportion of the original population had been removed (N. Marlow, pers. comm.).

In some cases, it may be possible to use track counts as an absolute measure of abundance. One way in which this can be done is to capture some individuals and to mark or measure them in such a way that their tracks become individually recognizable – for example, by toe-clipping (Skalski, 1991). A survey recording the number of tracks in an area, and the proportion of those produced by marked animals, can then be used to derive an estimate of population size.

Recent advances in the identification of individuals from small amounts of DNA via the polymerase chain reaction are likely to revolutionize population estimation using traces. It is now possible (although still not straightforward) to identify individuals from trace amounts of DNA in shed hairs and faeces (Höss *et al.*, 1992; Kohn & Wayne, 1997). The number of distinct genotypes detectable in a given area provides the equivalent of the 'minimum known to be alive' sometimes used as the simplest possible population estimate in mark–recapture studies. More sophisticated analysis should also be possible. If it is possible to ensure that apparently identical genotypes have been left by the same individual, and samples deposited over a defined period of time only are analysed (either by using hair collectors, or swept areas for faeces collection), then all the standard methods for mark–recapture analysis would be applicable.

Population estimation via DNA traces is likely to be most applicable to small populations of highly endangered species that either are hard to capture or cannot be captured without substantial risk to the animals. It is, for example, being implemented for the northern hairy-nosed wombat (*Lasiorhinus kreffti*), which exists only in a single population of less than 70 individuals.

In this case, there is strong evidence (Hoyle *et al.*, 1995) that trapping has a damaging effect on individuals. Unfortunately, rare animals persisting in very small isolated populations are precisely those that are likely to have very low levels of genetic diversity and high levels of inbreeding. It will be necessary to be very careful to ensure that genetic samples that cannot be distinguished do indeed belong to the same individual. In the case of the hairy-nosed wombat, virtually the entire population has been captured, marked and sampled for DNA. Polymerase chain reaction analysis of hairs will permit the continuation of intensive study without further disturbance, but ideally one would want not to need to interfere with the animals to validate the method.

Summary and recommendations

1   Ecologists often assume that using an index of population size is less problematical than attempting to calculate an estimate of actual population size. However, at best, indices are useful only for comparisons between places or times. To use them for this purpose, it is necessary to assume that the same constant connects the index to the actual population size at all times and places being prepared. This assumption cannot be tested without estimates of the actual population size.

2   For organisms that are easily visible, and that can be counted before they move, subsampling methods based on quadrats or transects are usually the best way to estimate population size.

3   Determining the best sampling strategy for a quadrat- or transect-based survey is not easy. Simple random sampling is rarely the optimal strategy. Table 3.1 should be consulted for guidance.

4   In general, strip transects perform better for population estimation than do square quadrats.

5   For surveying large vertebrates, line transect methods (in which the strip width is determined from the sighting data) should be considered.

6   Mark–recapture or mark–resight methods are needed to estimate the population size of small, mobile or cryptic animals. Methods using multiple capture occasions and individual marks are preferable compared to the simple Lincoln–Petersen method, as they allow some of the assumptions of the methods to be tested. Some departures from the basic assumptions can be corrected for, but variability in catchability between individuals is always a problem.

7   Indirect indices are not ideal, but are necessary in some cases.

CHAPTER 4

Vital statistics: birth, death and growth rates

Introduction

The most elementary equation in population dynamics is

$$N_{t+\Delta t} = N_t + B - D + I - E, \tag{4.1}$$

where N_t is the population size at some time t, $N_{t+\Delta t}$ is the population size some time interval Δt later, and B, D, I and E are the numbers of births, deaths, immigrants and emigrants in the time period. Having considered problems of estimating population size in the previous chapter, I now move on to the problem of estimating the numbers of births and deaths in a given time interval. Immigration and emigration are considered later, in Chapter 7, on estimating parameters for spatial models.

The estimation problems encountered depend very much on the type of animal under study. In terrestrial vertebrates, the principal problems are likely to be small sample size and lifespans that may exceed that of the study itself. On the positive side, however, it is often possible to follow the fate of particular individuals. In studies of insects and other invertebrates, sample size is not usually a problem, and the lifespan of the study organism is often very short. However, it is frequently impossible to trace particular individuals through time. Fish fall somewhere between these two extremes.

If the parameter estimates are required for a model structured by size or stage, it is essential also to estimate growth and development rates. Fecundity is often related to size, and the mortality through a particular size or stage class depends on how long an individual remains in that class.

Fecundity

There is considerable confusion about the most appropriate definition of fecundity in the ecological literature (Clobert & Lebreton, 1991). The most usual definition is the number of female offspring produced per adult female per unit of time, but there is no consistency on whether 'females' means female zygotes, or numbers surviving from the zygote stage to some other life-history stage at which the offspring are counted. If the sex ratio is uneven, it will be necessary to include it explicitly to estimate production of males and females separately. Usually, 'fecundity' will include some juvenile mortality. The most appropriate definition will depend on the context, and on how fecundity is to be used in a model. For example, the particular case of estimating fecundity for the Euler–Lotka equation is discussed in Chapter 5.

In general, fecundity is one of the easier demographic parameters to estimate in birds and mammals. Captive breeding data will usually provide estimates of gestation period and clutch or litter size that can be applied to wild populations without too much error. One adjustment that does need to be made, however, is to estimate the proportion of adult females that are breeding in a given population at a particular time. This is not a straightforward exercise (see Clobert *et al.*, 1994). In the fisheries literature, much attention is given to estimating fecundity, usually by estimating the number of eggs carried by gravid females. This is valuable for determining the relative contribution of various age or size classes to the overall reproductive output of the population. Without an estimate of larval mortality, however, it sheds limited light on the reproductive output itself.

Fecundity will always be age-dependent. It is frequently also size-dependent, particularly in ectothermic animals. Juveniles, by definition, do not reproduce, and in almost all organisms the pre-reproductive stage is not such a small fraction of the total lifespan that it can be neglected. This is especially so because the intrinsic growth rate of a population is particularly sensitive to early reproduction (see Chapter 5). Nevertheless, some very abstract models (for example, the simple host–parasite models of Anderson and May, 1978) assume a single birth-rate parameter and require it to be separated from the death rate. Attempting to estimate this as simply the rate of production of offspring by mature females will always cause a gross overestimate of the intrinsic growth rate of the population, to an extent that the qualitative behaviour of the system will probably be misrepresented.

There is no entirely satisfactory way to replace an age-specific fecundity estimate with a simple, unstructured rate. Probably the best approach is to use the full age-specific fecundity and mortality schedules to estimate the intrinsic rate of growth *r* (see Chapter 5), and then to subtract the death rate from this to yield a single birth-rate parameter.

Survival

Survival is inevitably a much more difficult process to quantify in wild populations than is fecundity. Survival of captive animals will always be a very poor indicator of survival in the wild, except that it may give an indication of the absolute maximum age the animals can reach. Estimating the rate of survival in the field ideally involves following the fate of individuals, preferably of known age. Given restrictive assumptions, some idea of the age-specific death rate can be gained from the age distribution of animals, but this requires a means of ageing individuals, a problem that is often technically difficult.

The most appropriate ways to describe survival for a model and then to estimate parameters to quantify the chosen form of survival depend critically on the type of organism under study and on the objectives of the study. In any but the most general models, the death rate will be a function of age. The death rate at a given age may also be dependent on time and any number of other variables.

Broadly speaking, the pattern of age-dependent survival in a population can be described in two ways. A life table breaks the life history up into a number of discrete ages or stages, and estimates a survival rate through each stage. A survival curve represents the death rate as a function of age. From the perspective of parameter estimation, the difference between the two approaches is similar to the difference between analysis of variance and regression. A life table requires the estimation of as many survival rates as there are stages in the table, but is completely flexible in the way that it allows survival to change with age. A survival curve will generally require the estimation of two or three parameters only, depending on the functional form fitted, but the way in which survival changes with age is constrained. Alternatively, a life table can be viewed as expressing survival as a nonparametric function of age, in contrast to the parametric approach of a survival curve (Cox & Oakes, 1984, p. 48).

Studies of insects generally have few problems with sample size, and often use life-table approaches to advantage. In many studies of vertebrate populations, however, sample size is a major constraint, and estimating many parameters simultaneously is impractical. An approach based on a two- or three-parameter survival curve may thus be preferable to a life table, even if the resulting mortality estimates are subsequently used in a model with discrete age classes.

Particular problems arise with models that are structured by size or developmental stage. For a given death rate per unit of time, the proportion of individuals surviving through a life-history stage will obviously depend on the time spent in that stage. Unless the development time through a stage is fixed, parameterization of stage-specific survival requires simultaneous estimation of the development and mortality rates.

The most appropriate method of estimating survival also depends on a number of practical considerations. Some methods can only be used if animals can be individually marked and then followed reliably until either death or the end of the study. If individuals are marked, but will not necessarily be detected on every sampling occasion, even if alive and in the study area, then other methods must be used. It may be possible to mark or identify members of, or a sample from, a cohort, but not to follow particular individuals through time. Again, different estimation methods must be used. Finally, in many situations, it is not possible to mark animals in any way, and attempts must be made to estimate survival from the age or size distribution of the animals.

Faced with this diversity of problems and possible approaches, I will first develop some general principles and concepts, and then deal with the actual estimation problems in a series of subsequent sections. Table 4.1 provides a guide to the possible approaches that can be taken to estimate survival or mortality rates.

Survival curves, failure times and hazard functions

A survival curve or survivor function is simply the proportion of individuals surviving as a function of age or time from commencement of the study. In the general ecological literature, it has become conventional to divide survival curves into three types, I, II and III (Krebs, 1985; Begon *et al.*, 1996a) (Fig. 4.1), depending on whether the relationship between log survivors and age is concave, straight or convex. These types correspond to death rates that are decreasing, constant or increasing with age. Almost all survival curves for real organisms, however, will include segments with very different survival rates, corresponding to different ontogenetic stages.

It is tempting to estimate a survivor function by fitting a line or curve through a plot of the proportion of individuals surviving versus time. This approach is invalid statistically because successive error terms are not independent: the survivors at any time have also survived to each previous time step. A better approach is to use the time until death of each individual (its 'failure time' in the statistical literature) as the response variable. A problem which then arises is that it is likely that some individuals will either survive to the end of the study or otherwise be removed. A minimum failure time can be assigned to these individuals, but the actual failure time is unknown. Such observations are 'censored'.

Survival of individuals throughout the course of a study is the most obvious way that censoring can occur, and will result in observations censored at a value equal to the total study duration. Censoring may, however, occur at other values. In many studies, subjects may enter the study at different times. For example, animals may be caught and radio-collared at times other than the start of the study. This is known as 'staggered entry' (Pollock *et al.*, 1989), and individuals surviving at the conclusion of the study period will have different censoring times, depending on their time of entry (Fig. 4.2). Alternatively, individuals may be removed from the study before its conclusion, due to factors other than natural mortality. Possibilities include trap mortality, permanent emigration, or radio-collar failure.

Statisticians usually describe survival data in terms of the hazard function, or the probability density of death as a function of time, conditional on survival to that time (Cox & Oakes, 1984). To ecologists, this concept will be more familiar as the instantaneous death rate of individuals, as a function

Table 4.1 A key for choosing a method of survival analysis. Use this like a standard dichotomous key for identifying organisms. Start at 1, on the left-hand side, and decide which of the alternatives best applies to your problem. Then go to the number associated with that alternative, and repeat the process. Continue until you reach a suggested method, or you reach a ×, which indicates that there is no satisfactory method

---

1   Do you want a continuous survival function or a life table (survival in discrete stages or intervals)?

Survival function   2
Life table   7

2   Can particular individuals be followed through time?

Yes   3
No   6

3   Can all individuals being followed in the population be detected at each sampling occasion, if they are present, or may some be missed?

All detected if present   4
Some may be missed   5

4   **Parametric survival analysis** (see p. 108)

5   **Mark–resight and mark–recapture methods** (see p. 119)

6   Can the number of survivors from an initial cohort of known size be determined?

Yes   4
No   ×

7   Can you follow an initial cohort through time?

Yes   8
No   13

8   Are the intervals of the life table of fixed (but not necessarily equal) duration, and can individuals be assigned to an interval unequivocally?

Yes   9
No   12

9   Do you need to use statistical inference, or is simple description sufficient?

Description only   10
Statistical inference required   11

10   **Cohort life-table analysis** (see p. 116)

11   **Kaplan–Meier method** (see p. 118)

12   **Stage–frequency analysis** (see p. 126)

13   Can individuals be aged?

Yes   14
No   18

14   Can the proportions of each age class surviving one time step be determined?

Yes   15
No   16

15   **Standard life-table techniques possible, but yields a *current* or *time-specific* life table** (see p. 116)

16   Is the rate of population increase known (or can it be assumed to be 0)?

Yes   17
No   ×

17   **Use 'methods of last resort'** (see p. 130)

18   **Try stage–frequency analysis** (see p. 126)

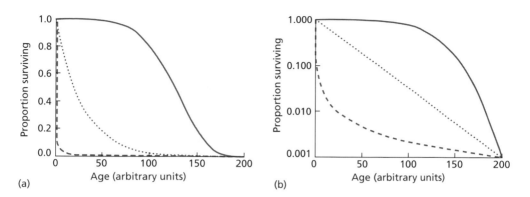

Fig. 4.1 Survival curves of types I, II, and III, with survival: (a) on a linear scale; (b) on a log scale. The solid line is a type I survival curve, the dotted line is a type II curve, and the dashed line is a type III curve. These were generated using a Weibull survival function (see Table 4.3), with: $\kappa = 5$ and $\rho = 0.007\ 36$ (type I), $\kappa = 1$ and $\rho = 0.034\ 54$ (type II); and $\kappa = 0.2$ and $\rho = 78.642$ (type III). I chose values of ρ so that survival to age 200 was standardized at 0.001 for each value of κ.

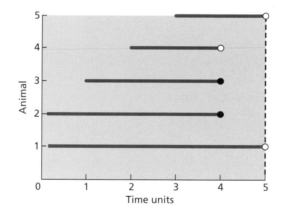

Fig. 4.2 Staggered entry and censoring in a survival study. This diagram represents the fates of five individuals in a hypothetical survival experiment running from times 0 to 5. The horizontal lines represent times over which animals were part of the study. Deaths are marked by solid dots, and open dots indicate *right-censored* observations: it is known that the animal survived to at least the time shown, but its actual time of death is unknown. Animal 1 was followed from the beginning of the study, and survived through to the end, so its survival time is right-censored at 5 time units. Animal 2 similarly was followed from the beginning, but was dead at time 4. Animal 3 did not enter the study until time 1, and was dead at time 4. Animal 4 entered the study at time 2, but at time 4 its radio-collar had failed, so it was not known if it was alive or dead. Its survival time is therefore censored at 2 time units. Finally, animal 5 entered the study at time 3, and survived until the end. Its survival time is thus also censored at 2 time units.

Table 4.2 Ways of representing the probability of survival

Function	Name	Meaning	Relationship to other representations
$S(t)$	Survivorship function	Proportion of cohort surviving to time t	$S(t) = \exp\left(-\int_0^t \mu(u)\,du\right)$
$\mu(t)$	Age-specific death rate, or hazard function	Probability density of failure at time t, conditional on survival to time t	$\mu(t) = f(t)/S(t)$
$f(t)$		Probability density of failure at time t	$f(t) = -\dfrac{dS(t)}{dt}$

of age. Survival can also be described as an unconditional probability density of failure as a function of age. This is different from an age-specific death rate, because it is the probability density of death at a given age of an individual at the start of the study. The unconditional probability density of failure is important, as it is this quantity that is necessary for maximum likelihood estimation of survival rate parameters. Table 4.2 summarizes these ways of describing age-specific survival, and the relationships between them.

Parametric survival analysis

The approaches described in this section produce an estimate of survival as a continuous function of age or time from the entry of the subject into the study. Survival may also depend on other variables, either continuous or categorical. These approaches can be applied in two situations: if animals are marked individually, and their continued survival at each sampling occasion can be determined unequivocally; or if animals are not followed individually, but the numbers from an initial cohort surviving to each sampling occasion can be counted. If animals are marked individually, but may not necessarily be detected on each sampling occasion, the mark–recapture methods discussed later in the chapter should be used instead.

It is first necessary to select an explicit functional form for the hazard function (the death rate, as a function of age), and the objective is then to estimate the parameters of that function. There are many forms of the hazard function in the statistical literature (see Cox & Oakes, 1984, p. 17). For most ecological purposes, two are adequate: a constant hazard or death rate, which yields an exponential survival function; and the Weibull distribution, which allows the hazard to either increase or decrease with time depending on the value of an adjustable parameter. Table 4.3 gives the hazard, probability density and

Table 4.3 Exponential and Weibull survival curves

Name	Hazard function	Density function	Survivorship function
Exponential	ρ	$\rho e^{-\rho t}$	$e^{-\rho t}$
Weibull	$\kappa \rho (\rho t)^{\kappa - 1}$	$\kappa \rho (\rho t)^{\kappa - 1} \exp[-(\rho t)^{\kappa}]$	$\exp[-(\rho t)^{\kappa}]$

survival functions of these two models. The exponential distribution has a single parameter ρ, which is the instantaneous death rate per unit of time. The Weibull distribution also has a rate parameter ρ, and, in addition, a dimensionless parameter κ, which determines whether the hazard increases or decreases with time. From Table 4.3, it can be seen that $\kappa > 1$ generates a death rate increasing with time (a type III survival curve) and that if $\kappa < 1$, the death rate decreases with time (a type I curve). If $\kappa = 1$, the Weibull reduces to the exponential distribution. Figure 4.1 shows examples of Weibull and exponential survival functions.

In many cases, it may be necessary to describe survival as a function of explanatory variables. These might be continuous, such as parasite burden, or discrete, such as sex. It is relatively straightforward to make ρ a function of such explanatory variables. For example, if a continuous variable X affects survival, ρ can be represented as

$$\rho(X) = \exp(\alpha + \beta X), \tag{4.2}$$

where α and β are parameters to be estimated. The exponential ensures that the death rate remains positive for all values of α, β and X. It is usual to assume that the index κ remains constant and does not depend on the value of the explanatory variable, although there is no good biological, as distinct from mathematical, reason why this should be so.

Most ecologists will use a statistical package to fit a survival function, although it is not difficult to do using a spreadsheet from the basic principles of maximum likelihood. Some statistical packages (for example, SAS) use a rather different parameterization from the one I have described here, which follows Cox and Oakes (1984). It is important to plot out the observed and predicted survival, both to assess the fit of the model visually, and to ensure that the parameter estimates have been interpreted correctly.

The survival rate of black mollie fish infected with *Ichthyophthirius multifiliis* (Table 4.4), an example first discussed in Chapter 2, is modelled using the Weibull distribution in Box 4.1, with the results presented graphically in Fig. 4.3. It is clear from the figure that the Weibull curve, using parasite burden as a continuous linear covariate, does not completely capture all the differences in survival between the different levels of parasite infection. In particular, the predicted mortality rate is too high for the lowest infection class. The fit can be improved by estimating a separate ρ for each infection class. This example emphasizes that procedures such as checking for linearity and lack of fit, which are standard practice for linear models, are equally important for survival analysis.

If very few data are available, the simplest and crudest way to estimate a death rate is as the inverse of life expectancy. If the death rate μ is constant through time, then the following differential equation will describe $N(t)$, the number of survivors at time t:

$$\frac{dN}{dt} = -\mu N(t). \tag{4.3}$$

This has a solution

$$N(t) = N(0) \exp(-\mu t). \tag{4.4}$$

Table 4.4 Black mollie survival data. The following table shows, as a function of parasite burden, the number of fish N that had died at the given times (in days) after the initial assessment of infection. The experiment was terminated at day 8, so that it is not known when fish alive on that day would have died. The survival time of those fish is thus censored at day 8

Burden	fail time	N	Burden	fail time	N
10	6	1	90	4	1
10	8*	10	90	5	2
30	3	1	90	6	1
30	7	2	90	7	1
30	8	4	90	8*	2
30	8*	3	125	3	1
50	3	2	125	4	4
50	4	1	125	5	1
50	5	3	125	6	2
50	6	1	125	8	1
50	7	3	175	2	2
50	8	5	175	3	4
50	8*	3	175	4	3
70	4	2	175	5	3
70	5	2	175	8	1
70	6	1	225	2	4
70	7	1	225	3	4
70	8	4	225	8	1
70	8*	4	300	2	1
			300	3	1
			300	4	4
			300	6	2
			300	7	1

*Censored observations.

Box 4.1 Fitting a Weibull survival function using maximum likelihood

The data used in this example are from an experiment investigating the survival of black mollie fish infected by the protozoan *Ichthyophthirius*. The experimental procedure is outlined in the legend to Table 2.1, and the data are presented in Table 4.4. The time of death of individual fish, up to a maximum of eight days after assessment of the infection level (which in turn was two days after infection) was recorded as a function of intensity of parasite infection. The objective is to use maximum likelihood to fit a Weibull survival curve with parasite infection intensity X as a covariate, and the rate parameter of the Weibull distribution given by

$$\rho = \exp(\alpha + \beta X). \tag{1}$$

There are two types of observation in the data set: uncensored observations, for which the actual time until death is known; and censored observations, for which it is known that the fish survived at least eight days, but the actual time of death is unknown. The probability density of death at time t, where $t \le 8$, is given in Table 4.3:

$$f(t) = \kappa \rho(\rho t)^{\kappa-1} \exp[-(\rho t)^{\kappa}]. \tag{2}$$

The probability that a fish survives beyond eight days is given by the survivorship function, evaluated at $t = 8$:

$$S(8) = \exp[-(8\rho)^{\kappa}]. \tag{3}$$

The log-likelihood function has two components, one for uncensored observations, and a second for the censored observations. Following the usual approach of maximum likelihood estimation, the log-likelihood function of the three unknown parameters α, β and κ can then be written as a function of the observed data:

$$l(\alpha,\beta,\kappa) = \sum_u \ln(f(t)) + \sum_c \ln(S(t)). \tag{4}$$

Here, the u represents a sum over the uncensored observations, and the c a sum over the censored observations. (Note that in this case, the time at censoring entered into $S(t)$ is 8 for every censored observation.) The maximum likelihood estimates of the parameters are those which maximize the value of eqn (4). Most ecologists would probably use a statistical package to perform the calculation. It is, however, quite straightforward to do using a spreadsheet. Substituting eqns (2) and (3) into eqn (4), the problem is to maximize

$$l(\rho,\kappa) = \sum_u (\ln \kappa + \kappa \ln \rho + (\kappa - 1)\ln(t) - (\rho t)^{\kappa}) - \sum_c (\rho t)^{\kappa}. \tag{5}$$

continued on p. 112

Box 4.1 *contd*

Substituting the expression for ρ from eqn (1), and further simplifying, the objective is to find estimates of α, β and κ to maximize

$$l(\alpha,\beta,\kappa) = \sum_u [\ln \kappa + \kappa(\alpha + \beta X) + (\kappa - 1)\ln(t)] - \sum_{u,c} (t \exp(\alpha + \beta X))^\kappa. \tag{6}$$

This can be done using the Solver facility in Excel, as is shown for a simpler example in Box 2.3. In this case, all that is necessary is to enter each component of the sums in eqn (6) into separate rows of a spreadsheet, and to maximize their sum. The solutions are:

α	β	κ
−2.211 56	0.002 789	2.892 263

As would be expected, β is positive, meaning that the death rate increases with parasite burden. As $\kappa > 1$, the death rate for a given parasite burden also increases strongly with time since infection. This is also to be expected, given that this is a parasite which grows rapidly in size over its 10-day lifetime on the host.

The predicted survival curves are shown in Fig. 4.3. Whilst the fit seems fairly good for intermediate parasite burdens, there is evidence of systematic differences between observed and predicted survivals, particularly at the lowest infection intensity. This subjective impression is confirmed if separate values of ρ are fitted for each value of the parasite burden. Again, this can be done quite easily on a spreadsheet. The estimated parameters are shown below.

Burden	10	30	50	70	90	125	175	225	300
$\ln(\rho)$	−2.860 88	−2.139 94	−1.987 8	−2.073 74	−1.979 54	−1.670 69	−1.509 17	−1.427 54	−1.59 146

The estimated value of κ is 3.00.

The change in deviance (twice the difference in the log-likelihood between the two models) is 23.58 at a cost of 7 additional parameters, a highly significant difference. The predicted survival using this categorical model is also shown in Fig. 4.3.

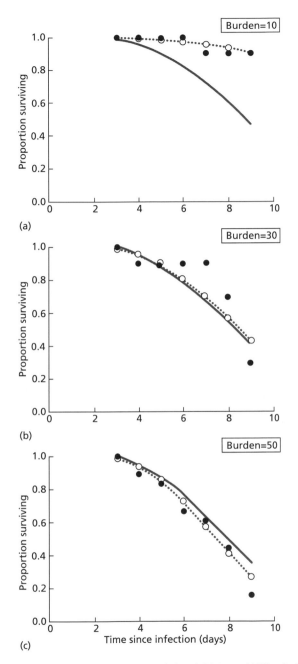

(a)

(b)

(c)

Fig. 4.3 Survival of black mollies infected with *Ichthyophthirius multifiliis*. Black dots show the observed proportions of fish surviving as a function of time after infection. The proportion surviving as predicted by a Weibull survival function, using burden as a continuous linear predictor, is shown as a solid line. The dotted line and open circles show the predicted proportion surviving if a Weibull with burden treated as a categorical variable is used.

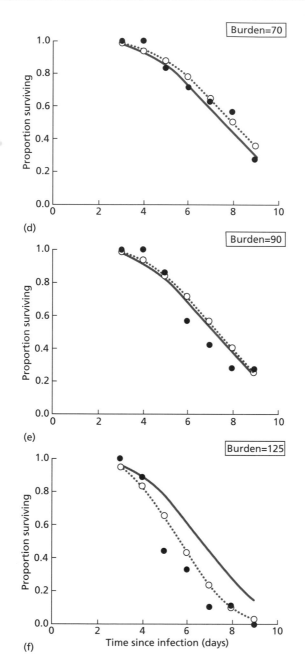

Fig. 4.3 *contd*

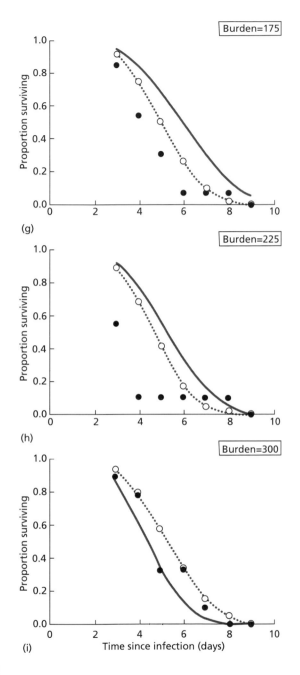

(g)

(h)

(i)

Fig. 4.3 *contd*

The definition of life expectancy, $E(t)$, is the average age each individual survives from time 0, and hence

$$E(t) = \frac{\int_{t=0}^{\infty} t\exp(-\mu t)dt}{\int_{t=0}^{\infty} \exp(-\mu t)dt}. \tag{4.5}$$

Integrating by parts, the solution to this is simply $1/\mu$.

This straightforward relationship is useful where a single death-rate parameter needs to be plugged into a model, as long as you remember that the life expectancy of an organism or life-history stage is usually much shorter than the maximum lifespan.

Life tables

A life table is simply a record showing the number of survivors remaining in successive age classes, together with a number of additional statistics derived from these basic data (Table 4.5). The method has its basis in human demography, and in particular, the life insurance industry (see Keyfitz, 1977). It is the first stage in constructing many age- or stage-structured models. The simplest life tables are single-decrement life tables, in which individuals dying in each age class or stage are recorded, without any attempt to identify the cause of death. A multiple-decrement life table (Carey, 1989) seeks further to attribute deaths to a range of possible causes, permitting the impact of various mortality factors on the population to be determined.

In developed countries, a full register of births and deaths for many years makes obtaining the data for human populations a straightforward exercise. Many animal and plant populations are a very different matter. With long-lived organisms, in particular, it is usually impossible to follow an entire cohort from birth to death. A wide variety of methods has been developed to deal, as far as is possible, with problems of limited sample size, limited study duration, organisms with unknown fate, difficulties with ageing and other sampling difficulties.

There are two basic forms of life table, whether single- or multiple-decrement tables are considered. The standard form is the cohort life table, in which a number of organisms, born at the same time, are followed from birth, through specified stage or age classes, until the last individual dies. The age- or stage-specific survival rates obtained are therefore for a number of different times, as the cohort ages. The second form of life table, a current life table, takes a snapshot of the population at a particular time, and then determines the survival over one time period for each age or stage class in the population.

Table 4.5 Structure of a basic single decrement life table. The simplest life table is a complete cohort life table, which follows an initial cohort of animals from birth until the death of the last individual. Two fundamental columns only are required to calculate the table, the age class (conventionally represented as x) and the number of individuals alive at the commencement of the age class (conventionally represented as K_x). From these, a further seven columns can be derived. Note that a life table *sensu stricto* is concerned with mortality only. Age-specific fertility is often added, and is of course essential for answering ecological questions about population growth rate, reproductive value and fitness. After Carey (1993)

Column	Definition	Calculation
x	Age at which interval begins, relative to initial cohort	Basic input
K_x	Number alive at the beginning of age class x	Basic input
D_x	Number dying in interval x to $x + 1$	$D_x = K_{x+1} - K_x$
l_x	Fraction of initial cohort alive at beginning of interval x	$l_x = K_x/K_0$
p_x	Proportion of individuals alive at age x that survive until age $x + 1$	$p_x = l_{x+1}/l_x$
q_x	Proportion of those individuals alive at age x that die in the interval x to $x + 1$	$q_x = 1 - p_x$
d_x	Fraction of the original cohort that dies in the interval x to $x + 1$	$d_x = l_x - l_{x+1}$ or $d_x = D_x/K_0$
L_x	Per capita fraction of age interval lived in the interval from x to $x + 1$	$L_x = l_x - \dfrac{d_x}{2}$
T_x	Total number of time intervals lived beyond age x	$T_x = \displaystyle\sum_{z=x}^{\infty} L_z$
e_x	Further expectation of life at age x	$e_x = T_x/l_x$

An *abridged life table* is simply one that has been amalgamated into a number of life-history stages, not necessarily of equal length. In the ecological literature, it would more often be called a stage-structured life table. The components of such a table are identical to the complete table, with the exception that L_x needs to be adjusted to take account of the varying length of each interval. Hence

$$L_x = n\left(l_x - \frac{d_x}{2}\right),$$

where n is the duration in time units of the life-history stage from x to $x + 1$. The idea here is simply that a proportion l_x of the individuals present initially enter the life-history stage, and those that die in the interval (a fraction d_x of the initial cohort) survive, on average, half the interval.

Fairly obviously, both methods will only be equivalent if death rates remain constant through time.

Cohort life tables

The simplest way conceptually of obtaining a life table is to identify or mark a large number of individuals at birth, and then to follow those individuals until the last individual dies. This approach is straightforward for univoltine insects, in which all of a single cohort is present at any given time, and the number of individuals followed can be very large indeed. There is a massive literature on the use of such life tables in applied entomology (see Carey, 1993, for a review). In many such cases, the life table is a set of descriptive statistics for the particular cohort under examination, so that estimation problems and determination of standard errors, etc. are not relevant. In other circumstances, particularly in studies of vertebrate populations, a sample of individuals in a single cohort may be identified and these may be intended to act as a basis for making inferences about the entire cohort. If this is the objective, statistical inference is involved. This can be done using the Kaplan–Meier or product-moment estimator (Cox & Oakes, 1984, p. 48).

Kaplan–Meier method

The idea behind this approach is simply that the K_x individuals entering age class x can be considered as independent binomial trials, and that the objective is to estimate the probability of failure in the stage. Using the notation of Table 4.5,

$$\hat{q}_x = \frac{D_x}{K_x}. \tag{4.6}$$

From this, it is easy to calculate an estimated survivorship function,

$$\hat{S}(x) = \prod_{j<x}\left(1 - \frac{D_j}{K_j}\right). \tag{4.7}$$

Equation (4.7) simply states that the proportion of individuals surviving from birth through to a given life-history stage can be obtained by multiplying the proportions that survive through each previous life-history stage. If there are no censored observations, eqn (4.7) will be exactly the same thing as l_x in the standard life table notation of Table 4.5. However, if observations are censored before the end of the study, the standard life-table formulation cannot be used. If K_x is considered to be the number of individuals at risk at time x (so that any censoring is assumed to act exactly at x, as would be the case if censoring occurred through trap mortality), then eqn (4.7) permits the survivorship to be estimated.

Approximate confidence intervals for the survivorship function can be calculated using Greenwood's formula (Cox & Oakes, 1984, p. 51),

$$\text{var}[\hat{S}(x)] = [\hat{S}(x)]^2 \sum_{j<x} \frac{D_j}{K_j(K_j - D_j)}. \tag{4.8}$$

This can be used to generate a confidence limit using the normal approximation. A more elaborate, but more accurate approach, particularly for the tails of the distribution, is to use profile likelihood-based confidence intervals (see Chapter 2; and Cox & Oakes, 1984, p. 51).

Whichever approach is used, any method based on binomial survivorship rests on the assumption that the probability of survival for each individual is independent. This will not be the case in many real situations. For example, one might expect heterogeneity in insect survival at a variety of scales, from individual plants through fields to regions. The use of standard binomial theory to calculate survival in such cases will result in the underestimation of confidence bounds, and such underestimation may be gross. The only solution is to measure survival rates separately for each subpopulation and to use the observed variation as a basis for estimation of the standard error in survival across subpopulations.

For many populations, following a set of individuals from birth is impractical. If detecting survivors relies on trapping to resighting, some animals may be missed on each sampling occasion, and may even never be recorded again, although they continue to live in the population. If cohorts overlap, it is necessary to mark and follow particular individuals to apply the standard approach, but this is often impossible. Other species may be so long-lived that no study of reasonable duration can follow a single cohort. A variety of less direct approaches must therefore be used.

Mark–resight and mark–recapture methods

I discussed methods for estimating population size using mark–recapture and mark–resight data in Chapter 3, and mentioned the fact that open-population methods also allow the estimation of recruitment and survival rates. In general, mark–resight or mark–recapture methods are rather more robust at estimating survival rates than population size. Most methods for estimating population size from mark–recapture data depend critically on the basic assumption that, on a given sampling occasion, the capture probability of all individuals is identical. This assumption is not as important for estimating survival rates. It is merely necessary that the survival rate of marked and unmarked animals should be the same, because the survival estimates are based on marked animals only. Reviews of the use of mark–resight and mark–recapture methods for estimating survival are provided by Lebreton *et al.* (1992) and Cooch and White (1998). The basic idea behind these approaches is discussed in Chapter 3, and follows theory first developed by Cormack (1964).

Mark–recapture methods for estimating survival require estimation of two sets of parameters: capture probabilities and survival probabilities. These parameters may depend on both the time at which an animal is recaptured,

Table 4.6 Parameters in an open mark–recapture study for a standard Cormack–Jolly–Seber model with six capture occasions. (a) Survival probabilities ϕ: the survival probability ϕ_t is the probability of survival from time t until time $t + 1$

Time of first capture (cohort) i	Capture time t					
	1	2	3	4	5	6
1	ϕ_1	ϕ_2	ϕ_3	ϕ_4	ϕ_5	
2		ϕ_2	ϕ_3	ϕ_4	ϕ_5	
3			ϕ_3	ϕ_4	ϕ_5	
4				ϕ_4	ϕ_5	
5					ϕ_5	

Table 4.6 *continued.* (b) Capture probabilities: the capture probability p_t is the probability that a marked animal, which is present in the population at time t, will be caught at that time. The final p_6 and ϕ_5 cannot separately be estimated, although their product $p_6\phi_5$ can be. Modified from Lebreton *et al.* (1992) and Cooch and White (1998)

Time of first capture (cohort) i	Capture time t					
	1	2	3	4	5	6
1		p_2	p_3	p_4	p_5	p_6
2			p_3	p_4	p_5	p_6
3				p_4	p_5	p_6
4					p_5	p_6
5						

and on the time at which it was first captured. The recapture probabilities p_{it} give the probability of an animal, which was first captured at time i, being recaptured at time t. The survival probabilities ϕ_{it} give the probability of an animal, which was first captured at time i, surviving over the period from t to $t + 1$. As with the Jolly–Seber method, discussed in Chapter 3, it is not possible, in general, to distinguish marked animals that did not survive the last recapture interval from those that survived, but were not captured at the final capture time. In the most general case, there will be $(k - 1)^2$ separate parameters that can be estimated if there are k capture occasions. Usually, however, a model simpler than the full one can be used.

The Cormack–Jolly–Seber model, first discussed in Chapter 3, has the parameter structure shown in Table 4.6. Survival and recapture probabilities vary between sampling times, but do not depend on when the animal was first caught. Often, a still simpler model with fewer parameters will be adequate. Recapture probabilities or survivals may be constant through time, or it may be possible to predict survival using a logistic regression with a covariate such as rainfall. If this can be done, survival over any number of periods can be modelled with two parameters only.

Table 4.7 Survival probabilities for an age-structured model. This matrix of survival probabilities is for a model in which there are three age classes – juveniles, subadults and adults – and six capture times. At first capture, all animals caught are juveniles. If they have survived one capture interval, they become subadults, and if they survive two capture intervals, they become adults. Survival depends only on age, and not time. The probability of survival through the juvenile stage is ϕ_j, the probability of survival through the subadult stage is ϕ_s, and the survival through each sampling period for adults is ϕ_a

Time of first capture (cohort) i	Capture time t					
	1	2	3	4	5	6
1	ϕ_j	ϕ_s	ϕ_a	ϕ_a	ϕ_a	
2		ϕ_j	ϕ_s	ϕ_a	ϕ_a	
3			ϕ_j	ϕ_s	ϕ_a	
4				ϕ_j	ϕ_s	
5					ϕ_j	

If animals are always first caught as juveniles, it is straightforward to add age-dependent survival or recapture probability to this model. For example, Table 4.7 shows the survival parameter structure for a model in which survival of a given age class does not vary through time, but is different for juveniles, subadults and adults. It would also be possible to make these survival parameters time-dependent as well, or to make capture probabilities age-dependent. Things are a little more complex if some animals are originally caught as subadults or adults. It is necessary to treat animals first caught as subadults or adults as two additional groups, each with their own matrix of capture and survival probabilities, and then to fit a model constraining adult and subadult survival to be the same at a given time in each of the three groups.

As raw data, mark–recapture methods use capture histories, which are sequences of 0s and 1s, representing whether or not a given animal was captured or not on any sampling occasion. I first discussed capture histories in the previous chapter. Given a set of recapture and survival probabilities, it is possible to write down the probability that an animal first captured at time t has any particular subsequent capture history. For example, in the model shown in Table 4.6, the probability that an animal first caught at time 2 having a subsequent capture history 0101 (missed at time 3, caught at 4, missed at 5, caught at 6) is $\phi_2(1 - p_3)\phi_3 p_4 \phi_4(1 - p_5)\phi_5 p_6$. By using a maximum likelihood method, closely related to logistic regression (see Chapter 2), it is possible to use a set of observed frequencies of occurrence of capture histories to estimate the parameters that best predict these capture history frequencies. I will not go into the technical details of the estimation procedure. Packaged programs to perform the analysis – for example, MARK (Cooch & White 1998) – are freely available on the World Wide Web, and few working ecologists would attempt to write their own algorithms from first principles.

The most important issue in using these models is how to go about choosing an appropriate model. As with any statistical model, if you fit too many parameters, the resulting estimates are imprecise, simply because relatively few data points contribute to the estimation of each parameter. Conversely, too few parameters can fail to represent the true pattern of survival and recapture adequately. Mark–recapture methods generate a quite bewildering number of models, each with a very large number of parameters, even if the range of underlying biological assumptions being explored is quite modest. There is an extensive literature on model selection for mark–recapture methods, and some controversy about the most appropriate ways to select the 'best' model (see, for example, Burnham & Anderson, 1992; Lebreton *et al.*, 1992; Burnham *et al.*, 1995; Buckland *et al.*, 1997; Cooch & White, 1998). The first, key, point is that models must be grounded in ecological reality. Using knowledge about the factors that might affect catchability and survival, you should select candidate models that you then can examine to see if they fit the data. Models should be compared to the data, and not simply derived from it. Model selection and development is an iterative process, and no single-step method is likely to work well. Having selected a suitable suite of candidate models, some objective methods for model comparison are helpful, but these will never replace sensible biological judgement.

There are two main 'objective' methods that have been used to guide model selection in mark–recapture studies: likelihood ratio tests and the Akaike information criterion. I discussed both of these in Chapter 2. You should recall that likelihood ratio tests can only be used on nested models. That is, they can be used to compare a complex model with a simpler one, in which one or more of the parameters of the complex model are set to a specific value (often 0). For example, you could compare a model in which survival was different for each sex with one in which survival for each sex was the same. Models that are not nested can be compared using the AIC, which is based on the goodness of fit of the model, adjusted by a 'penalty' for the number of parameters that have been fitted.

Likelihood ratio based model comparison has its roots in formal hypothesis testing, which will be only too familiar to anyone who has done a basic biometrics course. The underlying philosophy is that a simple model should be used, unless there is strong evidence to justify using a more complex one. In some contexts, applying a strict hypothesis-testing approach to model selection is clearly appropriate. For example, consider a mark–recapture study to see if animals treated with a drug to remove parasites differed in their survival from control animals. A likelihood ratio test comparing the complex model, in which survival differed between the two groups, versus the simple model, in which survival was equal, is the appropriate way to proceed. In this case, the simple model has logical primacy over the complex model. Consider, however, another scenario, in which a mark–recapture experiment is being used to estimate survival parameters for an endangered species, so that the future population trajectory can be modelled. Common sense might suggest that survival probably will be different in adults and juveniles. It might not be sensible to use a model with constant survival for both age classes, purely because a likelihood ratio test, using $p = 0.05$, was unable to reject the null hypothesis of equal survival. In this case, the survival model most strongly supported by the data should be used. One model does not have logical primacy over the other.

In algebra,

$$LR = \Delta Dev, \tag{4.9}$$

where ΔDev is the difference in the deviance between the simple and the complex models. To perform a likelihood ratio test, LR is compared with a χ^2 table, with degrees of freedom equal to the number of additional parameters in the complex model. If LR exceeds the tabulated value at a certain upper tail probability, the null hypothesis that the simple model is adequate can be rejected at that significance level.

AIC-based approaches do not give logical primacy to any particular model, but the criterion attempts to identify the model most strongly supported by the data. In algebra,

$$AIC = Dev + 2p, \tag{4.10}$$

where Dev is the deviance of the model in question, and p is the number of parameters fitted in the model. The model with the lowest AIC is most strongly supported by the data. Blind selection of the model with the minimum AIC should not be relied upon. One method that has been suggested (Buckland *et al.*, 1997; Cooch & White, 1998) to assess the relative support (loosely, the relative likelihood of each model) is to use normalized Akaike weights,

$$w_i = \frac{\exp\left(\dfrac{-\Delta AIC_i}{2}\right)}{\sum\left\{\exp\left(\dfrac{-\Delta AIC}{2}\right)\right\}}. \tag{4.11}$$

Here, ΔAIC_i is the difference in the value of AIC between candidate model i and the model with the lowest AIC, and the sum is over all models being compared. Given the form of eqn (4.11), the model with the highest Akaike weight will be the same model as the one with the lowest AIC. However, calculating Akaike weights will help you assess whether there are several models with roughly the same support, or whether the best model is considerably more likely than any of the others.

Box 4.2 shows an example of mark–recapture analysis using these model selection approaches.

Box 4.2 Analysis of survival in a ghost bat population by mark–recapture

The ghost bat *Macroderma gigas* is a large, carnivorous bat, endemic to Australia. The data in this box come from a long-term study (S.D. Hoyle *et al.*, unpublished) conducted on a population near Rockhampton, Queensland, between 1975 and 1981. At three-monthly intervals, bats were captured by mist nets at the mouths of caves. Approximately 40 different bats were captured in each three-monthly sampling interval. There is marked seasonality, both in the breeding of the bats (young are born from mid-October to late November, and are weaned by March) and in the climate (summer wet season). Many of the models investigated therefore include a seasonal component. I have selected this data set as an example, as it has an amount of messiness and ambiguity typical of real problems. I will consider only adult bats here, although the original paper includes age-structured survival analysis.

The following table shows results from fitting a series of models to these data using SURGE (Pradel & Lebreton, 1993), although MARK (Cooch & White, 1998) should produce almost identical results. Models are represented with factors connected by + for additive effects, and * for models including interaction. For example, if survival is modelled as sex*time, this means that separate survival parameters were estimated for each combination of sex (male or female) and time (the intervals between the 19 three-monthly sampling times). In addition to the self-explanatory factors (year, season, sex and time) a covariate, rain, which is six-monthly rainfall, and a dummy variable, d80, which separately fits the last two time intervals in 1980, were also used in models. Column headings show the model number; the survival model; the recapture model; the number of parameters in the model, *np*; the deviance of the model; *AIC*, the value of Akaike's information criterion; and *w*, the Akaike weights.

The first two models are a full model, in which survival and recapture probabilities are different for each combination of sex and time, and a simple model with time dependence only. Next, the table compares a series of models in which survival is kept with full sex and time dependence, but various simpler models for recapture are explored. Provided it makes biological sense, the model with the lowest value of *AIC* is the best model. Here, model (9), shown in bold, is selected. Recapture probability depends on an interaction between sex and season, with an overall year effect. Once an adequate representation for capture probability has been obtained, survival is then explored using this recapture model. The best model, as indicated by the lowest *AIC*, and an Akaike weight more than twice the next best model, is highlighted in bold. It suggests that survival varies between the sexes and seasons, is affected by rainfall, and is different for the last three sampling times. The resulting estimates of survival and recapture probability are shown below, with 95% confidence intervals. Females are

continued on p. 125

Box 4.2 *contd*

shown with closed squares, and males with open circles. The results are rather disappointing. The survival rates have substantial standard errors, and are not precise enough to be useful for population projection. This is despite the fact that most of the captured animals were marked, once the experiment was under way (close to 80%), and the fact that the total population size was only about three times the average number of animals captured per session (Jolly–Seber estimates over the period averaged about 120). Something quite odd is also happening to the winter survival rates.

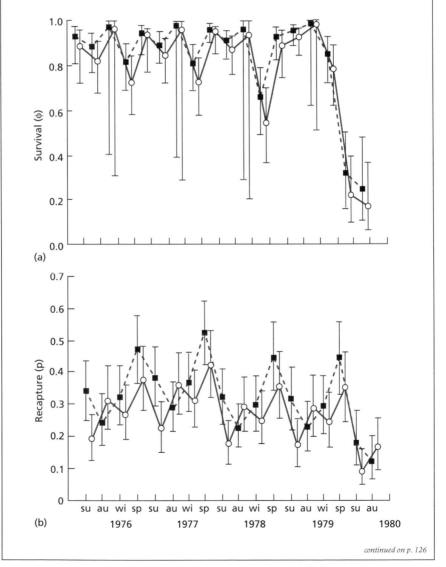

continued on p. 126

Box 4.2 *contd*

No.	Survival (ϕ)	Recapture (p)	np	Deviance	*AIC*	*w*
I Basic models						
(1)	sex*time	sex*time	70	1981.5	2121.5	
(2)	time	time	31	2027.3	2089.3	
II Modelling recapture						
(3)	sex*time	sex + time	55	2003.2	2113.2	
(4)	sex*time	sex	38	2044.6	2120.6	
(5)	sex*time	time	53	2005.6	2111.6	
(6)	sex*time	constant	37	2046.9	2120.9	
(7)	sex*time	sex*season	44	2017.7	2105.7	
(8)	sex*time	sex + season	41	2024.8	2106.8	
(9)	**sex*time**	**sex*season + year**	**48**	**2007.9**	**2103.9**	
(10)	sex*time	sex*season + d80	45	2013.4	2103.4	
III Modelling survival						
(11)	sex + time	sex*season + year	31	2020.1	2082.1	0.00
(12)	sex	sex*season + year	14	2059.9	2087.9	0.00
(13)	time	sex*season + year	30	2026.9	2086.9	0.00
(14)	constant	sex*season + year	13	2065.1	2091.1	0.00
(15)	sex*season	sex*season + year	20	2049.7	2089.7	0.00
(16)	sex + season	sex*season + year	17	2050.5	2084.5	0.00
(17)	sex + season + year	sex*season + year	21	2031.2	2073.2	0.02
(18)	sex*season + year	sex*season + year	24	2029.3	2077.3	0.00
(19)	sex*year	sex*season + year	22	2032.0	2076.0	0.01
(20)	sex + year	sex*season + year	18	2038.6	2074.6	0.01
(21)	sex + season + d80	sex*season + year	18	2034.7	2070.7	0.08
(22)	**sex + season + rain + d80**	**sex*season + year**	**19**	**2028.6**	**2066.6**	**0.59**
(23)	sex*season + rain + d80	sex*season + year	22	2027.2	2071.2	0.06
(24)	sex*rain + season + d80	sex*season + year	20	2028.4	2068.4	0.24

Stage–frequency data

Many populations are most sensibly modelled using life-history stages of differing and possibly varying duration, rather than with fixed age classes (see, for example Nisbet & Gurney, 1983; Caswell, 1989).

Parameterizing such models is not straightforward. The available data will usually consist of records of the number of individuals currently in each stage, as a function of time throughout a season or period of study (see Fig. 4.4). In many cases, recruitment to the first stage occurs over a limited period, at the beginning of the 'season', but in others it may continue for most or all of the study period. From such stage–frequency data, the usual objective is to estimate some or all of the following (Manly, 1989; 1990):

1   The total number of individuals entering each stage.
2   The average time spent in each stage.

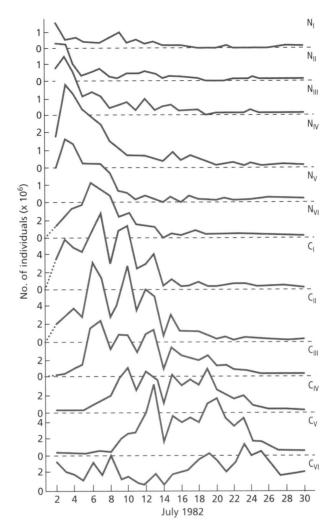

Fig. 4.4 Cohort analysis for one generation of the copepod *Diaptomus sanguineus*. Each panel in the figure is for one instar, labelled from N_I through to C_{VI} at the right of the figure. Samples were collected daily from a temporary pond, using triplicate traps. From Hairston and Twombly (1985).

3   The probability of surviving through each stage.
4   The mean time of entry into each stage.
5   Survival per unit time in each stage.

The estimation problems involved are tackled in parallel in the entomological and limnological literatures. A full discussion of the possible approaches and problems is beyond the scope of this book. Here I will outline the nature of the problem, discuss why some intuitively appealing methods are flawed, and describe in very broad terms the possible ways forward. Anyone attempting to use these methods in practice should consult Manly (1989) for entomological examples, and Hairston and Twombly (1985) or Caswell and Twombly (1989) for limnological examples. More recently, Wood (1993; 1994) has developed mathematically sophisticated approaches that should be considered. To date, however, they have not often been applied in practice.

The basic problem in attempting to analyse data of this sort is well summarized by Wood (1994): 'There were two bears yesterday and there are three bears today. Does this mean that one bear has been born, or that 101 have been born and 100 have died?' There is insufficient information in a simple record of numbers of individuals in various stages through time to define birth and death rates uniquely. If our bears are marked, we have no problem, but if they are not, it is necessary to make assumptions of some sort about the processes that are operating.

In the limnological literature (for example, Rigler & Cooley, 1974), it has been conventional to estimate the number of individuals entering an instar by dividing the area under a histogram of abundance through time by the stage duration, i.e.

$$N_i = \frac{A_i}{T_i}, \tag{4.12}$$

where N_i is the number of individuals entering the ith stage, T_i is the stage duration and A_i is the area under the histogram. This would be perfectly sound if the stage duration were known, all individuals entered the stage simultaneously and they suffered no mortality. Equation (4.12) then would express the elementary fact that the height of a rectangle is its area divided by its width. Less trivially, eqn (4.12) is also valid if entry to the stage is staggered, provided again there is no mortality and the stage duration is known, because the total area of the histogram is simply the sum of a series of components $n_{it}T_i$, where n_{it} is the number of individuals entering stage i at time t. Thus,

$$A_i = \sum_t n_{it}T_i = T_i \sum_t n_{it} = T_i N_i. \tag{4.13}$$

However, Hairston and Twombly (1985) show that eqn (4.12) may grossly underestimate N_i if there is mortality during life-history stage i. If α_i, the rate of mortality in the stage i, is known, they show that it can be corrected for, to yield

$$N_i = \frac{A_i \alpha_i}{[1 - \exp(-\alpha_i T_i)]}. \tag{4.14}$$

The utility of eqn (4.14) is limited, beyond showing that eqn (4.12) produces an underestimate, because stage duration and stage-specific survival are unknown, and are usually the principal parameters of interest, as well as being harder to estimate than N_i itself.

An intuitively obvious way of estimating stage duration is the time between the 'centres of gravity' of successive cohorts (Rigler & Cooley, 1974), with the 'centre of gravity' or temporal mean M_i of cohort i being defined as

$$M_i = \frac{\sum tn_i(t)}{\sum n_i},$$ (4.15)

where t is time and $n_i(t)$ is the number of individuals in cohort i present at time t. Unfortunately, this simple approach is also invalid if either development times or mortality rates differ between instars (Hairston & Twombly, 1985).

Manly (1990) describes a series of methods that overcome the above difficulties. However, the overparameterization problem remains: to make the methods work, it is necessary to place restrictions on the possible parameter values (for example, it may be necessary to assume that the mortality rate is equal in each stage). Other, more elaborate methods may experience numerical problems in the face of sampling variation, which is a ubiquitous property of all biological data (Caswell & Twombly, 1989).

A method promising to overcome many of these difficulties has recently been developed by Wood (1993; 1994). In principle, the approach is relatively straightforward, but it does involve some quite complicated computation. Here, I will outline the principle, and indicate what sorts of data might be approachable in this way. For computational details, Wood's original papers should be consulted.

The method is based on the McKendrick–von Forster equation, first presented in Chapter 1,

$$\frac{\partial n(a,t)}{\partial t} + \frac{\partial n(a,t)}{\partial a} + n(a,t)\mu(a,t) = 0.$$ (4.16)

Here $n(a,t)$ is the number of individuals of age a at time t, and $\mu(a,t)$ is the age-specific death rate at time t. A straightforward rearrangement of eqn (4.16) results in an equation for the age-specific death rate,

$$\mu(a,t) = \left(\frac{1}{n(a,t)}\right)\left(-\frac{\partial n(a,t)}{\partial t} - \frac{\partial n(a,t)}{\partial a}\right).$$ (4.17)

Thus, if the shape of the population surface $n(a,t)$ can be estimated, eqn (4.17) will allow the estimation of the age-specific death rate, and the birth rate as a function of time will simply be $n(a,t)$ evaluated at $a = 0$.

Unfortunately, estimating the shape of the population surface is a far from trivial exercise. The data available are likely to be estimates of the number of individuals in each of several age classes, at a number of sampling times. This obviously bears some relationship to the value of the population surface at that time, but will always be noisy (that is, subject to sampling error), and will group age ranges together into stages. More precisely, p_{ij}, the number of individuals in stage j on the ith sampling occasion, will be

$$p_{ij} = \int_{a_{j-1}(t_i)}^{a_j(t_i)} n(x,t_i)\mathrm{d}x + \varepsilon_{ij}. \tag{4.18}$$

Here, $a_j(t)$ is the age at which individuals mature out of stage j at t_i, the time of the ith sample, and ε_{ij} represents the noise.

Wood (1994) proposes a complex numerical procedure that uses eqn (4.18) to extract an estimated population surface $\hat{n}(a,t)$ from a set of observed p_{ij}s. There are two important limitations in this procedure. First, the mortality rate $\mu(a,t)$ must change smoothly across stage boundaries. This means that if different life-history stages have very different ecology (as is the case with many insects), the method may fail. Second, the procedure requires that the ages at which stage transitions occur should be known, and that all individuals of a given age make the transition to the next stage after the same maturation time. The method therefore cannot deal with distributed maturation times, although it can handle stage durations that vary through time.

Whether this approach is useful for a wide variety of problems remains to be seen. At this stage, it seems that the computational complexity has prevented its general application.

Methods of last resort

This section describes a series of methods for estimating survival rates that are not ideal. For the estimates to be valid, they require fairly restrictive assumptions about either the population itself or the way in which data are collected from it. Nevertheless, it is often necessary to use models for practical purposes, in the absence of ideal data. Provided the restrictions are kept in mind, these methods may at least provide an estimate of some sort for natural mortality. Unless attributed specifically elsewhere, these methods are taken from Caughley (1977a, pp. 90–5).

Two of the methods rely on the population having a stable age distribution. This concept is explained in more detail in the following chapter. For the purposes of this chapter, it is sufficient to note that if age-specific mortality and fecundity rates remain constant, then, whatever the initial age composition, a population will eventually either grow or decline exponentially at a rate fixed by those mortality and fecundity rates. When it does so, the proportions of individuals in each age class are a function of the mortality and fecundity schedules alone, and do not depend on the initial age distribution. This age distribution that is determined by the mortality and fecundity schedules is called the stable age distribution. In the special case that the population size is constant ($r = 0$), the stable age distribution is called the stationary age distribution (see Chapter 5 for a more detailed discussion).

It should be clear that reliance on a stable or stationary age distribution is a very restrictive assumption. Not only do the mortality and fecundity schedules need to be constant through time, but they must also have been so for a considerable period of time. The time taken to converge to a stable age distribution is dependent both on the nature of the distributions of the life-history parameters and the initial age distribution, but a reasonable rule of thumb is that it will require at least two or three generations.

Comparison of age structure at successive times

If it is possible to obtain an accurate representation of the age structure of a population at successive time intervals, together with an overall estimate of population size, then survival over the time interval can be estimated, assuming the population is closed. It is obvious that, if the number of individuals in age class $x + 1$ at time $t + 1$ is 90% of the size of age class x at time t, then survival of this age class over the interval x to $x + 1$ must have been 90%. The difficulty with this intuitive approach is that it critically dependent on the age structure of the collected sample being properly representative of the population as a whole. Many, if not most, ways of sampling animal populations are likely to be age- or size-selective. In addition, it is necessary to assume that the population is closed, and that the individuals in the population can accurately be aged.

'Pick-up' sample

This method can be used if ages at death have been recorded from a population, usually by 'picking up' skulls or skeletal remains (Caughley, 1977a, p. 90). If the remains are an unbiased sample of all individuals that have died (and this is a big 'if') and provided the population has a stable age distribution and the rate of increase is known and constant, age-specific survival can be determined. The method has usually been applied in cases where r is assumed to be 0. In this case, the number of carcasses D_x of each age x will be proportional to the age-specific death rate, so that

$$d_x = \frac{D_x}{\sum D_x}.$$ (4.19)

If r is not constant, but is known, these death rates need to be corrected via

$$d'_x = d_x \exp(rx).$$ (4.20)

Standing age distribution

If r is known, and the population has a stable age distribution, then a single unbiased sample from the population at the time of breeding will enable calculation of the age-specific survival curve.

Following the definitions for life-table parameters in Table 4.5 (Caughley's 1977a notation is slightly different), the number of individuals in the population of age x is K_x. Thus, the proportions of individuals in each age class relative to age class 0 are

$$k_x = \frac{K_x}{K_0}. \tag{4.21}$$

This means that an estimate of k_x, $\hat{k}_x$, can be obtained from estimates of the numbers of individuals in each age class, divided by an estimate of the total number of offspring produced in the birth pulse, K_0. Then, provided the $\hat{k}_x$ can be considered estimates from the stable age distribution,

$$\hat{l}_x = \hat{k}_x e^{rx}. \tag{4.22}$$

Two points should be made about this estimate. First, l_x and k_x are identical if the rate of increase is 0. Second, it is important that the method be applied only to birth-pulse populations sampled at the time of breeding. If this is not the case, then the number of juveniles (0 + individuals) is not K_0: it is K_0 times survival from birth to the sampling time. This survival would need to be measured separately.

In practice, there are often problems in estimating survival rates from a sampled age distribution, because the number of individuals in successive age classes may not decrease monotonically. This would lead to some estimated mortality rates being negative. If this problem occurred when an entire population was sampled, it would indicate that the underlying assumption of a stable age distribution was violated. However, in a finite sample, it may merely be an artefact of sampling. Caughley (1977a) suggests smoothing the age distribution so that it is always monotonically decreasing. This approach has been used several times since (e.g. Skogland, 1985; Choquenot, 1993). The smoothing should be applied to counts K_x corrected for the rate of increase in the same fashion as eqn (4.22),

$$\hat{K}_x = K_x e^{rx}. \tag{4.23}$$

Caughley suggests two methods for smoothing these corrected frequencies: a polynomial regression of $\log(\hat{K}_x)$ versus x, and fitting a curve of the form

$$\text{Probit}\left(\frac{\hat{K}_x}{\hat{K}_1}\right) = a + bx^i, \tag{4.24}$$

where a, b and i are constants to be estimated, and Probit(p) is the value of a standard normal deviate that corresponds to the probability p in a cumulative normal distribution. Caughley recommends using eqn (4.24) on post first year survival only. Hence, $\hat{K}_1$ and not $\hat{K}_0$ appears in the denominator. Equation (4.24) is more flexible than a polynomial regression, but rather messier to fit. Some statistics packages provide routines for performing probit regression, using maximum likelihood. A logit transformation $(\ln[p/(1-p)])$ in place of the probit would probably produce very similar results, and any standard logistic regression routine could be used.

Growth rates

Any model of a size-structured population requires modelling of the growth rate of individuals through time. This problem has been tackled in great detail by fisheries biologists, largely because 'growth overfishing' (fishing too hard so that fish are caught too soon when they are still too small) is one of the major practical considerations in fisheries management (Hilborn & Walters, 1992).

The first problem in parameterizing a growth model is to decide on a suitable functional form. There is a plethora of possible functional forms in the fisheries literature (Ricker, 1979; Kaufmann, 1981; Schnute, 1981). Some of these are summarized in Table 4.8. The most commonly used growth model, by a considerable margin, is the von Bertalanffy model (von Bertalanffy, 1938),

$$l_t = l_\infty[1 - e^{-K(t-t_0)}]. \tag{4.25}$$

Here, l_t is the length of an individual at time t, l_∞ is the asymptotic length, K is the age-specific growth rate, and t_0 is a shift parameter to allow the extrapolated size at age 0 to be nonzero. An additional allometric parameter b is also frequently added, particularly if weight w, rather than length, is modelled:

$$w_t = w_\infty[1 - e^{-K(t-t_0)}]^b. \tag{4.26}$$

The general shape of von Bertalanffy growth curves is shown in Figure 4.5. Note that the form in eqn (4.25) cannot generate a sigmoidal curve, whereas eqn (4.26) can do so.

Two quite different sorts of data may be used to estimate a growth function for a population (Francis, 1988). Age–size data consist of records of animal size, together with their age. Recapture data are obtained by measuring, marking, releasing and then recapturing animals, so that the growth over the period at liberty can be measured.

Age–size data can obviously only be used if a means of ageing animals is available. The technical difficulties in ageing wild animals are substantial, but are beyond the scope of this book. Maximum likelihood estimation for any of the models in Table 4.8 is relatively straightforward, provided that the ages are accurate, and that the errors can be assumed to be additive (Kimura, 1980). As I explain in Chapter 2, any nonlinear least squares regression routine will

Table 4.8 Growth functions. In each equation, S represents size; t represents age or time; t_0 is a constant of integration; S_∞ is the asymptotic size, if it exists, and a, b, c and n are parameters to be estimated. Equations describing indeterminate growth do not reach an asymptotic adult size, whereas those describing determinate growth do

Function	Differential equation form	Name, if any, and comments	Source
$S = \exp(b[t + t_0])$	$\dfrac{dS}{dt} = bS$	Exponential. Indeterminate growth	Kaufmann (1981)
$S = [ab(t + t_0)]^{1/a}$	$\dfrac{dS}{dt} = bS^{1-a}$	Power. Indeterminate growth	Kaufmann (1981)
$S = S_\infty \exp[-\exp(-a[t - t_0])]$	$\dfrac{dS}{dt} = aS(\ln S_\infty - \ln S)$	Gompertz. Sigmoid, determinate growth; inflexion point is at t_0	Ricker (1979)
$S = S_\infty[1 + \exp(-a[t + t_0])]^{-1}$	$\dfrac{dS}{dt} = aS\left(1 - \dfrac{S}{S_\infty}\right)$	Logistic. Sigmoid, determinate growth. Inflexion point is at $S_\infty/2$	Kaufmann (1981)
$S = S_\infty[1 - \exp(-a[t - t_0])]^b$	$\dfrac{dS}{dt} = abS\left(\left[\dfrac{S_\infty}{S}\right]^{\frac{1}{b}} - 1\right)$	Generalized von Bertalanffy. Determinate growth. If S is length, b is usually set to 1, and growth is not sigmoid. If S is weight, $b \approx 3$, and growth is sigmoid	Modified from Kaufmann (1981). Called the *Pütter* equation by Ricker (1979)
$n > 1, c_1 > 0, c_2 > 0$ $S^{1-n} = \dfrac{c_2}{c_1} + K\exp(c_1(1 - n)t)$	$\dfrac{dS}{dt} = c_1S - c_2S^n$	Richards. Determinate growth with $S_\infty = (c_2/c_1)^{1/(1-n)}$. The positive parameter K is a constant of integration. Inflexion point is at $S = S_\infty n^{1/(1-n)}$. More flexible sigmoid curve than those above. This form is equivalent to exponential if $c_2 = 0$ and equivalent to logistic if $n = 2$	Ricker (1979)
$0 < n < 1, c_1 > 0, c_2 > 0$ $S^{1-n} = \dfrac{c_1}{c_2} - K\exp(c_1(n - 1)t)$	$\dfrac{dS}{dt} = c_2S^n - c_1S$	Richards. See above. This form is equivalent to the generalized von Bertalanffy if $n = 1 - 1/b$	

produce maximum likelihood estimates of the parameters in a nonlinear model, provided the errors are additive, independent and normally distributed with a constant variance. A problem that is likely to occur in growth data is that larger errors may be associated with larger sizes, and thus older animals. A transformation to stabilize the error variance should be applied, if necessary, before using the nonlinear regression procedure. The parameters estimated by this approach are the means of the parameters for all members of the sampled population. In particular, if an asymptotic size is estimated, it is the mean size of 'old' individuals in the population, and some animals would be expected to reach a larger size.

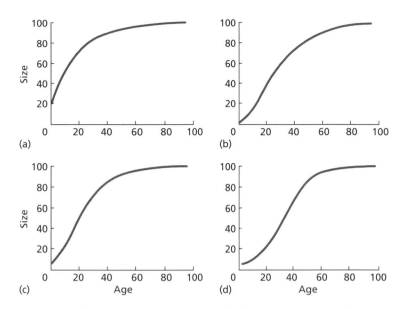

Fig. 4.5 Examples of growth curves: (a) von Bertalanffy: $S_\infty = 100$, $a = 0.05$, $b = 1$, $t_0 = -5.0$; (b) von Bertalanffy: $S_\infty = 100$, $a = 0.05$, $b = 3$, $t_0 = -5.0$; (c) Gompertz: $S_\infty = 100$, $a = 0.07$, $t_0 = 15$; (d) Logistic: $S_\infty = 100$, $a = 0.10$, $t_0 = -30$. Units of size and age are arbitrary.

Parameters estimated by recapture methods are not the same as those estimated from age–size data (Francis, 1988). Without an estimate of actual age, it is not possible to place a growth curve on the time axis. What can be done is to estimate the expected growth rate of an individual with an observed size S. It is easy to rearrange the von Bertalanffy equation for length (eqn (4.25)) into the form

$$\Delta l = (l_\infty - l)(1 - \exp(-k\Delta t)), \tag{4.27}$$

where Δt is the time at liberty, Δl is the increase in length, and l is the length at first capture. Given a data set including l, Δl and Δt for individual animals, it is also easy to fit eqn (4.27) by nonlinear least squares to produce estimates of l_∞ and k. The relationship between these estimates and those obtained from fitting eqn (4.25) to age–length data is not straightforward. It has been the topic of continuing debate in the fisheries literature (Sainsbury, 1980; Francis, 1988; James, 1991; Smith *et al.*, 1997). The problem is that fitting eqn (4.27) estimates the expected growth increment of an animal over a period Δt *given* that it is of length l at the start of the interval. This is not the same statistical problem as that of estimating the mean growth parameters over all individuals in the population. This point can be illustrated by considering the problem of estimating the asymptotic length l_∞. Unless any animals shrink, or unless there is measurement error, Δl in eqn (4.27) will always be positive, with the result that the estimated l_∞ will be larger than any observed l in the entire data set. This is not the case with estimates of l_∞ based on age–length data. (One would expect about half the 'old' animals to be larger than l_∞.) The solution to this problem depends on the assumptions that are made about the nature of the variability in the growth parameters between individuals in the population. Consult one of the above references for specific advice: the key point is not to expect estimates derived from age–length estimation and recapture data to be directly comparable.

If you want to fit a growth equation other than eqn (4.27) to recapture data, things are more complicated still. The general approach is to solve the differential equation form of the growth equation over a period Δt, subject to an initial condition that $S(0) = S_0$. For example, consider the form of the Richards equation for $0 < n < 1$ in Table 4.8,

$$\frac{dS}{dt} = c_2 S^n - c_1 S. \tag{4.28}$$

The solution of eqn (4.28) over a period t is also given in Table 4.8, with an integration constant K. If, at time 0, the value of S is S_0, then K must solve the equation

$$S_0^{1-n} = \frac{c_1}{c_2} - Ke^0 \tag{4.29}$$

$$\Rightarrow K = \frac{c_1}{c_2} - S_0^{1-n}.$$

Hence, when an animal that was initially caught at size S_0 was recaptured a period Δt later, its expected size $S_{\Delta t}$ would be the solution of

$$S_{\Delta t}^{1-n} = \frac{c_1}{c_2}[1 - (1 - S_0^{1-n})\exp(c_1[n-1]\Delta t)]. \tag{4.30}$$

Equation (4.30) is rather messier than eqn (4.27), but, given data on initial and final sizes, together with time intervals, its parameters can be estimated by nonlinear least squares. They will not have the same meaning as parameters fitted to the Richards equation using age–size data.

Stochasticity

Stochastic models, such as population viability analysis models, require an estimate of the probability distribution of survival and fecundity parameters, as well as point estimates. The interval estimates produced by most estimation procedures are based on the sampling distribution of the parameter. In other words, they are a measure of the precision by which the parameter is known, rather than a description of how the value of the parameter varies between individuals through time or across space. In the simple case of a parameter which is normally distributed, there is indeed a straightforward relationship between the estimate of the standard deviation of an estimated mean (its standard error) and the standard deviation of the quantity itself in the general population. This simple relationship does not exist for most of the parameter estimation procedures discussed in this chapter.

A common way to approximate the probability distribution of a parameter is to guess an appropriate value for its coefficient of variation (the standard deviation divided by the mean, and thus a measure of proportional variation), and then to assume that the distribution is of a standard form (usually normal or lognormal). This is, for example, the approach recommended in Burgman *et al.* (1993). As a first approximation, such an approach is defensible. However, it relies on a usually subjective estimate of the coefficient of variation, and furthermore assumes that a standard probability distribution is adequate. As extinction is often the product of extreme events (Ferson *et al.*, 1989), the tails of distributions are likely to be important, limiting the applicability of standard models.

A second approach is to identify an environmental variable, for which a long time series is available, as the major driver of stochasticity, and then to sample from the time series of that variable to generate stochasticity. This approach has been used for populations in arid areas, in which it can reasonably be assumed that rainfall is the fundamental driver of survival and fecundity (see, for example Bayliss, 1987; Caughley, 1987; McCallum *et al.*, 1995). There are two potential problems here. First, the relationship between rainfall and the parameters of interest may not be well known for endangered species. Second, and more fundamentally, this method assumes that the random variation is white, or uncorrelated variation. Annual rainfall itself may be serially correlated, due to large-scale phenomena such as El Niño (the southern oscillation). More fundamentally, rainfall probably acts on most animal populations via their food supply, introducing both lags and autocorrelation.

Summary and recommendations

1 Fecundity is one of the easier demographic processes to quantify. Data from captive or laboratory populations can be used to estimate clutch or litter size, and age at first breeding. However, the proportion of adult females that breed successfully also determines the mean fecundity of a population. This needs to be estimated from field data. In most modelling contexts, 'fecundity' also includes some juvenile mortality.

2 Survival is a difficult process to quantify adequately. Survival of captive animals rarely is a good indicator of survival in the field. Methods to quantify survival can broadly be clustered into life-table methods, which describe

survival through each of a series of life-history stages, and methods using parametric survival curves, which describe survival as a function of age. Consult Table 4.1 for guidance on selecting an appropriate method of analysis.

3 In the absence of reliable data, the inverse of the future life expectancy of an animal of a given age can be used as an estimate of its death rate. Remember, however, that the life expectancy is not the maximum lifespan: it is the mean lifespan.

4 Models of size-structured populations require the modelling of the growth rate of individuals through time. A wide range of growth curves is available in the fisheries literature, as summarized in Table 4.8. Parameter estimates based on age–size data are not directly comparable with those derived from growth increment data.

5 Stochastic models require estimates of the probability distributions of demographic parameters, as well as an estimate of their mean value. A common approach is to estimate the coefficient of variation of the parameter's distribution, using estimates from several different times or places, and then to use this in a standard probability distribution, such as the lognormal. An alternative approach is to estimate the parameter as a function of an environmental variable, and then to generate the parameter's distribution by sampling from an existing time series for the environmental variable.

CHAPTER 5

Rate of increase of a population

Introduction

The rate of increase of a population is a parameter basic to a great many ecological models. It is, however, an abstract quantity, being a combination and summary of age-specific mortality and fecundity. Perhaps more than any other parameter, the value appropriate for use in a model is context-dependent, varying both with the structure of the model and the model's purpose. In broad terms, estimates of r are required in two sorts of models: deterministic models, often of interspecific interactions or harvesting; and stochastic models, often used for population viability analysis of single species. Most deterministic models aim for a qualitative and general representation of population behaviour so that statistically rigorous estimation methods are not necessary. The main problem is to ensure that the rate being estimated is appropriate to the context of the model. Stochastic models, usually of single species, are likely to be directed at answering quantitative questions about specific populations and will thus require estimates that are as accurate as possible.

Definitions

The rate of increase r is defined from the most basic of population models, that for exponential growth,

$$\frac{dN}{dt} = rN,$$
(5.1)

with the solution

$$N(t) = N(0)e^{rt}.$$
(5.2)

Following this definition, r is an empirical quantity, and can be estimated from a time series of population size and time, without any need to consider the processes underlying population growth or decline.

In most deterministic models, however, the appropriate value of r is the maximum rate of population growth possible for the given species in the particular environment, excluding density-dependent constraints and any other factors that are explicitly included in the model. This quantity is often called r_{max} or r_m (for example, Caughley, 1977a) and referred to as the intrinsic rate of increase. An important point about this definition is that r_{max} is a property

139

of both the species and the environment the particular population inhabits. It is not a characteristic of the species alone.

Caughley (1977a) also defines two other rates of increase. These are r_S, the rate of increase at which a population would increase given the current fecundity and mortality schedules, once a stable age distribution is reached, and $\bar{r}$, the observed rate of increase of a particular population over a particular time period. Caughley's notation is useful, eliminating potential sources of confusion, but has not entered general usage. Most authors simply use r, and which of the different rates actually being referred to must be determined from the context.

In most circumstances, deriving r from a simple time series of population size will underestimate r_{max}, as various density-dependent factors are likely to be acting. The exception is when the population is increasing from a low base, in which case an estimate derived from a time series may well be appropriate.

The *finite rate of increase* λ_t is the proportional increase in total population size over a specified time interval $(t, t + \Delta t)$, i.e.

$$\lambda_t = \frac{N_{t+\Delta t}}{N_t} \tag{5.3}$$

or

$$\ln(\lambda_t) = \ln(N_{t+\Delta t}) - \ln(N_t). \tag{5.4}$$

Assuming that r has remained constant over the time interval Δt, λ and r are related by the simple equation

$$\ln(\lambda_t) = r\Delta t. \tag{5.5}$$

Estimating r from life-history information

The Euler–Lotka equation

Given a set of age-specific fecundities and mortalities, the potential rate of increase of a population can be determined from the solution of the Euler–Lotka equation:

$$\sum_{x=1}^{\infty} e^{-rx} F_x l_x = 1. \tag{5.6}$$

Here, x is the age of individuals, l_x is the age-specific survivorship (the proportion of individuals surviving from the first census to age x), and F_x is the age-specific fertility, or the number of individuals entering the first year class per individual of age x.

Defining parameters for the Euler–Lotka equation

There is considerable confusion in the literature concerning correct definitions of the parameters for the Euler–Lotka equation, and several textbooks are in error (Jenkins, 1988). To avoid these problems, a good solution is to index the age classes from $x = 1$, rather than from $x = 0$, that is, call the first age class age 1, not age 0. If this is done, it is best also to define l_x as the proportion of individuals surviving from the first census to age x, not from birth to age x as many authors do. Finally, it necessary to be very careful with the definition of age-specific fertility. In general, the number of newborn females produced per female of age x per unit of time (often called m_x) is *not* the appropriate definition of F_x required.

The difficulty arises because F_x is the number of individuals born since the previous census, alive at the current census, per individual of age class x present at the previous census. This means that the timing of the census relative to the timing of reproduction is important. Caswell (1989) has a particularly good discussion of the problems involved. You need first to decide whether your population breeds continuously (a birth-flow population) or whether births occur in a short, well-defined breeding season (a birth-pulse population).

In a birth-flow population, newborns will, on average, have to survive half a census interval to be counted in the first census and not all individuals of a particular age present at the previous census will survive long enough to give birth. Defining P_x as the probability that an individual will survive from age x to age $x + 1$, and $P_{J/2}$ as the probability of a newborn surviving half the interval between censuses, Caswell shows that

$$F_x = P_{J/2}\left(\frac{m_x + P_x m_{x+1}}{2}\right). \tag{5.7}$$

The formal derivation of eqn (5.7) is given in Caswell (1989). The first term is straightforward to understand: on average, newborns must survive half a census interval to be counted. The second term is trickier. The easiest way to understand it intuitively is that it is the average of m_x, the fecundity of individuals of age class x immediately after one census (when they are all still alive) and the reproductive output immediately before the next census when only the proportion P_x of the age class remains alive, but their fecundity is m_{x+1}.

In principle, birth-pulse populations could be censused at any time relative to the breeding season. It is convenient, however, to restrict discussion to either of two extremes: a census immediately preceding breeding, or a census immediately following breeding. For a prebreeding census, juveniles will only be counted if they have survived the entire census interval, but all individuals in age class x in the previous census will have survived to be able to breed. Thus

$$F_x = m_x P_J,\tag{5.8}$$

where P_J is the survival rate of juveniles in their first census interval. Conversely, for a post-breeding census, all juveniles born since the previous census will survive to be counted, but only a proportion P_x of individuals in age class x in the previous census will survive to reproduce. Hence

$$F_x = m_x P_x.\tag{5.9}$$

The problem of estimating survival and fecundity is considered in Chapter 4.

Solving the equation

The Euler–Lotka equation is an implicit equation for r. This means that (except for particular functional forms of l_x and F_x), it cannot be manipulated into a form $r = f(l_x, F_x)$. It must instead be solved by a process of successive approximation, in which the value of r is varied until the left-hand side of the equation equals 1. This is very easy to do using most spreadsheets (see Box 5.1). The r estimated is r_S: the rate at which the population would grow once a stable age distribution is reached.

Cole (1954) derived the following approximate form of the Euler–Lotka equation:

$$1 = e^{-r} + be^{-ra} - be^{-r(m+1)},\tag{5.10}$$

Box 5.1 Using a spreadsheet to solve the Euler–Lotka equation

This example uses the schedules shown in Table 5.1, and Excel notation.
Step 1. Enter age, fecundity and survival schedules:

	A	B	C	D	E	F	G	H	I
1	Age	1	2	3	4	5	6	7	8
2	Fx	0	0.25	0.4	0.8	0.8	0.6	0.5	0.2
3	Px	0.5	0.8	0.9	0.9	0.9	0.8	0.7	0

Step 2. Enter formulae to calculate l_x for ages 1 and 2 into cells B4 and C4 and for $e^{-rx}l_x F_x$ for age 1 into cell B5. Note that J2 ensures that cell J2 is referenced even when the formula is copied to other columns. A trial value of r is placed in cell J2.

continued on p. 143

Box 5.1 *contd*

	A	B	C	D	E	F	G	H	I	J
1	Age	1	2	3	4	5	6	7	8	r
2	Fx	0	0.25	0.4	0.8	0.8	0.6	0.5	0.2	0.05
3	Px	0.5	0.8	0.9	0.9	0.9	0.8	0.7	0	
4	lx		1	= B4*B3						
5	exp		= exp(−B1*J2)*B4*B2							

Step 3. Copy these across the other ages and sum row 5 in cell J5.

	A	B	C	D	E	F	G	H	I	J	K	L
1	Age	1	2	3	4	5	6	7	8	r		
2	Fx	0	0.25	0.4	0.8	0.8	0.6	0.5	0.2	0.05		
3	Px	0.5	0.8	0.9	0.9	0.9	0.8	0.7	0			
4	lx	1	0.5	0.4	0.36	0.32	0.29	0.23	0.16			
5	exp	0	0.11	0.14	0.24	0.2	0.13	0.08	0.02	=Sum(B5:15)		

Step 4. Use 'Goal Seek' to set cell J5 to 1 by altering cell J2.
'Goal Seek' is an Excel feature that uses successive approximations to set a certain cell in the spreadsheet to a specified value by altering the value of a number in another cell of the spreadsheet. It needs to be given an initial guess of the value of *r* to commence the process.

	A	B	C	D	E	F	G	H	I	J	K	L
1	Age	1	2	3	4	5	6	7	8	r		
2	Fx	0	0.25	0.4	0.8	0.8	0.6	0.5	0.2	0.03		
3	Px	0.5	0.8	0.9	0.9	0.9	0.8	0.7	0			
4	lx	1	0.5	0.4	0.36	0.32	0.29	0.23	0.16			
5	exp	0	0.11	0.14	0.24	0.2	0.13	0.08	0.02	1		

Table 5.1 A hypothetical fertility and survivorship schedule

Age	1	2	3	4	5	6	7	8
F_x	0	0.25	0.4	0.8	0.8	0.6	0.5	0.2
P_x	0.5	0.8	0.9	0.9	0.9	0.8	0.7	0

where b is the average number of offspring produced, a is the age at first reproduction and m is the age at last reproduction. Cole's equation assumes that reproduction occurs once per time interval, and that all females survive until after age m. It has been used in at least two recent studies (Hennemann, 1983; Thompson, 1987), but will always overestimate r because of the simple step function assumed for survivorship. Given the ease with which the Euler–Lotka equation can be solved on spreadsheets, there is now little need for approximate forms.

Depending on the shape of the initial age distribution, r_S may be very different from the r calculated by applying eqn (5.5) to a few years of growth in total population size. Consider the hypothetical fertility and survivorship schedules in Table 5.1, chosen to represent a population of macropods in good conditions immediately following the breaking of a drought. The value of r_S predicted from this schedule using eqn (5.6) is $r_S = 0.032\ 0$. Just before a drought breaks, it is likely that there will be very few juveniles in the population, and the age distribution might resemble that shown in Fig. 5.1(a). Figure 5.1(b) shows the total population size projected from this initial age distribution using the schedules in Table 5.1, and Fig. 5.1(c) shows r calculated from applying eqn (5.5) from year 1 forward to each subsequent year. Finally, Fig. 5.1(d) shows the stable age distribution reached after 20 years.

It is obvious that, for the first few years, the population grows at an average rate much greater than that predicted by the Euler equation. In fact, in the first year, the growth rate is 10 times higher than r_S. The reason in this particular case is that the initial age distribution contains a far higher proportion of animals in the age range of maximum fecundity than does the stable age distribution. One would expect a similar pattern in a translocation. It is usual to commence translocations with young adults, which will be individuals with particularly high survivorship and fecundity. Even without changes in the underlying age-specific survivorship and fecundity, one would not expect the initial rate of increase of a translocated population to be maintained.

Estimating r_{max} from the Euler–Lotka equation

Assuming that the parameters of the life table have been defined correctly, there is still the substantial problem of choosing estimates of these to estimate

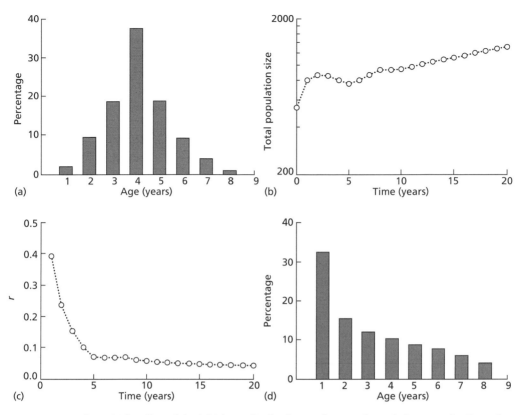

Fig. 5.1 The effect of the initial age distribution on the rate of population growth. Given the initial age distribution in (a), the fecundity and survivorship schedules in Table 5.1 will lead to the population growing as shown in (b). The estimated growth rate from time 0 to each time is shown in (c). The stable age distribution for these schedules is shown in (d). When this distribution is reached, the population will grow at the constant rate $r_s = 0.032\ 0$.

r_{max}. Data from a field population will not be appropriate unless that population either is increasing at r_{max}, or at least would do so if current conditions continued long enough for a stable age distribution to be reached. Suitable values of m_x can often be obtained from laboratory or captive populations: age at first breeding, gestation time, interval between breeding and litter or clutch size of captive populations are probably reasonable indications of what might be expected in the field under good conditions. As I discuss in the previous chapter, survival of captive animals gives a much poorer indication of survival in the wild. One possible approach is that used by Tanner (1975), who used the highest rates of juvenile and adult survival he could find in the literature to calculate r_{max} for 13 species of mammals. His estimates were, however, criticized by Caughley and Krebs (1983) as being too low. For example, Caughley and Krebs (1983) derive r_{max} for white-tailed deer as being at least 0.55 on the basis that an introduction of six animals had increased to 160 in six years, whereas Tanner's estimate is 0.30. The example in Fig. 5.1 cautions, however, that the initial rate of increase of a population of adult animals may greatly exceed r_S for that same population and environment.

Estimating r for stage-structured populations

The population dynamics of many species can be better described by dividing the population into stages, which may be of differing or variable duration, rather than ages, where each age class has an equal and fixed duration. Stages may be distinct developmental stages (egg, larva, pupa, etc.) or may be size classes. The critical difference from age-structured populations is that surviving individuals do not necessarily progress into the next stage each time step, but may also remain in the same stage, possibly jump two or more steps forward, or even drop back one or more stages.

There is an outstanding discussion of stage-structured discrete models in Caswell (1989), which I will summarize very briefly here. In principle, r can be estimated by a straightforward generalization of the approaches used for age-structured models. A square projection matrix is constructed, the elements of which contain transition coefficients. For example, a_{ij}, the coefficient in the ith row and jth column, gives the number of individuals in stage i next time step, per individual in stage j at the previous time. Hence,

$$n_i(t+1) = \sum_{j=1}^{s} a_{ij} n_j(t). \tag{5.11}$$

Having constructed the matrix, the long-term finite rate of increase, once a stable stage distribution is reached, is given by the dominant eigenvalue λ of the transition matrix. By rearranging eqn (5.5),

$$r_s = \frac{\ln(\lambda)}{\Delta t}, \tag{5.12}$$

where Δt is the length of the time step. Various computer packages for mathematical analysis can extract eigenvalues from matrices – for example, Mathematica (Wolfram, 1991). Methods to calculate eigenvalues can also be found in Press *et al.* (1994). In general, an $n \times n$ matrix has n eigenvalues, and these are often complex numbers. The dominant eigenvalue is the one with the largest magnitude of the real part.

As with age-structured models, it is necessary to be careful about defining fecundity terms in the matrix, particularly with respect to the timing of reproduction relative to the timing of the census. The crucial point remains that the first row of the matrix must contain the number of members of the initial stage *at the current census* per individual of each stage *present at the previous census.*

Estimating *r* from time series

The most obvious way to estimate *r* from a time series is from the slope of a least squares linear regression of ln(*N*) versus *t*, as suggested, for example, by Caughley and Sinclair (1994). A simple linear regression in this situation is a rather dubious exercise, however, as the errors associated with successive data points are likely to be related (autocorrelated) because random variations in any given sampling occasion will probably influence population size over a number of subsequent samples. Whether or not this statistical problem matters depends on what the estimate is required for and the relative importance of sources of variation in the time series.

There are at least three sources of 'error' or variation in a series of population size values versus time. As is discussed in Chapter 2, there is observation error: most such data sets are a series of estimates of population size or density rather than complete enumerations of the population. There also is process error of two sorts: some is associated with the fact that population size is finite and must be a whole number. Equation (5.1) is a deterministic equation, whereas actual population growth will be stochastic, or random. Birth and death are discrete processes that either do or do not occur to particular individuals in a certain time interval with particular probabilities. This form of randomness is often called 'demographic stochasticity'. The other source of process error is environmental stochasticity: temporal changes in the underlying rates of birth and death caused by environmental variability such as good or bad seasons.

There is no reason to expect observation errors in adjacent recording times to be related, unless the recording interval is so short that weather-related sightability or catchability influences more than one census. An exception may occur if population estimates are based on mark–resight or mark–recapture methods (see Chapter 3), although autocorrelation of such estimates does not seem to have formally been studied. If the population under study is large, it exists in a fairly constant environment and it is not subject to significant density dependence, then its population dynamics may be approximated well by eqn (5.1). In that case, variation in the estimated population size will predominantly be a result of sampling error, and a simple regression of ln(*N*) versus time will be the best way to estimate *r*.

In many (probably most) cases, there will be substantial variation in population size and population growth rate as a result of environmental stochasticity. This will introduce autocorrelation into the error terms of the time series, producing problems, particularly in the estimation of the standard error of the growth rate. A standard method in time-series analysis to reduce autocorrelation is to use differences between successive observations as the analysis variable. It is possible to show that this process will eliminate autocorrelation from a time series of population size generated from age- or stage-structured models such as those described by eqn (5.11), provided the transition probabilities are not correlated (Dennis *et al.*, 1991). In ecological terms, this means that age-dependent survival or fertility can vary according to environmental conditions, but such variation must be independent between observation periods.

If differences in log population size are calculated between observation periods with an initial population size observation N_0, and n subsequent observations $N_1, \ldots, N_n$ one time unit apart, then the mean growth rate can be estimated as

$$\bar{r} = \left[\sum_{i=1}^{n} \ln\left(\frac{N_i}{N_{i-1}} \right) \right] \bigg/ n.$$

(5.13)

This follows in a straightforward way from eqn (5.4). There are, however, complications. First, eqn (5.13) can easily be arranged into the form

$$\bar{r} = [\ln(N_n) - \ln(N_0)]/n,$$

(5.14)

meaning that the estimate of r is obtained only from the first and last values in the series. Intermediate values do not contribute at all to the point estimate, although they will influence its standard error. This is not unreasonable if the N_i are true counts, because both values are ones the population has in fact taken, and intermediate values have contributed to N_n in the sense that each population size builds on previous values. If the N_i are only estimates of population size, this estimation method has the undesirable feature that a single aberrant observation at the beginning or end will greatly effect the estimate of r. It is intriguing that this statistically rigorous approach produces the same point estimate as the very elementary approach that has been used by several authors (for example, Caughley & Krebs, 1983; Hanski & Korpimäki, 1995) of estimating r from two records only of population size over time.

A second difficulty is that eqns (5.13) and (5.14) measure the average growth rate of the population. This is not the same as the growth rate of the expected, or average, population size (Caswell, 1989). The difference is that the expected or average growth rate of a population is determined by the average of log population size, whereas the growth rate of the average population size is determined by the log of average population size. The growth rate of the average population size can be estimated by

$$\hat{r} = \bar{r} + \frac{\tilde{\sigma}^2}{2}.$$

(5.15)

Here is $\tilde{\sigma}^2$ the sample variance of $\ln(N_i/N_{i-1})$ (Dennis *et al.*, 1991), given by

$$\tilde{\sigma}^2 = \frac{\sum(r_i - \bar{r})}{(n-1)},$$

(5.16)

where $r_i = \ln(N_i/N_{i-1})$.

If the interval between observations is not constant, the problem becomes more complex. Defining τ_i as the time interval between recording N_{i-1} and N_i, Dennis *et al.* (1991) show that $\bar{r}$ and $\tilde{\sigma}^2$ can be estimated by a linear regression without intercept between

$$y_i = \ln\left[\frac{N_i}{N_{i-1}}\right]\bigg/\sqrt{\tau_i} \qquad (5.17)$$

as the dependent variable and $\sqrt{\tau_i}$ as the independent variable. The slope of the relationship is $\bar{r}$, and $\tilde{\sigma}^2$ is the mean square error for the regression. Provided n, the number of terms in the series, is not too small, $\hat{r}$ is approximately normally distributed with standard error

$$se_{\hat{r}} = \sqrt{\tilde{\sigma}^2\left[\frac{1}{t_n} + \frac{\tilde{\sigma}^2}{2(n-1)}\right]}, \qquad (5.18)$$

where t_n is the length of the series in time units.

An example of the application of this method to the classic data set of reindeer population growth after their introduction to St Paul Island in the Bering Sea (Scheffer, 1951) is given in Box 5.2.

It should be emphasized that, strictly speaking, this approach is only applicable if population size is known exactly. The question of how best to analyse data when both environmental stochasticity and sampling error are important has not satisfactorily been answered, but the problem is considered in Barker and Sauer (1992) and Sauer *et al.* (1994). (Also see Chapter 11.)

Box 5.2 Analysis of the growth rate of the St Paul Island reindeer population

In 1911, four male and 21 female reindeer were introduced on to St Paul Island (106 km^2) in the Bering Sea. By 1938, the population had increased to about 2000 animals in what has come to be regarded as a classic example of exponential population growth (Krebs, 1985). The population then crashed to eight animals in 1950. In this box, some of the methods of estimating r in Chapter 5 are applied to this data set using the data from 1911 through to 1940.

The original data are shown in the figure, plotted on a log scale with the least squares regression line from 1911 to 1940 superimposed. It appears that the residuals are highly autocorrelated, as there are long runs or values

continued on p. 150

Box 5.2 *contd*

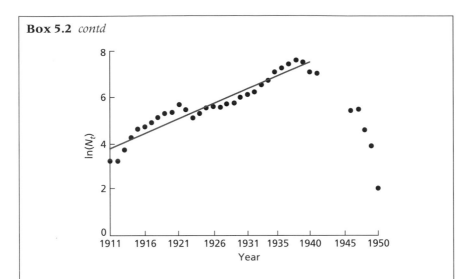

above and below the linear regression line. A standard way of testing for autocorrelation is to use a Durbin–Watson test (Kendall, 1976). The test statistic d is around 2 if a time series is not autocorrelated. The value of d obtained from analysis of $\ln(N)$ versus time up to 1940 is 0.388, which is much smaller than the critical value at $p = 0.01$ of 1.13. The gradient of the relationship, which is often used as an estimate of r, is 0.130. The standard error of the estimated value of r will be an underestimate of the true standard error because of the autocorrelation. It is 0.007 1.

The alternative approach is to use $X_t = \ln(N_t/N_{t-1})$ as the analysis variable. In this case, the data are still autocorrelated ($d = 1.127$, $p = 0.01$). The result suggests that there may be serial correlation in age-specific survival and mortality, as Dennis *et al.* (1991) have shown that X_t should be independent if fecundity and mortality are independent. The significant autocorrelation is not surprising, as the animals use a slow-growing resource. Together with some unknown observation errors, the autocorrelation in $X(t)$ should cause one to interpret the following results with caution.

Using equations (5.15) and (5.18) and data from 1911 to 1940, the estimated rate of increase and its standard error are $\hat{r} = 0.167\,33$ and $se_{\hat{r}} = 0.094\,1$. These are obviously different from the results obtained by linear regression. In particular, the standard error is an order of magnitude greater. In general, the point estimates of r obtained by the two methods will coincide only if all the data points fall on a single line. Here, $\hat{r} > r_{\text{regression}}$ for two reasons. First, the first data point is below the line of best fit and the last is above it, meaning that $\bar{r} > r_{\text{regression}}$ (see eqn (5.14)). Second, $\hat{r} = \bar{r} + \hat{\sigma}/2$ (see eqn (5.15)).

continued on p. 151

Box 5.2 *contd*

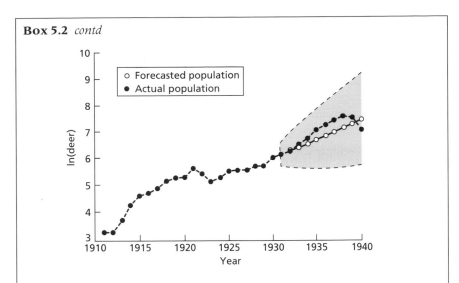

To see how well the method of Dennis *et al.* (1991) forecasts population size, I derived r from the first 19 years of data, and then used it to predict $N(t)$ for the next 10 years. Defining s as the number of years forward the forecast is required for, and n as the number of years from which the estimate of r is obtained, the standard error of $N(s)$ is

$$se_{N(t)} = \sqrt{\tilde{\sigma}^2 s(1 + s/n)}.$$

The forecast value of $N(t)$ is shown in the second figure, together with the observed $N(t)$ and a 95% confidence interval. As the observed data keep within the confidence band, the projection is satisfactory up to 1940.

Allometric approaches

It is obvious that r_{max} is higher for bacteria than mice, and that in turn r_{max} is greater in mice than elephants. This allometric relationship between body size and the intrinsic rate of increase appears first to have been formally described by Smith (1954) and has since been investigated by a number of authors. The relationship has the potential to be used for estimating r in cases where reliable life-history or time-series data are unavailable.

Table 5.2 shows details of some of the allometric relationships reported in the literature, standardized to units of grams and days. There is no doubt that there is a highly significant relationship between $\log(r_{max})$ and $\log(\text{mass})$ both between taxa and within taxa, and that the relationship is close to linear. On theoretical grounds, the slope of the relationship can be predicted to be close to -0.25 (Peters, 1983). In most cases, the estimated slope is remarkably

Table 5.2 Allometric relationships between r_{max} (d^{-1}) and mass (g). Standard errors (where available) are given in brackets after the parameter estimate

Taxon	Intercept	Slope	Sample size	Source
'Unicellular organisms'	−1.936 7	−0.28	26	Fenchel (1974)
Heterothermic metazoa	−1.693 1	0.273 8	11	Fenchel (1974)
Homoiothermic metazoa	−1.4	−0.275	5	Fenchel (1974)
Viruses to mammals	−1.6	−0.26 (0.012 6)	49?	Blueweiss *et al.* (1978)
Mammals	−1.872 1	−0.262 2	44	Hennemann (1983)
Mammals	−1.31	−0.36	9	Caughley and Krebs (1983)
Mammals	−1.436 0 (0.085 6)	−0.362 0 (0.026 6)	84	Thompson (1987)
Insects	−1.736 0 (0.075)	−0.267 (0.097)	35	Gaston (1988)

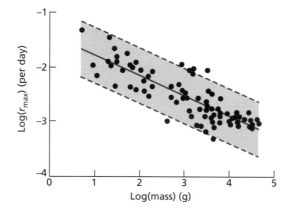

Fig. 5.2 Regression of $\log_{10}(r_{max})$ versus $\log_{10}$(mass) using data from Thompson (1987). The dashed lines are 95% prediction intervals for estimation of $\log(r_{max})$ for an individual species given its mass.

close to this value, the exceptions being the relationship derived by Caughley and Krebs (1983) (based on nine data points only) and Thompson (1987). Thompson's method of calculation could be expected to overestimate the slope, as he used the equation of Cole (1954) (eqn (5.10)) to estimate r. Because this equation omits mortality before last breeding, it will overestimate r in all cases, but could be expected to do so particularly for mammals with small body sizes which have high juvenile mortality compared to larger mammals. Somewhat surprisingly, Hennemann (1983) estimates a slope in accordance with other studies, despite also using Cole's method.

Despite the strength of these allometric relationships, they do not allow for accurate estimation of r. For example, Figure 5.2 shows the largest data set, that of Thompson (1987), together with a 95% confidence interval for the predicted r. At any mass, the interval spans almost an order of magnitude, an error that will make substantial differences even to the qualitative results of most models. Allometric relationships should therefore be used as methods of last resort for predicting r, and only when life-history information is totally unavailable.

The ratio of production to biomass, P/B, is closely related to r_{max}. Paloheimo (1988) shows that, in an increasing population, P/B is equal to the sum of the rate of increase (in terms of biomass) and the biomass-weighted average mortality rate. In cases where population size is better described by biomass than numbers (for example, fisheries models) P/B is a more appropriate parameter to describe the potential rate of population increase. Allometric relationships between mass and P/B have been investigated by several authors (Banse & Mosher, 1980; Paloheimo, 1988; Boudreau & Dickie, 1989), particularly in marine organisms. In general, patterns are similar to those reported for r_{max}.

Two features of the relationships are worth commenting on. First, there is apparently a stronger relationship between mass and P/B within taxonomic groups of marine organisms than there is overall (Boudreau & Dickie, 1989). Gaston (1988) reported the opposite pattern for r_{max} in insects, but this may be because insects within taxonomic groups covered relatively smaller size ranges. Second, somewhat surprizingly, P/B does not appear to depend on temperature (Banse & Mosher, 1980), at least over the range 5–20°C. This pattern was also reported by Gaston (1988). There appears to be little prospect of improving predictions of r_{max} or P/B by inclusion of ambient temperature in a regression.

Stochasticity

If an estimate of r is required for a stochastic model, a point estimate (a single value) for the rate of increase is not sufficient. It is necessary to quantify the variability of the rate of increase through time. This has been touched upon previously in the chapter, but it is important to understand that the standard error of r (eqn (5.18)) describes the accuracy with which the point estimate of r is known, and does not in itself quantify the variation in r through time.

The stochastic version of eqn (5.1) is

$$dN(t) = rN(t)dt + \sigma N(t)dW(t). \qquad (5.19)$$

Loosely, σ is the standard deviation of the variation in r and $dW(t)$ is white noise, or uncorrelated random variation. This equation and its properties have been studied in considerable detail (Tuljapurkar, 1989; Dennis et al., 1991).

For practical purposes, however, it is much more common to use a discretized version of eqn (5.19),

$$N(t+1) = N(t) + rN(t) \qquad (5.20)$$

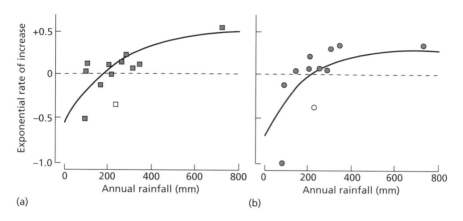

Fig. 5.3 The relationship between r and rainfall (lagged six months) for kangaroos. The figure shows this relationship for (a) red kangaroos, *Macropus rufus*, and (b) western grey kangaroos, *M. fuliginosus*, on Kinchega National Park, western New South Wales. The fitted curves are $r = -0.57 + 1.14(1 - e^{-0.004R})$ (red kangaroos) and $r = -0.69 + 1.04(1 - e^{-0.005R})$ (western grey kangaroos), where R is the annual rainfall. The relationship was calculated with a six-month lag: the annual rainfall in a calendar year was regressed against the rate of increase from 1 July of that year to 1 July of the next year. The open symbols were not used in calculating the fitted curves. From Bayliss (1987).

(see, for example Burgman *et al.*, 1993). This can be iterated repeatedly over the time period of interest, sampling from a probability distribution for r, and the resulting frequency distributions for $N(t)$ can be studied. The task is thus to find an appropriate probability distribution for r.

The most direct way this can be done is essentially identical to that described earlier. Given a time series of population sizes one time unit apart, the variance of r is estimated by the sample variance of $\ln(N_t/N_{t-1})$. The difficulty with this approach is that it requires a long time series to derive a good estimate, and that the variance is simply one parameter of the probability distribution of r. It would usually be assumed that the distribution was approximately normal (Burgman *et al.*, 1993).

A more indirect approach is to estimate the relationship between r and an environmental variable for which a long time series is available. This relationship can then be used to generate a probability distribution for r by sampling from the distribution of the environmental variable. In many terrestrial environments, rainfall is the obvious variable to choose. Even in a recently settled country like Australia, yearly records going back at least a century are available in most areas. For kangaroos, there is a very good relationship between rainfall and r (Bayliss, 1987; Cairns, 1989) (see Fig. 5.3). This has been used to generate the stochastic driving parameter in models of red kangaroos (Caughley, 1987) and adapted to generate stochastic variation in a model of bridled nailtail wallaby populations (McCallum, 1995a).

Why estimate r?

Models requiring an estimate of r are usually general and at a high level of abstraction. They are typically 'strategic' rather than 'tactical' models, intended to provide a simple and general representation of a range of ecological situations, rather than a precise representation of any particular situation. As the model will only approximately represent the ecological processes operating, there is no point in putting large amounts of effort into rigorous statistical estimation of the parameter. A far more important problem is to ensure that the value of r being estimated is appropriate for the problem being examined.

For example, an important issue in bioeconomics is whether it is in the economic interests of a sole owner of a population to conserve it, or whether there is more economic benefit in elimination of the population and investment of the profits elsewhere. Clark (1973a; 1973b) has shown that extinction of a population is optimal if

$$1 + i > (1 + r)^2 \tag{5.21}$$

and

$$p > g(0), \tag{5.22}$$

where i is the discount rate (the yearly rate at which financial managers discount future versus current profits), p is the unit price of the animals, and $g(0)$ is the cost of harvesting the last individual.

Clark's analysis was based on a simple logistic model of population growth without harvesting, with equally simple economic elements. The dynamics of the harvested population are represented as

$$\frac{dN}{dt} = rN\left(1 - \frac{N}{K}\right) - EqN, \tag{5.23}$$

where N is the size of the harvested population, K is the carrying capacity in the absence of harvesting, E is the harvesting effort and q is the catchability. The analysis was directed at baleen whale harvesting, but the same model has more recently been applied to elephant hunting (Caughley, 1993) and kangaroo harvesting (McCallum, 1995b).

Inspection of eqn (5.23) shows that r is being used to encompass all aspects of the population dynamics of the population other than resource limitation and harvesting. It is the long-term average rate of increase the population could sustain without resource limitation, not the rate of increase that could be sustained in unusually favourable environmental conditions.

There is limited value in estimating r for its own sake. The rate of increase usually is the output from a model, and there is little point in using it to forecast population growth using eqn (5.2). Almost always, the original age- or stage-structured model will do a better job of projection than will plugging the estimate of r the model provides into eqn (5.2). This presupposes that the age- or stage-class composition of the population is obtained at the same time as the stage-specific mortality and fertility rates.

Summary and recommendations

1   The most appropriate definition and estimate of the rate of increase r is dependent on the context in which it is to be used. In many contexts, r is intended to represent the per capita growth rate of the population in the absence of all the other factors that are included in the model explicitly.

2   It is important to distinguish between: $\bar{r}$, the observed rate of increase of a population; r_s, the rate at which it would grow, given the current mortality and fecundity schedules, once a stable age distribution is reached; and r_{max}, the maximum rate at which a population of a given species could grow, in a given environment, without density-dependent constraints.

3   Given estimates of age-specific fecundity and age-specific mortality, r_s can be estimated using the Euler–Lotka equation. Care needs to be taken in defining the fecundity terms correctly.

4   In the short term, a population may grow much faster than the rate r_s predicted from its age-specific fecundity and mortality schedules.

5   The observed rate of increase of a population, $\bar{r}$, can be estimated from time-series data. The most appropriate way to estimate this rate depends on whether the error in the time series is predominantly observation error or process error. If both are important, no currently available method is entirely satisfactory. Chapter 11 provides some suggestions on how to proceed.

6   Observation error occurs because the recorded values of population size are only estimates, and not the true population size. If observation error predominates, r can be estimated simply from the slope of a linear regression of $\ln(N)$ versus time (N is the estimated population size at time t).

7   Process error is variation in the actual size of the population, at a given time, from the size predicted by the model being used. If process error predominates, then r should be estimated from the mean of the change in $\ln(N)$ between censuses one time unit apart.

8   There is a strong allometric relationship between the size of an animal and r_{max}. However, estimating r_{max} from adult body size should be considered only as a method of last resort.

9   If a probability distribution of r is required, a useful approach is to find a relationship between r and an environmental variable such as rainfall.

CHAPTER 6

Density dependence

Introduction

The parameters considered up to this point in this book have been descriptive parameters of a single population. As such, they are most appropriately estimated by properly designed and statistically soundly based surveys or observations of the population in question. This will not always be possible, but is at least an ideal to be targeted. From here on, however, the parameters involve interactions, either between individuals in a single population, the topic of this chapter, or between individuals of different species. Interaction parameters often cannot be measured easily by simple observation of the interacting entities. The ideal way of measuring an interaction is with a manipulative experiment, in which one or more of the interacting entities is perturbed, and the response of the other entities measured. Unfortunately, ecological experiments in natural conditions, at appropriate scale, with controls, and with sufficient replication are very difficult and expensive to undertake, and in some cases are quite impossible. This and the following chapters therefore contain a number of *ad hoc* approaches that lack statistical rigour, but, provided they are interpreted with due caution, may permit at least some progress to be made with model building.

Importance of density dependence

There has been a long and acrimonious debate amongst ecologists concerning the importance of density-dependent processes in natural populations (e.g. Hassell *et al.*, 1989; Hanski *et al.*, 1993b; Holyoak & Lawton, 1993; Wolda & Dennis, 1993). This book is not the place to review that controversy. All modellers are convinced of the importance of density dependence, for the simple reason that population size in any model will tend rapidly to either extinction or infinity unless the parameters of the model are carefully adjusted so that net mortality exactly balances net fecundity. It seems implausible that this could happen in nature unless some factor (density dependence) acts to perform the adjustment in real populations. This is not to say, however, that density dependence must act on all populations all of the time, nor that real populations spend most of their time at some 'carrying capacity' set by density-dependent factors. What is necessary is that, at some times and at some population level, the population should decline *because* it is at high density. A good working definition of density dependence is provided by Murdoch and Walde (1989): 'Density dependence is a dependence of per capita growth rate on present and/or past population densities'.

Why do you need an estimate of density dependence?

There are two reasons why you might want to estimate one or more parameters to quantify density dependence. First, you might be trying to fit a model to a time series of population size or density. Such models may perhaps be used to draw indirect inferences about some of the mechanisms affecting population size, but the primary objective of the model-building process is either to describe the time series statistically or to be able to project the population size into the future, together with some idea about the possible range of future values. Such a projection, based only on the properties of the time series to date, will only be valid if the same processes that have determined the time series until the present continue to operate in essentially the same way into the future. The general principles that apply to the fitting of any statistical model should apply in this case: additional terms should only be included in the model if they improve the fit according to a significance level specified in advance. For the purposes of prediction, as distinct from hypothesis testing, one may be able to justify a significance level above the conventional $p = 0.05$. Nevertheless, in a purely statistical fitting exercise, the logical necessity for density dependence to exist at some density is not, in itself, a justification for including it in the model.

The second major reason for estimating density dependence is to include it in a mechanistic model. As discussed above, some density dependence is essential in many models to keep the results within finite bounds. To omit it because a time series or set of experiments failed to detect it will cause the model to fail.

The easiest way of including generic density dependence in a continuous-time model is via a simple logistic relationship:

$$\frac{dN}{dt} = rN\left(1 - \frac{N}{K}\right). \tag{6.1}$$

Here, r is the intrinsic rate of growth, and K is the carrying capacity. The form

$$\frac{dN}{dt} = rN - kN^2 \tag{6.2}$$

is identical mathematically; here k ($= r/K$) is a parameter describing the strength of the density dependence. Usually, there will be additional terms tacked on to the right-hand side to represent other factors (harvesting, predation, etc.) being modelled.

The same thing can be done in a discrete-time model, but a possible

Table 6.1 Forms for expressing density dependence in discrete time. Density-dependent survival can be represented in the following general form: $S = Nf(N)$, where S is the number of survivors, N is the number of individuals present initially, and $f(N)$ is the proportional survival as a function of the initial density. This table, modified from Bellows (1981), displays some forms that have been suggested for this function. The first three forms are single-parameter models, which inevitably are less flexible at representing data than are the final four two-parameter models. Bellows fitted the two-parameter models to two sorts of insect data: experimental studies which recorded numbers of survivors as a function of numbers present initially (set 1, 16 data sets) and long-term census data (set 2, 14 data sets, some field, some laboratory). The final two columns show the number of times each model fitted the data sets best, using nonlinear least squares

	$f(N)$	k value ($\log(N/S)$)	Set 1	Set 2
1	N^{-b}	$b \ln N$		
2	$\exp(-aN)$	aN		
3	$(1 + aN)^{-1}$	$\ln(1 + aN)$		
4	$\exp(-aN^b)$	aN^b	2	3
5	$(1 + (aN)^b)^{-1}$	$\ln(1 + (aN)^b)$	10	9
6	$(1 + aN)^{-b}$	$b \ln(1 + aN)$	1	0
7	$(1 + \exp(bN - a))^{-1}$	$\ln(1 + \exp(bN - a))$	3	2

problem is that high values of k may cause population size to become negative. Bellows (1981) provides an excellent review of the various functional forms with which density dependence can be represented in discrete time, particularly in insect populations (see Table 6.1).

It is worth emphasizing that logistic density dependence is a very stylized way of including density dependence in a model. Few real populations have been shown to follow such a growth curve. In fact, most basic textbooks use the same example of logistic growth in the field: that of sheep in Tasmania (Odum, 1971; Krebs, 1985; Begon *et al.*, 1990). That this highly unnatural population is used as the standard example of 'natural' logistic population growth is a good indication of the rarity of this mode of population growth. Nevertheless, the simplicity and generality of this model make it an appropriate way to represent density dependence when detail is either unavailable or unnecessary. The simple linear form the logistic assumes for density dependence can also be considered as the first term in a Taylor series expansion of a more general function representing density dependence. (A Taylor series is a representation of a function by a series of polynomial terms.)

Quick and dirty methods for dealing with density dependence in models

Omission to obtain conservative extinction risks

In some cases, it may be appropriate to omit density dependence from a mechanistic model, despite the logical necessity for its existence. An example is the use of stochastic models for population viability analysis (Ginzburg *et al.*, 1990). The problem in population viability analysis is to predict the risk of 'quasi-extinction', which is the probability that a population falls beneath some threshold over a specified time horizon. As populations for which this is an important question are always fairly rare, estimating all parameters, but especially those concerned with density dependence, is very difficult. The effect of direct density dependence, if it is not extreme, is to act to bring a population towards an equilibrium. Ginzburg *et al.* (1990) suggest that a model omitting density dependence will therefore produce estimates of quasi-extinction risks that are higher than they would be in the presence of density dependence. This will lead to conservative management actions.

To use this idea in practice, it is necessary to adjust the estimates of survival and fecundity used in the model so that the rate of increase r is exactly 0. This is effectively assuming that the initial population size is close to the carrying capacity, so that the long-run birth and death rates are equal. If the initial r is less than 0, extinction would be certain, although density dependence might 'rescue' the population through lower death rates or higher birth rates once the population had declined. Conversely, if the initial r is greater than 0, the extinction rate will be underestimated by omitting density dependence, because the population may become infinitely large.

This approach is potentially useful over a moderate time horizon. In the very long run, the random walk characteristics of a population model with $r = 0$ would cause it to depart more and more from the trajectory of a population that was even weakly regulated around a level close to the initial population size of the random walk.

There are substantial difficulties of estimating survival and fecundity in the wild (see Chapter 4). If a clear population trend is not evident, quasi-extinction probability predictions made by adjusting r to equal 0 are likely to be superior to those obtained by using the best point estimates of survival and fecundity in most cases, given the sensitivity of population trajectories to these parameters.

Scaling the problem away

In many deterministic models, the variable defining population size can be considered to be density per unit area. The unit of area is essentially arbitrary, and population size can therefore be rescaled in any unit that is convenient. Defining it in units of carrying capacity eliminates the need to estimate a separate parameter for density dependence. In many cases, such a scaling is helpful in interpreting the results of the model anyway. For example, if a model is developed to investigate the ability of a parasite or pathogen to depress a host population below its disease-free level (see, for example, Anderson, 1979; 1980; McCallum, 1994; McCallum and Dobson 1995), then it makes good sense to define the disease-free population size as unity.

This approach is much less likely to be useful in a stochastic model, because the actual population size, rather than some scaled version of it, is of central importance in determining the influence of stochasticity, particularly demographic stochasticity.

Estimating the 'carrying capacity' for a population

The problem of quantifying density dependence in a simple model with logistic density dependence is one of determining a plausible value for the carrying capacity of the population, as set by all factors other than those included explicitly in the formulation of the model. A little thought and examination of eqn (6.1) will show that the carrying capacity is *not* the maximum possible size of the population. It is rather the population density above which the unidentified density-dependent factors will, on average, cause the population to decline. This population level is unlikely to be identifiable easily, except in special cases. For example, suppose a model is being constructed of a harvested population. Assuming that the population was at some sort of steady state, the average population density before harvesting commenced would be a natural estimate of the carrying capacity. Unfortunately, for many exploited populations, the data needed to calculate a long-term mean density before the commencement of harvesting are unavailable.

In many cases, factors in the model other than the unspecified density dependence will be acting on the population, and the mean population density will not estimate K. Frequently *ad hoc* estimates, based on approximate estimates of peak numbers, have been used (e.g. Hanski and Korpimäki, 1995). For the reasons discussed in the previous paragraph, this method should be one of last resort.

Statistically rigorous ways of analysing density dependence

One of the primary reasons why the debate over density dependence persists is that it is remarkably difficult, for a variety of technical reasons, to detect density dependence in natural, unmanipulated ecological systems. As detecting the existence of density dependence is statistically equivalent to rejecting a null hypothesis that some parameter describing density dependence is equal to zero, it is correspondingly difficult to quantify the amount of density dependence operating in any real population.

Analysis of time series

Many approaches have been suggested for detecting density dependence in ecological time series. Most of these have been criticized on statistical grounds. Recent developments in the theory of nonlinear time-series analysis (see, for example, Tong, 1990) show considerable promise for quantifying and detecting density dependence. Although it is unlikely that there will ever be a single 'best test' for density dependence, we now seem to be at a point where reliable, statistically valid methods for examining density dependence in ecological time series have been established. For a recent review, see Turchin (1995).

Dennis and Taper (1994) propose the following model, described as a 'stochastic logistic' model:

$$N_{t+1} = N_t \exp(a + bN_t + \sigma Z_t). \tag{6.3}$$

Here, N_t is the population size, or some index derived from it, at time t, a and b are fitted constants, and σZ_t is a 'random shock' or environmental perturbation, with a normal distribution and variance σ^2. The model is thus a generalization of the Ricker model (see Box 2.1), with random variation added.

Before discussing the technical details of parameterizing this model, it is worth noting a few of its features that may limit its application. First, the variation is 'environmental stochasticity'. This is not a model that deals explicitly with the demographic stochasticity that may be important in very small populations. Second, the population size next year depends only on that in the current year, together with random, uncorrelated noise. This means that the model does not allow for the possibility of delayed effects, such as may occur if the density dependence acts through a slowly recovering resource or through parasitism or predation (see Chapters 9 and 10; or Turchin, 1990). Finally, the model assumes that all the variation is 'process error'. That is, the available data are records of the actual population size or index, with no substantial observation error. Handling both types of error simultaneously is difficult, but is possible, provided the observation error can be independently characterized (see Chapter 11).

Nevertheless, provided appropriate statistical checks are made (see below) the model is sufficiently simple to be a workable predictive model for population dynamics including density dependence.

The first step in analysis is to define $X_t = \ln N_t$ and $Y_t = X_{t+1} - X_t$. Then eqn (6.3) becomes:

$$Y_t = a + b \exp X_t + \sigma Z_t. \tag{6.4}$$

This looks as if it should be amenable to solution by a linear regression of Y_t versus X_t. Both can easily be calculated from a time-series population size N_t, and the 'random shock' term σZ_t looks essentially the same as the error term in a regression. Up to a point, ordinary linear regression can be used.

There are three possible models of increasing complexity. The simplest model is a random walk, with no density dependence and no trend. Thus, $a = 0$ and $b = 0$. The only quantity to be estimated is the variance σ^2. Next, there is

Table 6.2 Parameter estimates for the model of Dennis and Taper (1994)

Model	$\hat{a}$	$\hat{b}$	$\hat{\sigma}^2$
Random walk	0	0	$\hat{\sigma}_0^2 = \dfrac{1}{q}\sum\limits_{t=1}^{q} y_t^2$
Exponential growth	$\hat{a}_1 = \bar{y}$	0	$\hat{\sigma}_1^2 = \dfrac{1}{q}\sum\limits_{t=1}^{q}(y_t - \bar{y})^2$
Logistic	$\hat{a}_2 = \bar{y} - \hat{b}_2\bar{n}$	$\hat{b}_2 = \dfrac{\sum\limits_{t=1}^{q}(y_t - \bar{y})(n_{t-1} - \bar{n})}{\sum\limits_{t=1}^{q}(n_{t-1} - \bar{n})^2}$	$\hat{\sigma}_2^2 = \dfrac{1}{q}\sum\limits_{t=1}^{q}(y_t - \hat{a}_2 - \hat{b}_2 n_{t-1})^2$

Here, n_t is the population size at time t; $y_t = x_t - x_{t-1}$, and q is the number of transitions in the time series.

the possibility of exponential growth, with no density dependence ($b = 0$). Finally, there is the full model with all three parameters to be estimated. Formulae for the estimators are given in Table 6.2.

The estimators are simply as one would expect from elementary regression methods using least squares. If there is no trend or density dependence, the variance of the Y_t is simply the average of the squared values. If there is trend, the best estimate of the rate of increase is the mean of Y_t, and its variance is as generated from the usual formula. In both these cases, the standard errors, etc. given by any regression package will be perfectly valid.

In the full model, both a and b must be estimated, but the formulae are still simply those one would get from a conventional least squares regression with N_{t-1} as the predictor and Y_t as the response. The estimated variance $\hat{\sigma}_2^2$ is also simply the mean square error from a regression. The problem is that the full model violates the assumptions of conventional regression, because of its auto-regressive structure (X_t appears on both sides of eqn (6.4)). This means that the usual hypothesis-testing methods and ways of calculating standard errors are not valid.

Dennis and Taper (1994) suggest bootstrapping the estimates (see Chapter 2). In this particular case, the method involves a parametric bootstrap. The parameters a, b and σ^2 are estimated by the formulae above and then used to repeatedly generate artificial time series of the same length as the observed series, using eqn (6.4). From each artificial series, new bootstrapped estimates of the parameters can be generated. When a frequency distribution of these is produced, a 95% bootstrapped confidence interval is then just the range between the 2.5th and 97.5th percentiles. Box 6.1 shows an application of this approach to long-term census data for a population of birds.

Box 6.1 Density dependence in silvereyes from Heron Island, Great Barrier Reef

Silvereyes (*Zosterops lateralis*) are a small terrestrial bird common in eastern Australia and on many south Pacific islands. The population on Heron Island, Great Barrier Reef, has been studied for over 30 years by Jiro Kikkawa (University of Queensland). As the island is very small, the population essentially closed, and almost all individuals are banded, the population size used below can be considered to be a complete census, not an estimate.

The following graph, from Kikkawa, McCallum and Catterall (in preparation), shows the time series of the observed data from the period 1965–95, with cyclones marked as dashed lines:

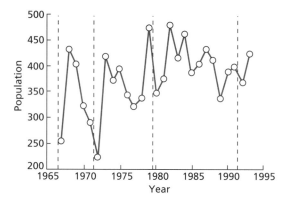

Following Dennis and Taper (1994), the following model was fitted to the data:

$$N_{t+1} = N_t \exp(a + bN_t + \sigma Z_t)$$

where N_{t+1} is the population size at time $t+1$, N_t is the population size at time t, a is a drift or trend parameter, b is a density-dependent parameter, and Z_t is random shock to the growth rate, assumed to have a normal distribution with a standard deviation σ.

Note that the above equation can be written as

$$\ln\left(\frac{N_{t+1}}{N_t}\right) = a + bN_t + \sigma Z_t,$$

meaning that the rate of increase between one year and the next is a linear function of the population size. The transformed data are shown below with declines associated with cyclones shown as solid points:

continued on p. 165

Box 6.1 *contd*

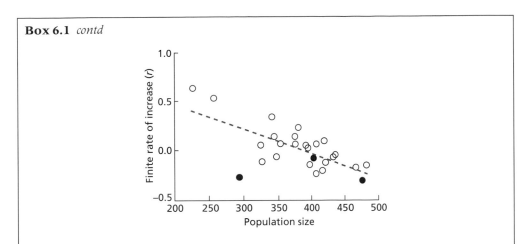

Least squares (and also maximum likelihood) estimates of these parameters can be calculated using a standard linear regression of $Y_{t+1} = \ln(N_{t+1}/N_t)$ versus N_t and are given below.

	Estimate	Standard error	*t* Stat	*p*-value	Lower 95%	Upper 95%
a	0.730 9	0.216 4	3.37	0.002 5	0.284 3	1.177 5
b	−0.003 12	0.000 932	−3.35	0.002 7	−0.005 05	−0.001 2
σ	0.204 8					

However, the usual standard errors are not valid because of the autoregressive structure of the model. Instead, 2000 time series of the same length as the observed series, and with the same starting point, were generated, and the three parameters estimated from each, generating a parametric bootstrap distribution. The bootstrapped estimates are shown below:

	Estimate	Standard error	Lower 95%	Upper 95%
a	0.970 6	0.175 2	0.640 07	1.315 8
b	−0.002 50	0.000 454	−0.003 400	−0.001 674
σ	0.176 2	0.025 40	0.128 6	0.226 70

Hence, there is clear evidence of density dependence in the time series, as the bootstrapped confidence interval for *b* does not include 0.

The mean population trajectory, with 90% confidence bounds (dashed) and maxima and minima of simulated series (dotted), is shown below, together with the actual trajectory (open squares).

continued on p. 166

Box 6.1 *contd*

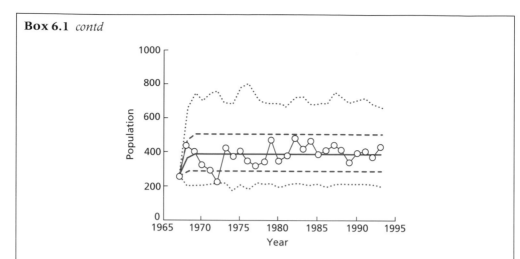

The time series thus shows very clear evidence of density dependence, around an equilibrium population size of about 380. There are two main caveats, however. First, the noise in this system is clearly not normal. Cyclones are the major perturbation, and these occur discretely in certain years. Second, inspection of the time series suggests that there may be some qualitative change in it occurring after about 1981. The population declined steadily after the two previous cyclones, but did not do so after the 1979 cyclone.

Density dependence in fledging success
The following table shows the number of individuals attempting to breed, the numbers banded as nestlings and their percentage survival from 1979 to 1992:

Year	No. attempting to breed	No. banded at nest	% surviving until 1st breeding	No. surviving until 1st breeding
1979	383	301	13.95	42
1980	314	287	13.94	40
1981	327	324	27.47	89
1982	370	259	22.39	58
1983	368	319	19.75	63
1984	348	281	16.73	47
1985	328	282	17.38	49
1986	317	231	31.17	72
1987	316	139	22.3	31
1988	365	301	15.28	46
1989	288	392	23.47	92
1990	317	369	20.87	77
1991	335	309	22.01	68
1992	315	237	20.68	49

If the percentage surviving to first breeding is plotted versus numbers attempting to breed, there is a suggestion of density-dependent survival (see below), but this tendency is not statistically significant ($t = 1.62$, 12 df, $p = 0.13$).

continued on p. 167

Box 6.1 *contd*

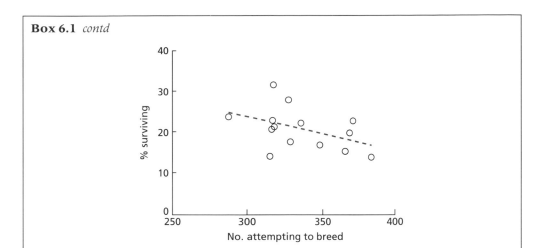

However, this approach ignores much of the information in the data, in particular that the numbers of individuals surviving can be considered as the results of a binomial trial. We therefore fitted the following logistic regression model to the data:

$$\ln\left(\frac{p}{1-p}\right) = \alpha + \beta X,$$

where p is the probability of a bird banded as a nestling surviving to first breeding, and α and β are constants. Using GLIM, the following results were obtained.

	Estimate	Standard error
α	0.178 5	0.480 2
β	−0.004 610	0.001 438

The analysis of deviance table was as follows:

Source	Deviance	df	p
Model	10.37	1	0.005
Residual	43.272	12	0.005
Total	53.637	13	

There is thus evidence of density dependence, although there is strong evidence of significant residual variation. Residuals for three years (1980, 1981 and 1986) were greater than 2.5, and thus probably outliers. If these are omitted there is still a highly significant effect of numbers attempting to breed on survival (Δ dev = 8.34, 1 df, $p < 0.005$) and there is no suggestion of significant residual deviance (dev = 11.58, 9 df).

Pollard *et al.* (1987) proposed a broadly similar approach. Their test assumes that the density dependence is a function of the logarithm of current population size, rather than N_t itself. Dennis and Taper (1994) suggest that this will weaken the power of the test to detect density dependence. There is no strong a priori reason to prefer one form of density-dependent function over another, but each could be compared to see how well they represented particular time series. Pollard *et al.* use a randomization method to test for density dependence. This does not lend itself to placing confidence bands around estimates of the parameters describing density dependence in the way that is permitted by the parametric bootstrap of Dennis and Taper. There is no reason, however, why one should not bootstrap estimates of the parameters from the model of Pollard *et al.*

These methods are not conceptually particularly difficult, but do involve some computer programming. Fox and Ridsdillsmith (1995) have suggested that the performance of these recent tests as means of detecting density dependence is not markedly better than that of Bulmer (1975). If the objective is to go further and parameterize a model, then Bulmer's test will not fulfil this need.

In an important series of papers, Turchin and co-workers (Turchin & Taylor, 1992; Ellner & Turchin, 1995; Turchin, 1996) have developed a model which is a generalization of Dennis and Taper's model. The basic idea is that the logarithm of the finite rate of increase of a population,

$$r_t = \ln\left(\frac{N_t}{N_{t-1}}\right), \tag{6.5}$$

should be a nonlinear function of population density at various lags or delays,

$$r_t = f(N_{t-1}, N_{t-2}, N_{t-d}, \varepsilon_t), \tag{6.6}$$

where d is the maximum lag in the model, and ε_t is random noise. The problem is then how best to represent the function f and to estimate its parameters. For moderately short time series (less than 50 data points), Ellner and Turchin (1995) recommend using a response surface methodology which approximates the general function f by a relatively simple polynomial. Specifically, they suggest first power-transforming the lagged population size:

$$X = N_{t-1}^{\theta_1}. \tag{6.7}$$

This transformation is known as a Box–Cox transformation (Box & Cox, 1964): $\theta_1 = {}^1\!/_2$ corresponds to a square root, $\theta_1 = -1$ to an inverse transformation, and $\theta_1 \to 0$ is a logarithmic transformation.

For a maximum lag of 2 and a second-order polynomial, eqn (6.6) can be written as

$$r_t = a_0 + a_1X + a_2Y + a_{11}X^2 + a_{12}XY + a_{22}Y^2 + \varepsilon_t, \tag{6.8}$$

where

$$Y = N_{t-2}^{\theta_2}. \tag{6.9}$$

and X is given by eqn (6.7). Higher-order polynomials and longer delays are possible, but require many more parameters to be estimated. Dennis and Taper's equation (eqn (6.4)) is a special case of eqn (6.8), with a_2, a_{11}, a_{12} and a_{22} all 0, and $\theta_1 = 1$. Pollard's model is similar, with $\theta_1 \to 0$.

The problem now is one of deciding how many terms should be in the model, and to estimate parameters for the terms that are included. As is the case with Dennis and Taper's method, ordinary least squares is adequate to estimate the a_{ij} parameters (but not the power coefficients θ). However, the standard errors, etc. thus generated are not suitable guides to the number of terms that should be included. Again, a computer-intensive method is required. Turchin (1996) suggests a form of cross-validation similar to a jackknife. Each data point is left out in turn, and the model is assessed by its ability to predict this missing data point. The a_{ij} parameters are fitted by ordinary least squares, and the values (-1, -0.5, 0, 0.5 and 1) are tried in turn for θ. Full details of the fitting procedure are given by Turchin (1996) and Ellner and Turchin (1995). Commercial software – RAMAS/time (Millstein and Turchin, 1994) – is also available. Box 6.2 shows results from using this approach on a time series for larch budmoth.

In summary, autoregressive approaches to parameterizing density dependence in time series are potentially very powerful. The most general model is eqn (6.6). Most current models are special cases of eqn (6.8). Standard least squares methods are adequate to generate point estimates of parameters, but a computer-intensive method, whether randomization (Pollard et al., 1987), bootstrapping (Dennis & Taper, 1994) or cross-validation (Turchin, 1996), must be used to guide model selection. All models assume white (uncorrelated) noise. This is a major problem (see below), although additional lagged terms may sometimes remove autocorrelation from the residual error. Further progress in this area can be expected in the next few years.

There are inherent problems in attempting to quantify density dependence in any unperturbed time series. All regression-based statistical methods rely on a range of values being available for the predictor variables: the wider the range, the better are the resulting estimates of the regression parameters. In methods that search for density dependence, the predictor variable is population density itself. If the system is tightly regulated, then an unmanipulated population will not vary in density much at all, making the relationship hard to detect. Conversely, if a system is weakly regulated, there will be a range of population densities available on which to base a regression, but the strength

Box 6.2 Analysis of larch budmoth time series

The following analysis is from Turchin and Taylor (1992), and is based on
a time series of the larch budmoth *Zeiraphera diniana*, a forest pest in
the European Alps, running from 1949 to 1986. The data shown below
are the population density in the upper Engadine Valley, Switzerland,
expressed as the logarithm to base 10 of the weighted mean number of
large larvae per kilogram of larch branches (digitized from Baltensweiler &
Fischlin, 1987).

Year	Population density	Year	Population density	Year	Population density
1949	−1.713 01	1962	1.404 39	1975	0.660 446
1950	−1.064 18	1963	2.373 45	1976	−1.905 19
1951	−0.324 71	1964	2.254 91	1977	−2.108 36
1952	0.674 519	1965	0.504 905	1978	−1.266 31
1953	1.867 09	1966	−1.728 38	1979	−0.708 14
1954	2.509 89	1967	−2.614 29	1980	0.991 958
1955	2.107 33	1968	−1.222 38	1981	2.190 57
1956	1.324 13	1969	−0.718 68	1982	2.289 5
1957	0.323 372	1970	0.032 79	1983	1.959 49
1958	−1.076 16	1971	1.019 89	1984	0.765 392
1959	−1.091 98	1972	2.236 68	1985	−0.918 11
1960	−0.406 9	1973	2.408 11	1986	−0.191 07
1961	0.260 11	1974	2.235 15		

Below are the observed data, log-transformed, together with the autocorrelation function (ACF). The ACF is the obtained by correlating the
data against itself, at lags shown on the horizontal axis. The maximum of
the function at a lag of 9 suggests that the series has a period of that length.

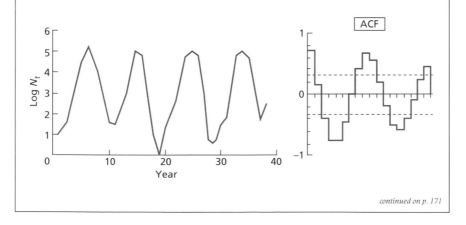

continued on p. 171

Box 6.2 *contd*

The parameters obtained by fitting eqn (6.8) to the data are:

θ_1	θ_2	a_0	a_1	a_2	a_{11}	a_{22}	a_{12}
0.5	0.0	−4.174	4.349	−1.790	−1.280	−0.124	0.437

The response surface generated by eqn (6.8) and the above parameters is shown in (a), together with the actual data points. Also shown (b) is the rather confusing relationship between r_t and $\ln N_{t-1}$. It is clear that this particular series requires the second lag term N_{t-2}.

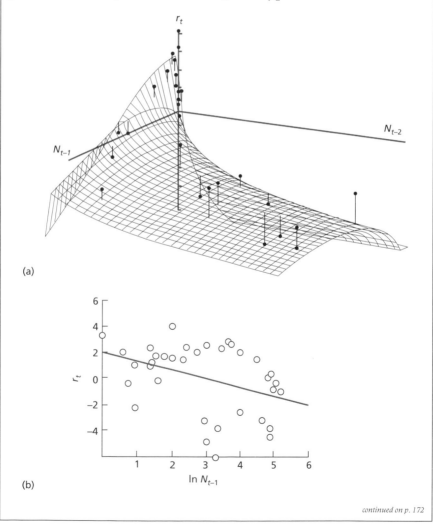

(a)

(b)

continued on p. 172

Box 6.2 *contd*

The reconstructed series is shown below, both without and with noise: (a) observed series; (b) reconstructed series without random noise; (c) reconstructed series with random process noise, standard deviation $\sigma = 0.2$.

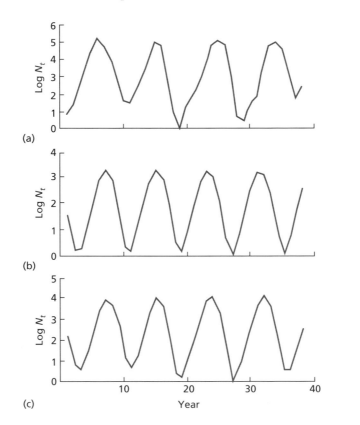

Note, first, the good correspondence between the observed and reconstructed series; and second, that the series without random noise does not exactly repeat itself. The analysis thus suggests that the series is quasi-periodic. That is, the dynamics are complex, and the period is not a simple integer.

of the relationship is, of course, weak. There will also be substantial problems with observation error bias, if population density is estimated with error. I discuss this bias in more detail in Chapter 11. A common effect of observation error bias is that a relationship that does exist is obscured. Nevertheless, in a study of nearly 6000 populations of moths and aphids caught in light or suction traps in Great Britain, Woiwod and Hanski (1992) found significant density dependence in most of them, provided sufficiently long time series were used: 'sufficiently long' means more than 20 years. It is always going to be difficult to detect and measure density dependence, although it is present, in unmanipulated time series only a few years in length.

A further problem with unmanipulated time series is that they rely on environmental stochasticity to generate the variation in population density. This is almost always assumed to be white noise. In reality, much stochasticity in ecological systems is 'red', or highly serially correlated, noise (Halley, 1996; Ripa and Lundberg, 1996). For example, if random, uncorrelated variation in rainfall is used to drive herbivore dynamics via rainfall-dependent vegetation growth, then variation in the resulting herbivore population size is strongly serially correlated – see Caughley and Gunn (1993) or Chapter 9 for more details. Serial correlation may have unpredictable effects on statistical tests for density dependence. For example, a high level of forage will produce high population growth and, if maintained, high population density, obscuring the actual density-dependent effect of food supply (see Chapter 9 for further details).

Demographic parameters as a function of density

An alternative to seeking density dependence in a time series directly is to look for a relationship between either survival or fecundity and density in observed data from a population. Clearly, this requires more information about the population than does a simple analysis of a time series. On the other hand, there are fewer statistical complications, as current population size no longer is part of both response and predictor variables, as it is for time-series analysis. However, the problems with lack of variation in population size in a tightly regulated population, observation error bias in the predictor variable, and auto-correlation in errors are still likely to be present in many cases, particularly if the data come from following a single population through time, rather than from examining a number of replicate populations at differing densities.

Density dependence in fecundity data can normally be looked for using standard regression techniques. An example is Shostak and Scott (1993), which examines the relationship between parasite burden and fecundity in helminths. Individual hosts can be considered as carrying replicate populations of parasites, avoiding the problems of time-series analysis. Survival data are likely to be analysed using logistic regression methods (see Chapter 2, and Box 6.2 for an example).

Experimental manipulation of population density

It should be clear from the previous discussion that there are unavoidable problems with any attempt to describe density dependence from purely observational data. Manipulative experiments, provided they are not too un-realistic, will almost always provide less equivocal results than purely observa-tional methods. In principle, density may be manipulated either by increasing it beyond normal levels or by experimentally reducing it. The latter approach is to be preferred, because any closed population at all will show either a reduction in fecundity or an increase in mortality or both, if the population is at sufficiently high density. This obvious result does not necessarily shed light on whether density-dependent factors are operating on the population when it is at the densities occurring in the field. On the other hand, an increase in fecundity or decrease in mortality when the population is reduced below normal levels is far more convincing evidence that density dependence oper-ates on natural populations.

Harrison and Cappuccino (1995) review recent experiments on population regulation in which density has been manipulated. Their overall conclusion is that 'surprisingly few such studies have been done'. Nevertheless, they were able to locate 60 such studies on animal taxa over the last 25 years, and of these 79% found direct density dependence. Most of these studies, however, were concerned solely with determining whether density dependence was present, and could not be used to parameterize a particular functional form to represent density dependence. To do this it is necessary to manipulate the population to several densities, rather than merely to show that there is some relationship between density and demographic parameters.

Experimental manipulation of resources

If density dependence is thought to act via resource limitation, an alternative to manipulating density itself is to manipulate the level of the resources. Several studies have examined the effect of supplementary feeding of wild populations. However, if the objective is to quantify density dependence through describing the relationship between resource level and demographic parameters, it is necessary to include resource dynamics in the model explic-itly. This requires the methods described in Chapter 9.

Stock–recruitment relationships

Obtaining a sustained harvest from a population that, without harvesting, is at steady state, relies on the existence of some form of density dependence operating on the population. Otherwise, any increased mortality would drive the population to extinction. It is not surprising, therefore, that detection and parameterization of relationships that quantify density dependence is a major part of fisheries management. Fisheries biologists have usually approached the problem through estimating the form of the stock–recruitment relationship for exploited species. 'Stock' is usually defined as the size of the adult population. Since this needs to be weighted according to the fecundity of each individual, a more precise definition of the average stock size over a short time interval is the total reproductive capacity of the population in terms of the number of eggs or newborn offspring produced over that period (Hilborn & Walters, 1992). Historically, recruits were the individuals reaching the size at which the fishing method could capture them, but 'recruitment' can be defined as the number of individuals still alive in a cohort at any specified time after the production of the cohort.

Any stock–recruitment relationship other than a linear relationship passing through the origin implies density dependence of some sort. The simplest possible model in fisheries is the Schaefer model, which assumes logistic density dependence. This is often parameterized by assuming that the population size of the unexploited population is the carrying capacity K. The maximum sustainable yield can then be obtained from the population by harvesting at a rate sufficient to maintain the population at a size of $K/2$ (see, for example, Ricker, 1975). This very simplistic model is frequently used, despite both theoretical and practical shortcomings (Hilborn & Walters, 1992).

Hilborn and Walters (1992) contains an excellent review of methods for fitting stock–recruitment relationships to fisheries data. The two most commonly used models of stock–recruitment relationships in fisheries are the Ricker model and the Beverton–Holt model. The Ricker model can be written as

$$R = Sa \exp(-bS), \tag{6.10}$$

where R is the number of recruits, S is stock size, the parameter a is the number of recruits per spawner at low stock sizes, and b describes the way in which recruitment declines as stock increases. The Beverton–Holt model can be written as

$$R = \frac{aS}{1 + \frac{a}{c}S} \tag{6.11}$$

where a has the same meaning as for the Ricker model, and c is the number of recruits at high stock levels. The key distinction between the two models is that the Ricker model allows for overcompensation, whereas the Beverton–Holt model does not.

Almost all stock–recruitment data are very variable (see Fig. 6.1), which makes fitting any relationship difficult. It is essential to recognize, however, that variation in the level of recruitment for a given stock size is an important property of the population dynamics, with major implications for management. Such variation is not simply 'noise' or error obscuring a 'true'

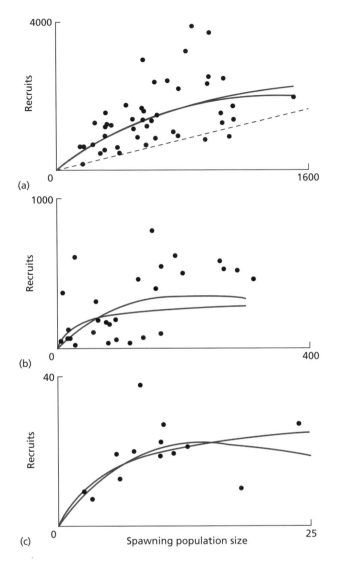

Fig. 6.1 Examples of Ricker and Beverton–Holt stock–recruitment relationships: (a) Skeena River sockeye salmon; (b) Icelandic summer spawning herring; (c) Exmouth Gulf prawns. In each case, the curve that decreases at high stock size is the Ricker curve, whereas the one that reaches an asymptote is the Beverton–Holt curve. All were fitted using nonlinear least squares on log-transformed data, although the plots shown are on arithmetic axes. The straight dashed line on the Skeena River relationship is the line along which stock equals recruitment. In a species which reproduces only once, such as this salmon, the intersection of the line with the stock–recruitment line gives the equilibrium population size. The line has no particular meaning for species that may reproduce more than once, so it is not shown on the other two figures. Note the amount of variation in each relationship. From Hilborn and Walters (1992).

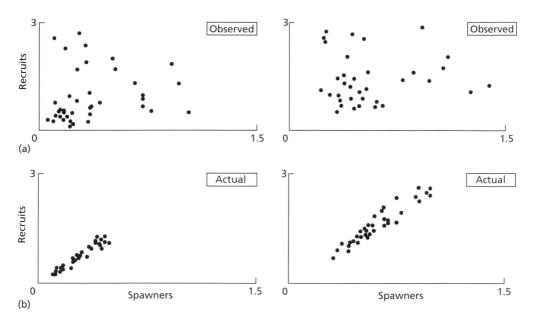

Fig. 6.2 Simulated relationships between estimated stock size S and estimated recruitment R with observation error: (a) observed relationship, and (b) actual relationship without observation error. These relationships were generated from the simple stock–recruitment equation $R_{t+1} = R_t(1 - H_t) \exp(\alpha + u_t)$, where R_t is the number of recruits in generation t, H_t is a harvest level (assumed to be fairly high), α is a measure of the intrinsic growth rate, and u_t is a random normal deviate, to simulate process noise. The number of recruits is measured with lognormal observation error, such that $O_t = R_t \exp(v_t)$, where O_t is the observed recruitment, and v_t is a random normal deviate. In both of these examples, $\sigma_u^2 = 0.1$ and $\sigma_v^2 = 0.25$. From Walters and Ludwig (1981).

relationship. It an essential part of the stock–recruitment relationship and needs to be quantified and taken into account, not eliminated. As well as such process error, there is also likely to be substantial observation error. The observed stock and recruitment levels are estimates only of the true levels, and are usually quite imprecise. Observation error is discussed in Chapter 2. It is particularly a problem when it occurs in the predictor variable, which in this case is the stock size. Unless the error in the estimation of stock size is much smaller than the amount of variation in actual stock size, it will be quite impossible to detect a stock–recruitment relationship. This point is illustrated using a simple simulation in Figure 6.2. In Chapter 11, I discuss some approaches that may prove useful in dealing with observation error in predictor variables.

In the face of these problems, Hilborn and Walters's (1992) advice is as follows:

1 Any fitted stock–recruitment curve is at best an estimate of the *mean* relationship between stock and recruitment, not an estimate of a *true* relationship. The true relationship includes variation around the mean.

2   The process error in recruitment size can be expected to have an approximately lognormal distribution. This means that to obtain maximum likelihood estimates of parameters, recruitment should be log-transformed before least squares methods are applied.

3   Asymptotic confidence intervals, as generated by standard nonlinear estimation packages, are likely to be quite unreliable, both because of the level of variation present, and because of the likelihood of strong correlations between parameter estimates. Computer-intensive methods such as bootstrapping or jackknifing (see Chapter 2) are needed.

4   Observation error is likely to obscure the relationship between stock and recruitment. If the size the error is known or estimated from independent data, it can be corrected for to some extent (see Chapter 11; or Ludwig & Walters, 1981). Otherwise, it is worth using simulation with plausible estimates to explore the possible consequences of observation error.

5   It is always worth looking at residuals.

More sophisticated methods of estimating stock–recruitment relationships can be considered as density-manipulation experiments, in which the varying harvest levels are used to alter density, although there are rarely controls or replications. This idea has been formalized as 'active adaptive management' (Smith & Walters, 1981), in which harvest levels are deliberately varied in order to obtain information about the nature of the stock–recruitment relationship.

Further complications

Stochasticity

It is very clear that density dependence does not always act in a predictable way on a given population. For example, where density dependence acts through food limitation, the amount of food available to a population will vary according to environmental conditions. Where density dependence acts through a limited number of shelter sites from which to escape harsh weather, the level of mortality imposed on individuals in inadequate shelter will depend on the harshness of the weather. This fairly obvious concept has been labelled 'density-vagueness' (Strong, 1986).

To include such stochastic density dependence in a mechanistic model, it is necessary not only to find parameters and a functional form capable of representing the mean or expected level of density dependence at any given population size, but also to describe the probability distribution of density dependence at any given population size. This will never be easy, given the proliferation of parameters that is required.

The most general and flexible approach, but the one most demanding of data, is to use a transition-matrix approach. Getz and Swartzman (1981) took

Table 6.3 A transition matrix for a stochastic stock–recruitment relationship. This example is for a purse-seine anchovy fishery off South Africa. The stock data were divided into eight levels, and the recruitment into seven levels. Each element in the table is the probability of a given stock level producing a particular recruitment level. From Getz and Swartzman (1981)

Recruitment level j		Low ←			Stock level i				→ High
		1	2	3	4	5	6	7	8
High	7	0.0	0.05	0.05	0.05	0.05	0.05	0.05	0.05
	6	0.0	0.05	0.1	0.1	0.15	0.2	0.25	0.25
	5	0.05	0.1	0.15	0.25	0.30	0.35	0.35	0.35
	4	0.15	0.2	0.35	0.3	0.25	0.2	0.2	0.2
	3	0.15	0.4	0.25	0.25	0.2	0.15	0.1	0.1
	2	0.45	0.15	0.1	0.05	0.05	0.05	0.05	0.05
Low	1	0.2	0.05	0.0	0.0	0.0	0.0	0.0	0.0

extensive data sets on stock size and resulting recruitment levels, and grouped both stock and recruitment data into six to eight levels. They then calculated the frequency with which stock level i gave rise to recruitment level j, producing transition matrices like the one shown in Table 6.3. This method of generating a stock–recruitment relationship is entirely flexible, allowing for any form of density dependence whatsoever, and allows the probability distribution of recruitment for a given stock size to take any form consistent with the data. However, it requires the estimation of as many parameters as there are elements in the transition matrix (at least 42 in the examples Getz and Swartzman discuss), and cannot be expected to be reliable without some hundreds of data points.

An approach less demanding of data is simply to use the variance estimate generated by a standard least squares regression to add stochasticity to a simulation. As is mentioned above, it is often reasonable to assume that 'abundance' has an approximately lognormal error. If this is the case, least squares regression on log-transformed data produces maximum likelihood estimates of the model parameters. This is the case whether the regression is linear or nonlinear. A maximum likelihood estimate of the variance of the error of the log-transformed data can then be obtained from

$$\hat{\sigma}^2 = \frac{SSE}{n-p}, \tag{6.12}$$

where SSE is the error (residual) sum of squares, n is the number of data points and p is the number of parameters fitted. This can then be used in simulations. If there is substantial observation error in either dependent or predictor variables or both, this approach may not work well.

A final approach is to model the process producing the density dependence, and to introduce stochasticity into that model. For example, Chapter 9 discusses a model in which a herbivore is limited by food supply, and plant growth is driven by stochastic rainfall. This approach probably produces the best representation of stochastic density dependence, but at a cost of adding a level of additional complexity.

Inverse density dependence

Most discussions of density dependence concentrate, as has this chapter, on direct density dependence. There is also the possibility of inverse density dependence, in which survival and/or fecundity increase as density increases. The phenomenon is often called an Allee effect. In the fisheries literature, inverse density dependence is usually called depensation. Inverse density dependence may be of considerable applied importance. If it occurs at low population densities, there may be a minimum population density below which the population will decline to extinction, because the mean death rate exceeds the mean birth rate. If it occurs at higher population densities, population outbreaks are likely to occur. It may be of particular importance when attempting to predict the rate of spread of an invading species. For example, Veit and Lewis (1996) found that it was necessary to include an Allee effect in a model of the house finch invasion of North America. Otherwise, the qualitative features of the pattern of spread could not be reproduced. However, they were unable to estimate the size of the effect from the available data, other than by choosing the value for which their predicted rate of population spread most closely approximated the observed rate of spread.

A variety of mechanisms have been suggested that may produce inverse density dependence. Allee's original idea (Allee *et al.*, 1949) was that, when a population becomes very sparse it is difficult for animals to locate mates. Other possibilities are that group-based defences against predators might be compromised at low population densities, or that foraging efficiency might be reduced (Sæther *et al.*, 1996).

Inevitably, inverse density dependence is difficult to detect or quantify in the field. Because it is expected to be a characteristic of rare populations, it will always be hard to obtain adequate data for reliable statistical analysis, and the effects of stochasticity will be considerable. Nevertheless, there have been several attempts to investigate it systematically.

Fowler and Baker (1991) reviewed large mammal data for evidence of inverse density dependence, including all cases they could find in which populations had been reduced to 10% or less of historical values. They found no evidence of inverse density dependence, but only six studies met their criterion for inclusion in the study of a 90% reduction in population size. Sæther *et al.* (1996) applied a similar approach to bird census data, obtaining 15 data sets satisfying the criterion that the minimum recorded density was 15% or less of the maximum density recorded. They then used a quadratic regression to predict recruitment rate, fecundity or clutch size as a function of population density. Their expectation was that an Allee effect would produce a concave relationship between the recruitment rate and population density, so that the coefficient on the quadratic term would be positive. In all cases, however, they found that a convex relationship better described the data. A limitation of this approach is that, assuming an Allee effect, one might expect a sigmoidal relationship between recruitment and population size over the full range of population size (direct density dependence would cause a convex relationship at high population densities). If most of the available data are at the higher end of the range of population densities, then the directly density-dependent component of the curve would dominate the relationship.

Myers *et al.* (1995) have produced the most complete analysis of inverse density dependence in time-series data. They analysed stock–recruitment relationships in 128 fish stocks, using a modified Beverton–Holt model,

$$R = \frac{\alpha S^\delta}{1 + S^\delta / K}. \tag{6.13}$$

Here, R is the number of recruits, S is the stock size, and α, K and δ are positive constants (compare with eqn (6.11)). If there is inverse density dependence, $\delta > 1$, as is shown in Fig. 6.3. Myers *et al.* (1995) fitted eqn (6.13) by nonlinear regression to log-transformed recruitment. (To reiterate, this process corresponds to using maximum likelihood estimation, if the variation is lognormal.) They found evidence of significant inverse density dependence in only three cases. This approach may lack power, if few data points are at low population densities. Myers *et al.* used a power analysis, which showed that in 26 of the 128 series, the power of the method to detect $\delta = 2$ was at least 0.95. The results suggest that depensation is not a common feature of commercially exploited fish stocks.

Liermann and Hilborn (1997) reanalysed these same data from a Bayesian perspective. They suggest that the δ of eqn (6.13) is not the best way to quantify depensation, because the amount of depensation to which a given δ corresponds depends on the values taken by the other parameters. As a better measure, they propose the ratio of the recruitment expected from a depensatory model to that expected from a standard model, evaluated at 10% of the maximum spawner biomass. Using this reparameterization, they show that the possibility of depensation in four major taxa of exploited fish cannot be rejected.

Comparing the conclusions reached by Liermann and Hilborn (1997) with those arrived at by Myers *et al.* (1995), from precisely the same data set makes two important and general points about parameter estimation. First, reparameterizing the same model and fitting it to the same data can produce

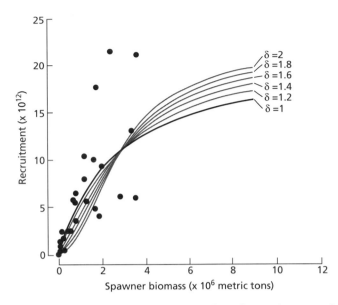

Fig. 6.3 Inverse density dependence in a Beverton–Holt stock–recruitment model. The dots are actual recruitment data for the California sardine fishery, and the heavy solid line is the standard Beverton–Holt model (eqn (6.11)) fitted to the data. The light solid lines show how this estimated curve is modified by using eqn (6.13), with varying values of δ. The parameters of each of these curves were chosen so that they have the same asymptotic recruitment as the fitted curve, and so that they have the same point of 50% asymptotic recruitment. From Myers *et al.* (1995).

different results: to make biological sense of modelling results, it is important that parameters should have direct biological or ecological interpretation. Second, reinforcing a point made in Chapter 2, conventional hypothesis-testing approaches give logical primacy to the null hypothesis. That is, they assume that a given effect is not present, in the absence of strong evidence that it is present. For management purposes, such as the prudent management of fish stocks, giving logical primacy to the null hypothesis may not be sensible. As Liermann and Hilborn suggest, a Bayesian approach may be more appropriate.

As with any other ecological process, manipulation is likely to be a more satisfactory way to detect an Allee effect than is observation. Kuusaari *et al.* (1998) report one of the few manipulative experiments on Allee effects in the literature. They introduced differing numbers of Glanville fritillary butterflies (*Melitaea cinxia*) to empty habitat patches, and found a significant inverse relationship between population density and emigration rate, even after patch area and flower abundance had been controlled for. They also found a positive relationship between the proportion of females mated and local population density in unmanipulated populations.

Summary and recommendations

1 Density dependence is difficult to detect and quantify in natural populations, but its omission will cause many mechanistic models to fail.

2 Adjusting survival and fecundity estimates so that $r = 0$ and omitting density dependence produces estimates of extinction risks that will lead to conservative management actions, provided that the omitted density dependence is direct. This approach is useful over moderate time horizons. However, it will not produce useful estimates of extinction risks if the objective is to investigate the impact of additional mortality, such as harvesting.

3 Models based on the logistic equation require an estimate of the 'carrying capacity' of a population. This is the population level beyond which unspecified density-dependent factors will cause the population growth rate to be negative. It is not the peak population size ever attained.

4 A variety of nonlinear time-series methods are now available to produce a statistically rigorous framework for studying density dependence in time series. Obtaining point estimates using these approaches is relatively straightforward, but standard regression packages cannot be used to perform tests of significance or to estimate standard errors. Computer-intensive methods such as bootstrapping need to be used. Even these methods are unreliable if there is substantial observation error, or if there is autocorrelated 'red' process error.

5 It is easier to quantify density-dependent effects on demographic parameters than it is to quantify density dependence in a time series itself. However, if the population density or population size is estimated imprecisely, this observation error will tend to obscure density dependent effects.

6 Manipulative experiments are usually a better way to quantify density dependence than is simple observation. If your objective is to show that density dependence is important in a natural situation, it is better to reduce the natural population density than to augment it artificially.

7 Density dependence or 'compensation' must occur if a population is to be harvested sustainably. Fisheries biologists usually quantify it by using a stock–recruitment relationship, fitting one of two standard models: the Ricker or Beverton–Holt. Specific advice on applying these is provided earlier in the chapter.

8 Stochasticity is an essential feature of density dependence, but is difficult to deal with.

9 Inverse density dependence at low population density may be very important, particularly in conservation biology. It is difficult to detect and quantify, because it occurs when animals are rare. It can potentially be quantified using a modified stock–recruitment relationship. A more powerful approach is to investigate inverse density-dependent effects on particular demographic parameters.

CHAPTER 7

Spatial parameters

Introduction

The importance of spatial processes in population dynamics has increasingly been recognized by ecologists in recent years (Levin, 1992). For some ecological questions, spatial dynamics are quite explicitly part of the problem. It is clearly nonsensical, for example, to attempt to predict the rate of spread of an invading species without a spatial component in the model. However, it has also become apparent recently that the qualitative behaviour of interspecific interactions (host–parasitoid, predator–prey and competitive) may be totally different in spatially heterogeneous environments than in the homogeneous environments implicitly assumed by most basic models (Hassell *et al.*, 1991a; 1994).

Models may represent spatial processes at several levels of abstraction. Spatially explicit models (sometimes called landscape models) seek to represent populations, or even individuals, within the actual heterogeneous landscape that they inhabit (Dunning *et al.*, 1995). Such models will usually represent the landscape as a series of cells that may be occupied by one or more individuals of the species under study, with the properties of the cells determined from a Geographic Information System (GIS). Movements and the various ecological processes acting on the occupants of the cells are then functions of the cell properties. Models of this type have been developed for the spotted owl (*Strix occidentalis*) in old growth forests of the western United States (Lamberson *et al.*, 1992) and also for the endangered Bachman's sparrow (*Aimophila aestivalis*) in the southeastern United States. Any such model will inevitably require large numbers of parameter estimates and it is unlikely that field data suitable for estimating all of these reliably will be available, even for the most well-known species (Conroy *et al.*, 1995; Dunning *et al.*, 1995).

Cellular automaton models (Caswell & Etter, 1993; Tilman *et al.*, 1997) are a more abstract version of spatially explicit models. Again, the environment is divided into a large number of cells, and interactions are more likely to occur between close neighbours than cells far apart (in standard cellular automaton models, only neighbouring cells contribute to transition rules for each cell). The state of each cell is described by a binary variable (usually presence or absence). Cells are considered to be identical or of a relatively small number of types, so that despite the very large number of cells, the number of parameters needed is much smaller than in a spatially explicit model. The abstract nature of these models means that they are valuable for exploring the qualitative influence of spatial structure on population dynamics, rather than for making accurate quantitative predictions about particular species in particular habitats.

Lattice models (e.g. Rhodes & Anderson, 1997) are a more complex version of a cellular automaton model. As with a cellular automaton model, interactions occur only (or primarily) between adjoining cells. In lattice models, however, each lattice point may contain more than one individual organism, and more than one interacting species.

Patch models incorporate a number of patches with particular properties and a specified spatial distribution, and then attempt to model dynamics within patches, together with the rate of movement between patches (for example, Lindenmayer & Lacy, 1995b). As with spatially explicit models, the number of parameters and thus the data requirements are substantial. More abstract versions of these models, with more modest numbers of parameters, include classical metapopulation models, in which only presence or absence on a patch is modelled, and patch models in which all patches are assumed to be equidistant.

Finally, reaction–diffusion models are based on models in physics that describe diffusion of heat across a surface or a solute through a liquid (Murray, 1989). Spatial spread is modelled in continuous space using partial differential equations. An excellent and succinct introduction to these models is provided by Tilman *et al.* (1997).

Corresponding to these levels of abstraction, spatial information will be needed with varying degrees of resolution. At the finest scale, the level of the individual organism, data may be required on home range, foraging range or territory or dispersal and migration distances and directions. At the level of the individual population, the numbers or proportions of individuals immigrating or emigrating per unit of time may be needed. At the level of the metapopulation, or group of populations, estimates of the extinction and colonization rates of each habitat patch or population may be required. At the highest level of abstraction, models operating in continuous space require estimates of the diffusion rate (or rate of spread through space) of the organisms.

Parameters for movement of individuals

Information on individual movements can be required in a model for several reasons. First, knowledge of home range or territory size will help determine the boundaries of the population under study. Is it appropriate to consider your population at the scale of square metres, hectares or square kilometres? Second, and more importantly, the range of an individual defines the environment in which it lives, in terms of resources, potential competitors and potential exploiters. In a spatially explicit model, the size of the home range will determine either the size of the cells (if each cell represents a territory) or the number of cells each individual occupies. Similarly, dispersal and migration information are necessary to define the appropriate scale for modelling and the environment of the target organisms.

Collecting data

Almost all methods for collecting data about movements of individuals rely on capturing animals, marking them and then either following them, or recapturing or resighting them at some time or times in the future. It is beyond the scope of this book to describe methods for marking, tracking or relocating animals in any detail. Tracking methods vary in sophistication from following footprints in snow (Messier & Crête, 1985) or attaching reels of cotton to animals' backs (Key & Woods, 1996) to radiotelemetry (see White & Garrott, 1990, for an excellent review), use of radioactive tracers (Stenseth & Lidicker, 1992) and satellite tracking (Grigg et al., 1995; Tchamba et al., 1995; Guinet et al., 1997). Do not assume that technologically sophisticated methods are necessarily the best. Given a constraint of fixed cost, there is always a trade-off between cost per sample and sample size.

Standard methods for survival analysis and population estimation based on mark–recapture or mark–resight are discussed in Chapter 4. If locations of initial capture and resight or recapture are recorded, then obviously some information on ranging behaviour is available. Equally obviously, a home range thus determined will usually be only a subset of the true range.

Analysis of range data

The home range of a species was originally defined as 'that area traversed by an individual in its normal activities of food gathering, mating and caring for young' (Burt, 1943). It is thus not necessarily a defended territory, and neither is it defined by the extreme bounds of the area within which the individual moves.

The simplest way of defining a home range is by a convex polygon, which is an envelope drawn around all available locations in such a way that all internal angles are less than 180° (see Fig. 7.1). As a means of defining the limits of the parts of the habitat used by an individual, a minimum convex polygon has several problems. First, it is determined by extreme observations, and will therefore be greatly affected by outliers. Second, it will be strongly influenced by sample size: the more points, the greater will be the minimum convex polygon. Third, if the habitat is patchy, the minimum convex polygon may include regions into which the animal never goes, if these areas are bounded by or surrounded by suitable habitat (White & Garrott, 1990).

Because of the limitations of the minimum convex polygon, various alternatives involving more sophisticated statistical analysis have been suggested. For example, Jennrich and Turner (1969) described a home range using a bivariate normal distribution. As this restricts the shape of the home range to an ellipse, its practical application is severely limited. More recent approaches have taken the approach of smoothing and interpolating the frequency distribution of locations, without imposing any particular shape upon them a priori. Two such approaches are the harmonic mean method (Dixon & Chapman, 1980) and kernel estimation methods (Seaman & Powell, 1996).

Migration and dispersal

A distinction is often drawn between dispersal, which is a spreading or movement of individuals away from others (diffusive movement), and migration, which is a mass movement of individuals from one place to another (advective movement) (Begon et al., 1996a, p. 173). However, the distinction cannot be applied too rigidly. Depending on the objectives of the model being constructed, the quantities that require estimation might include:

1   The probability distribution of distance moved by individuals, per unit of time. This might be considered either migration or dispersal, depending on whether movements were away from, or together with, conspecifics.

2   The probability distribution of the distance between the location of adults and the settlement point of their offspring, or equivalently, the probability distribution of the distance between the position of birth and the position of first reproduction. This would normally be considered to be dispersal.

3   For a given patch or subpopulation, the probability that an individual leaves it, per unit of time, or the number of individuals entering or leaving

Fig. 7.1 (on following pages) Home ranges for bridled nailtail wallabies (Onychogalea fraenata), as determined by various methods. These maps show an area in Idalia National Park, western Queensland, where a reintroduction of the endangered bridled nailtail wallaby is being attempted. The unshaded areas along drainage lines are potentially wallaby habitat, whereas the shaded areas are residual outcrops, unsuitable as habitat for these animals. The dots on each map are daytime sheltering locations obtained by radiotelemetry, over six months, for about 30 animals. The composite home ranges were calculated using the package CALHOME (Kie et al., 1996). For all but the convex polygon 50%, 75% and 95% utilization contours are plotted. From F. Veldman and A.R. Pople, University of Queensland (unpublished data). (a) Convex polygon. This encompasses all data points, but includes large amounts of habitat where animals were never found. (b) Jennrich–Turner bivariate normal distribution. This always will have an elliptical shape, whatever the actual pattern of space use. (c) Harmonic mean. This distribution is no longer constrained to have a particular shape, but substantial areas of unused habitat are still included. (d) Kernel estimator. This has done the best job of indicating where the animals actually were. The 'square' shape of some of the utilization contours is an artefact caused by the grid used to calculate the distribution.

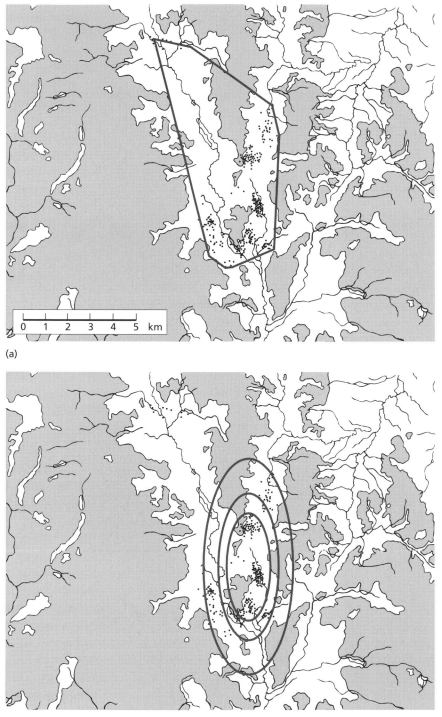

(a)

(b)

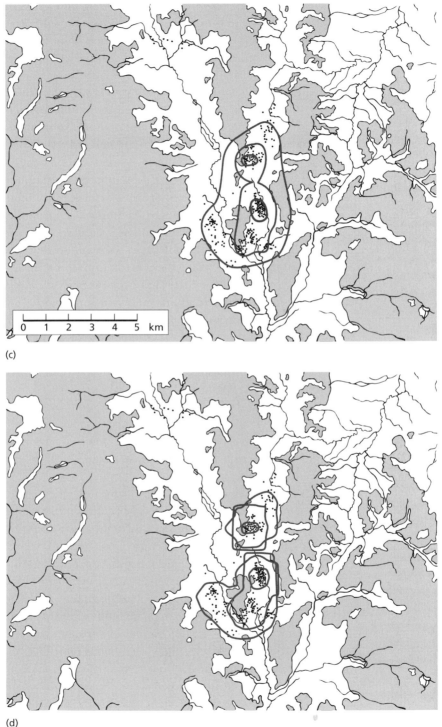

(c)

(d)

it, per unit of time. These patch-specific entering and leaving processes are usually called emigration and immigration.

4   For a network of patches or subpopulations, the probability of movement, or the numbers of individuals moving, between each pair of patches. These rates are usually called migration rates. Typically, they may be functions of the physical size of each patch, any other measure of 'quality' of patches, population density in patches, and distance between, or arrangement of, patches in space.

The first two types of quantity are likely to be estimated from raw data consisting of distances moved by individuals, either animals that have been followed through time, or animals that have been marked at a particular location and subsequently recaptured, recovered, or resighted at another. For almost all distance-based movement data, it is essential to describe a probability distribution of distance moved. Almost always, the probability distribution of distance moved is highly skewed, so that the mean dispersal distance is a poor indication of the distance that 'most' individuals move. On the other hand, many important biological processes (such as the rate of spread of the population) are dependent on the tails of the distribution (how far a few animals manage to disperse) rather than on the mean (Kot *et al.*, 1996).

Patch-related movement statistics, such as (**3**) and (**4**) above, are likely to be estimated from mark–recapture data, giving proportions of individuals marked on particular patches that are subsequently recovered on that patch, or on other patches. Genetic data are also increasingly being used to determine the extent to which subpopulations are isolated from each other.

Probability distributions of movement or dispersal distances

The probability distribution of dispersal distance is usually assumed to be negative exponential,

$$f(x) = ae^{-bx} \tag{7.1}$$

where $f(x)$ is the probability density of dispersal at distance x and a and b are constants; or sometimes is assumed to follow a power relationship,

$$f(x) = ax^{-b} \tag{7.2}$$

(Okubo & Levin, 1989). These relationships are phenomeological: that is, they are chosen because they seem to represent observed distributions of dispersal data quite well. Both distributions are monotonically decreasing, and will therefore fail to fit dispersal distributions which have a maximum at a distance greater than 0. Such distributions might be produced from the seed rain from tall trees (Okubo & Levin, 1989), or from larvae of marine species that have a fixed pre-patent period and are released into a net current flow. An example of the latter would be crown of thorns starfish larvae on the Great Barrier Reef

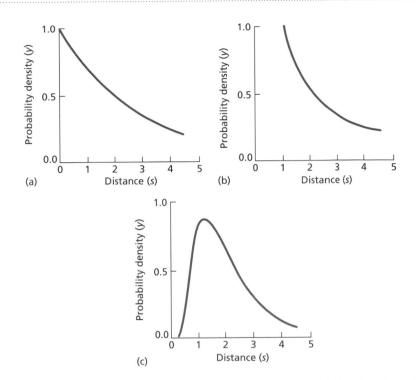

Fig. 7.2 Dispersal distributions: (a) negative exponential; (b) power; (c) Gaussian plume. In each case, the horizontal axis shows the distance from the source *s*, and the vertical axis is the probability density of settling propagules, *y*. From Okubo and Levin (1989).

(Scandol & James, 1992). Dispersal distributions of this type may be represented by Gaussian plume models (Okubo & Levin, 1989), which are based on a mechanistic representation of a plume of propagules that diffuses and is dispersed by wind or current. Figure 7.2 shows some theoretical dispersal distributions that can be generated by exponential, power and Gaussian plume models.

There are several ways in which data might be collected to determine a probability distribution for dispersal distance. Unfortunately, there are substantial biases associated with some methods. These biases are frequently not recognized (Koenig *et al.*, 1996). One approach, developed most fully for dispersal of seeds or spores (Fitt & McCartney, 1986), is to use collectors of a given surface area at various distances from the source. This approach might also be applied to dispersal of invertebrates, or to any other situation where settlement of very large numbers of dispersers or propagules from a single source can be recorded. The number of individuals recovered per collector, as a function of distance, is not an estimate of the probability distribution of dispersal distance of individual propagules. The problem is that almost all dispersal clouds expand as they disperse. The simplest model is that dispersal is uniformly distributed over a circle, or a segment of a circle. If this is the case, the area of settlement increases as the square of distance. The result is that the numbers of individuals settling per unit area would decline following an inverse square relationship, given a uniform probability distribution of dispersal distance.

Estimating the parameters a and b in eqn (7.1) or (7.2) for data of the above type is often done by log-transforming the cumulative proportion of individuals dispersing to particular distances, and then using conventional regression (e.g. Hill *et al.*, 1996). This approach is technically incorrect, as cumulative proportions are not independent observations. This is the same problem as occurs with estimating survival rates by a regression of log survivors against time (see Chapter 4). The solution is also the same. The dispersal distance of each individual should be considered as an independent observation, with a negative exponential distribution. This means that a standard survival analysis program with a constant hazard rate can be used, substituting 'distance' for time (for eqn (7.1)) or log(distance) for time (for eqn (7.2)). Alternatively a generalized linear model with a gamma errors, a scale parameter of 1, and a reciprocal link can be used (see Crawley, 1993).

In cases where the modal dispersal distance is greater than zero, survival analysis approaches using a Weibull distribution (see Chapter 4) should be considered. They will not give results identical to the Gaussian plume models of Okubo and Levin (1989), but should produce qualitatively similar results.

In many studies of vertebrate dispersal, an area of finite size is studied intensively. Within this area, individuals are marked, and their position at a later date is recorded. Only movements within the study area are recorded with high reliability, although there may be haphazard records from outside the main study area. Such a methodology will result in underestimates of the mean or median dispersal distance, unless dispersal distances are much smaller than the dimensions of the study area (Koenig *et al.*, 1996). The problem is simply that dispersal events that take the animal beyond the study boundary are not recorded. The consequence is not just a truncation of the frequency distribution of dispersal. Because some individuals initially quite close to the study boundary will disperse out of the study with only a small movement, the effect is to generate a distribution of dispersal distances that appears to decline with distance, even if the true distribution is uniform (see Fig. 7.3). If there is no preferred direction of dispersal, the study area is circular, and individuals are distributed randomly throughout the area, Barrowclough (1978) provides a means of correcting for this type of bias. These are, however, restrictive assumptions that limit the validity of this correction.

Radiotelemetry generally will provide better estimates of the distribution of dispersal distances than will direct observation (Koenig *et al.*, 1996), simply

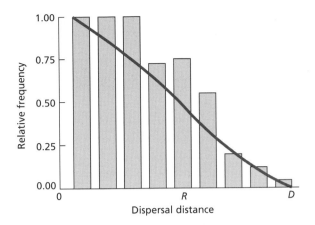

Fig. 7.3 The effect of a finite study area on observed dispersal distance. The frequency distribution of observed dispersal distances of 141 female acorn woodpeckers (*Melanerpes formicivorus*) within a finite study area is shown, overlaid with the probability of detection of a given dispersal distance, if dispersal directions are random and the study area is circular with radius *R*. Despite the apparent shape of the distribution, the observed distribution is quite consistent with a uniform distribution of dispersal distances over distances between 0 and *D*. From Koenig *et al.* (1996).

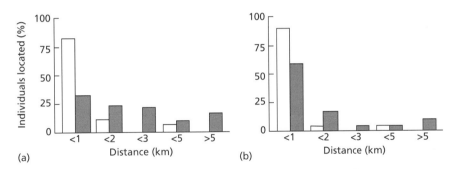

Fig. 7.4 Natal dispersal in yellow-bellied marmots (*Marmota flaviventris*): (a) Dispersal distances for males, based on intensive observation and trapping (unshaded bars) compared with those from radiotelemetry (shaded bars) (18 trapped, 52 radiotracked). (b) Dispersal distances for females, represented similarly to (a) (22 trapped, 38 radiotracked). From Koenig *et al.* (1996).

because fewer long dispersal distances will be missed. Figure 7.4 shows an example of natal dispersal distances in marmots that emphasizes this point.

Using mark–recapture methods to estimate migration rates

Applied to a single population, standard open-population methods for mark–recapture analysis (see Chapter 4) produce estimates of the number of new entrants to the population between sampling occasions. The methods

themselves cannot discriminate between offspring produced by the population under study and immigrants. In some cases, it will be obvious that the
new entrants cannot be recently recruited juveniles. It will rarely be possible
to distinguish between new recruits that have arisen from reproduction
within the population and new recruits that have migrated into the population, but have arisen from reproduction outside it. These methods also estimate 'survival': the probability that individuals present in the population on one
occasion are still present on the next trapping occasion. The methods cannot
distinguish between death and permanent emigration. Temporary emigration,
in which individuals leave the sampled population for one or more trapping
occasions, but subsequently return, is a problem for mark–recapture methods.
All such methods behave badly if the probability of capture varies substantially
between individuals. If animals temporarily leave the population, then their
probability of capture whilst they are absent is zero, and clearly this is different
from the capture probability of those animals that remain in the population.
Kendall *et al.* (1997) describe some possible ways of dealing with, and estimating the rate of, temporary emigration.

Tag–recovery studies, in which large numbers of animals are tagged, and
recoveries of tagged animals are later reported, are commonplace in fisheries
research and in ornithology. A major objective of such studies is to investigate
movements, but the data are usually only analysed qualitatively. Typically,
animals will be reported to have moved at least x kilometres, or $y\%$ of recoveries will be reported to have occurred within a certain distance of release.
Information of this type is adequate to determine how far some animals can
move, but not to quantify the amount of interchange between subpopulations, or to determine the frequency distribution of movements.

If animals are marked and recaptured or recovered from different subpopulations, it is possible to analyse recovery data more formally, to estimate
the probability of movement between subpopulations or patches. The approach
is an extension of standard mark–recapture methods for single populations.
Instead of estimating ϕ_i, a single parameter representing the probability of
survival from time period i to $i + 1$, the objective now is to estimate a matrix $\hat{\phi}_i$
of survival/migration probabilities. Each element ϕ_i^{rs} in the matrix is the
probability that an individual present in subpopulation r at capture occasion i
is alive and present in subpopulation s at capture time $i + 1$. The diagonal elements of the matrix are thus simply the probabilities that animals are alive and
have not moved from particular subpopulations, whereas the other elements
are the probabilities that the animal has both survived and moved from subpopulation r to subpopulation s. It is impossible to separate migration probability from survival probability, without making certain restrictive assumptions
(for example, that survival is identical in all subpopulations, or that movements
only occur immediately before recapture).

Not surprisingly, the technical details of fitting these models are not straightforward. The basic approach is an elaboration of the standard Cormack–Jolly–Seber model (see Chapter 4). The raw data will consist of a series of capture histories, giving the stratum in which the animal was captured on each sampling occasion. For example, if there are three subpopulations A, B and C, and six capture occasions, the history 0AA0B0 would represent an animal first captured on the second sampling occasion in subpopulation A, recaptured the next time still in A, missed on the next occasion, picked up again in subpopulation B on the fifth occasion, and then not seen again. (In a mark–recovery study, as distinct from a mark–recapture study, there will only ever be one recapture.)

From these histories, the objective is to estimate the survival/migration parameters ϕ_i^{rs} (as defined above), and also the capture probabilities p_i^s, (each p_i^s is the probability of recapture at sampling time i for an animal present in subpopulation s at that time). As is the case with a standard Cormack–Jolly–Seber model, the estimates for the survival/migration parameters and capture probabilities over the last time interval are confounded: it is not possible to separate animals that were not captured, but were present, in a given subpopulation on the final sampling occasion from those that were no longer present.

For even a fairly small number of subpopulations, the number of parameters to be estimated rapidly becomes prohibitive. Brownie *et al.* (1993) suggest investigating reduced models, in which either or both the transition probabilities ϕ or the recapture probabilities p are considered to be constant through time, but variable across subpopulations. However, they also suggest that an additional complication may be necessary. The model, as described above, is 'memoryless' or Markovian. The probability of an animal moving from one population to the next depends only on where it currently is, not where it has been in the past. If some animals are 'transients' and others are 'residents', this will not be a reasonable assumption.

In addition to the standard parameters ϕ and p, some derived parameters based on these are likely to be of ecological interest, perhaps more than the original parameters themselves. For example,

$$\phi_i^{r\cdot} = \sum_{j=1}^{s} \phi_i^{rj} \tag{7.3}$$

is the total probability of survival through the time period $(i, i+1)$ of the animals present at time i in subpopulation r, wherever they end up, and

$$\Psi_i^{rs} = \frac{\phi_i^{rs}}{\phi_i^{r\cdot}} \tag{7.4}$$

is the probability that an animal which was in subpopulation r at time i is present in subpopulation s at time $i+1$, given that it has survived. If it can be assumed that movement occurs just before recapture, or that survival rates within subpopulations are equal, eqns (7.3) and (7.4) enable survival and migration to be disentangled.

If independent data on population size in each subpopulation are available, the number of immigrants into or emigrants from each subpopulation can also be calculated (see Schwartz *et al.*, 1993).

These methods are likely to be unreliable if animals make multiple moves and recapture rates are relatively low. For example, if the capture history

A 0 0 0 B 0

is recorded, it is not possible to determine whether the animal has survived in subpopulation A for three intervals, but has been missed, and has then migrated to B, or whether it has moved to B in the first time interval, but then has been missed whilst in B until capture occasion 5. It may even have gone to B, returned to A and then gone back to B, or even have gone to B via C. In most studies, migration is likely to be relatively unusual, so that multiple migration is rarer still.

The details of the calculations to estimate these parameters are in Brownie *et al.* (1993) (multiple recaptures) or Schwartz *et al.* (1993) (tag–recovery). Few ecologists are likely to want to undertake the calculations from scratch. The program MSSURVIV is available to estimate them (Hines, 1993).

Genetic approaches

The methods of molecular genetics are increasingly providing useful data for estimating ecological parameters, particularly for spatial models. At the coarsest level of resolution, strong genetic differentiation between two or more subpopulations can be used to show that they function as independent breeding stocks. At a somewhat finer level of resolution, genetic data may provide information on the nature of the structuring of subpopulations. For example, some patterns of genetic variation suggest that subpopulations are connected according to an 'island' model (similar to a classical metapopulation or patch model), whereas other patterns suggest that subpopulations are connected according to a 'stepping-stone' model (similar to a lattice model). These models and the associated patterns of genetic variation are shown in Figure 7.5. The extent of genetic differentiation between populations is sometimes used to estimate the number of migrants exchanged per generation, but, as will be seen below, this relies on simplifying assumptions, which are liable to make such estimates inaccurate. A very recent and exciting development is to use high-resolution information about individuals' genotypes to identify migrant individuals directly. This should enable much more accurate estimation of migration rates than has been possible in the past.

This book is not the place to go into the details of collecting and processing

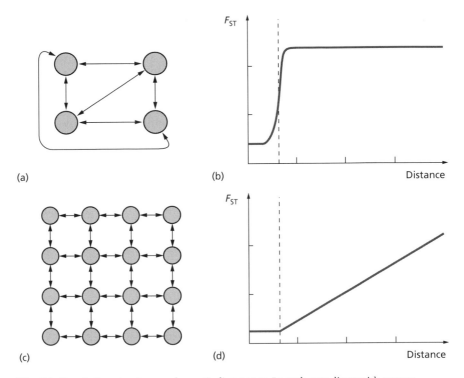

Fig. 7.5 Population structure and genetic divergence. In each case, lines with arrows represent equal rates of migrant exchange between patches (circles). Mating is assumed to be random within patches. (a) An island model. This is similar to an ecologist's classic metapopulation model. (b) The expected relationship between the distance apart samples are taken and F_{ST} for an island model. The vertical dashed line is the patch size. (c) A stepping-stone model (in this case, two-dimensional). This is similar to an ecologist's lattice model. (d) The expected relationship between the distance apart samples are taken and F_{ST} for a stepping-stone model. The vertical dashed line is the patch size. Note that it may be difficult to discriminate between the relationships between genetic divergence and geographic distance in (a) and (b), given sampling error in the estimation of genetic divergence. This will be particularly so if the patch size is not known a priori.

genetic samples. Excellent reviews of the range of techniques available can be found in Hillis *et al.* (1996) or Parker *et al.* (1998). A summary of some of the available techniques and their potential applications is provided in Table 7.1.

Genetic data can potentially provide information on gene flow between subpopulations or through continuous space. This may not necessarily be the same thing as migration or movement of individuals. The classic method of estimating gene flow (Slatkin, 1985; 1987; Slatkin & Barton, 1989) is via the statistic F_{ST}. F_{ST} was originally defined by Wright (1951) as the component of the overall inbreeding coefficient F in a subdivided population that can be attributed to population subdivision. A simple conceptual definition is provided by Slatkin and Barton (1989):

Table 7.1 Molecular genetic methods and spatial parameters. For all but the first two methods, it is possible to use PCR, enabling data to be obtained from very small amounts of genetic material. Modified from Moritz and Lavery (1996), Jarne and Lagoda (1996) and Parker et al. (1998)

Method	Genotype freqs	Allele freqs	Allele genealogy	Ease of use/cost	Applications	Limitations
Allozyme electrophoresis	Yes	Yes	No	Straightforward, cheap	Good for population structure, if enough polymorphism	Low levels of polymorphism in many populations. May be subject to selection
Multilocus minisatellite VNTRs 'DNA fingerprints'	No	No	No	Cheap for a DNA technique. Analysis is tricky	Excellent for identifying individuals	Relatively poor for population structure. As many loci scored simultaneously, genotypes cannot be determined.
Mitochondrial RFLPs	Yes	Yes	Yes	Relatively straightforward	Good for population structure. Mutation rate (and hence polymorphisms) greater than most nuclear DNA loci	Entire MtDNA functions as a single locus. Only maternally inherited. Cannot identify individuals
Nuclear RFLPs	Yes	Yes	Yes	Somewhat more difficult than MtDNA	Good for overall population structure	Amount of available polymorphism may be limited
Single locus microsatellite VNTRs	Yes	Yes	Yes	Relatively expensive. Primers must often be specially developed	Excellent for identifying individuals. Sufficient resolution for assignment tests	High initial setting-up costs for working on a new species
RAPDs	No	Yes	No	Method straightforward, analysis tricky	Potentially good both for population structure and individual identification	Methods of analysis and interpretation are not yet clearly established
Sequences	Yes	Yes	Yes	Currently expensive and time-consuming	Excellent for historical ecology	Costs and resulting small sample sizes possibly make method not optimal for current population structure or for identifying individuals

MtDNA, mitochondrial DNA; RAPD, random amplified polymorphic DNA; RFLP, restriction fragment length polymorphism; VNTR, variable number of tandem repeats.

$$F_{ST} = \frac{f_0 - \bar{f}}{1 - \bar{f}}, \tag{7.5}$$

where f_0 is the probability of identity by descent of two alleles chosen at random from a single subpopulation (deme) and $\bar{f}$ is the probability of identity by descent of two alleles chosen at random from the entire population. Thus, if a population is not subdivided, $F_{ST} = 0$, and F_{ST} increases with population subdivision. In an infinite island model without mutation or selection, Wright (1951) showed that

$$F_{ST} \approx \frac{1}{1 + 4Nm}, \tag{7.6}$$

where Nm is the number of gametes (not individuals!) migrating into each patch per generation, made up from the probability that each gamete is a migrant m and the effective population size N. The first and most obvious point to emerge from eqn (7.6) is that, under this model, it does not take much migration (about one individual per generation) to prevent significant genetic differentiation. Further, it appears that, by using eqn (7.6) it may be possible to estimate the number of migrants per generation between populations, given an estimate of F_{ST}.

Equation (7.6) is based on a highly idealized model, which has an infinite number of subpopulations of equal size, exchanging migrants equally with every other subpopulation, and with no selection or mutation occurring. Quite clearly, these are unrealistic assumptions, but Slatkin and Barton (1989) review simulations suggesting that the result in eqn (7.6) applies reasonably well in the face of mutation rates less than the migration rate, in cases where there are several (rather than infinite) subpopulations, where populations are connected via a stepping-stone model rather than an island model, and even where there is selection, provided the selection coefficient is less than the migration rate. Some other authors, however (e.g. Weir, 1996; or Bossart & Prowell, 1998) are less confident about the applicability of the equation in real situations.

A second issue is that eqn (7.5) is a conceptual definition, and cannot be used to estimate F_{ST} directly from data. There are several ways that F_{ST} can be estimated, and somewhat confusingly, these are often represented with symbols other than F_{ST}. There is a review of these methods in Weir (1996). Most people are likely to use a packaged program such as FSTAT (Goudet, 1995) to calculate them.

Other approaches to estimating Nm from genetic information do not use F_{ST}. One approach is to assume that the distribution of allele frequencies follows a beta distribution, one of the parameters of which is Nm, and then to estimate Nm by maximum likelihood (Slatkin & Barton, 1989). This approach, however, appears to overestimate Nm unless a very large number of subpopulations is sampled. The rare alleles or 'private alleles' method uses the mean frequency of alleles that are present in only one population to estimate gene flow, following the intuitive idea that there will be more such alleles if there is little gene flow than if gene flow is substantial. However, it appears that this method is probably not as useful as those based on F_{ST} (Slatkin & Barton, 1989).

An entirely different approach to the use of genetic data for estimating migration rates has recently been developed. The idea is to use high-resolution genetic methods to identify, as probable migrants, individuals whose genotype is more typical of a neighbouring population than the population in which they were recorded. The raw data necessary are simply a set of genotypes, based on a number of loci, from individuals in more than one population. It is not essential that each genotype is unique, but the more different genotypes there are in the data set, the more powerful the method will be. The genetic data have usually been derived from single-locus microsatellites, but allozymes, RAPDs, or combinations of all these can also be used (Waser & Strobeck, 1998).

The method is based on an extremely simple 'assignment test', which is used to assign an individual to the population to which it is most similar, on the basis of its genotype. The test follows these steps (Waser & Strobeck, 1998):

1 Remove the test individual's genotype from the population in which it was found, and then estimate allele frequencies at each locus using the remaining individuals sampled from that population. These frequencies can be represented as p_{il}, p_{jl}, ... for alleles i, j, ... at locus l.

2 At any particular locus l, the likelihood of the test individual having the genotype it was observed to possess, given the population's allele frequencies, is then p_{il}^2 if it was homozygous for allele i, and $2p_{il}p_{jl}$ if it was heterozygous for alleles i and j.

3 The log-likelihood of the test individual possessing its overall genotype, given the population's observed allele frequencies, is then simply the sum of the log-transformed individual likelihoods.

4 Repeat the above steps for each of the putative populations to which the individual may belong.

5 Assign the individual to the population in which it has the highest likelihood of occurrence.

It is implicit in step 2 that each population is in Hardy–Weinberg equilibrium, and implicit in step 3 that there is no linkage disequilibrium. There is a problem if an allele possessed by the individual being assigned is not present in a particular sample from a test population. This would cause the probability of the individual being assigned to that population to be calculated as zero, making the log-likelihood infinitely small. If the allele were indeed absent from the test population, this would be correct. However, as the data are samples from the populations, and not entire populations, there is a finite probability that a particular allele is present in the population, but not the sample. An appropriate work-around is to assign the missing allele a very low frequency such as $1/2N$, where N is the number of individuals sampled from the test population. Figure 7.6 shows examples applying this method.

It should be obvious that this method may detect animals as 'immigrants' that are not immigrants themselves, but the descendants of immigrants. If there is random mating within each population, as is assumed by the method, immigrant genotypes should get diluted fairly rapidly. However, the same basic approach has been used, with reasonable power, to detect immigrant human genotypes in populations up to about the second generation (Rannala & Mountain, 1997).

There is great potential in these microsatellite-based methods to investigate dispersal over ecological time scales, rather than over evolutionary time scales, which most other genetic methods operate on. Further research is required to determine the extent to which this potential may be realized. Individuals can only be unequivocally identified as migrants if they are clearly genetically distinct from the population in which they are found, but it may be that even modest levels of gene flow are sufficient to prevent enough genetic differentiation occurring to be able to make genetic distinctions between individuals from different populations. For example, the polar bear populations in Fig. 7.6(b) are separated by several thousand kilometres, yet several individuals lie very close to the assignment boundary between the populations.

Even if individual dispersers cannot clearly be identified, assignment-based methods may still be valuable in making inferences about dispersal. For example, using microsatellite data, Favre et al. (1997) found that female shrews were less likely to be assigned to the population in which they were found than were males. This matched the results of an extensive mark–recapture program showing that dispersal was strongly female-biased.

Metapopulations

A metapopulation is a population of subpopulations, separated in space from each other, and with limited movement between subpopulations. Such a broad definition could apply to almost all species. The 'classic' metapopulation model, developed by Levins (1969; 1970), is, however, much more restrictive. It assumes a large number of identical, discrete habitat patches, connected by migration. Patches are scored as either occupied or vacant, with the population density on each patch not being considered. The single variable modelled is the proportion of occupied patches P, and this is assumed to be influenced by two parameters only: c, the colonization rate; and e, the extinction rate. Colonization is assumed to occur at a rate proportional to the product of the proportion of occupied patches (sources of colonists) and the proportion of

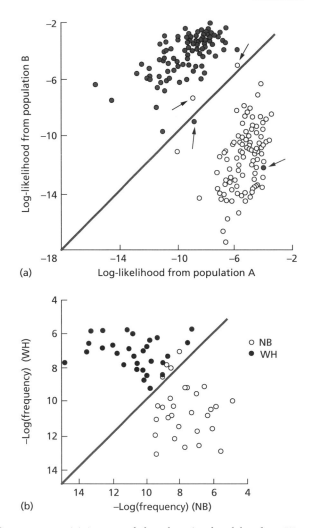

Fig. 7.6 Assignment tests. (a) An example based on simulated data from Waser and Strobeck (1998). The simulated populations have eight loci with 12 alleles each. Each population consists of 100 individuals, and the migration rate is 0.005 (one individual every second generation). Genotypes actually from population A are shown with open circles, whereas those from B are shown with closed circles. Genotypes above the 45° line have a higher likelihood of belonging to population B than to population A. Four individuals (arrowed) are misaligned. The B individual at the right-hand side of the A cluster is in fact an immigrant. (b) An example using real data on polar bears from Paetkau and Strobeck (1998). The likelihood of bears' genotypes coming from Western Hudson Bay (WH) or the North Beaufort Sea (NB) is shown. There is clear genetic separation between the two populations, although four North Beaufort Sea animals are assigned wrongly to Western Hudson Bay. As these are very close to the 45° line, they do not have a much greater likelihood of belonging to one population than the other, and they cannot be inferred to be immigrants. Based on eight microsatellite loci with four to nine alleles per locus (Paetkau & Strobeck, 1998). Note that polar bears have particularly low levels of genetic variation when assayed with allozyme electrophoresis or mitochondrial DNA.

vacant patches (available targets). Extinction occurs at a constant rate per occupied patch. These assumptions lead to the following simple equation to describe the rate of change in the proportion of occupied patches:

$$\frac{\mathrm{d}P}{\mathrm{d}t} = cP(1-P) - eP \tag{7.7}$$

This equation is identical in structure to the logistic equation. Its extreme simplicity makes it very valuable for drawing very general conclusions about populations divided into discrete patches (see, for example, Hanski, 1997a). There is little value, however, in attempting to apply such a general model in any specific case, and therefore little point in discussing estimation of the colonization and extinction parameters.

Some authors have questioned how common metapopulations are in the real world (e.g. Harrison, 1994). Certainly, no real population will meet the assumptions of this highly stylized model exactly. There is little value in applying the model to species that move between patches so frequently that 'extinction' on a patch is a meaningless term, and there is equally little point in applying it in situations where movement between patches is so rare that it has no appreciable impact on population dynamics.

The crucial property of a system that makes the metapopulation paradigm useful is that the proportion of occupied patches should be an adequate descriptor of the overall state of the population. If the population density per patch and its frequency distribution are needed to describe the population adequately, then it is a spatially subdivided population, but not one to which the classic metapopulation paradigm applies easily. Hanski (1997a) provides four conditions that should be satisfied if a metapopulation approach is likely to prove useful:

1 The suitable habitat occurs in discrete patches which may be occupied by local breeding populations.
2 Even the largest local populations have a substantial risk of extinction.
3 Habitat patches must not be too isolated to prevent recolonization.
4 Local populations do not have completely synchronous dynamics.

If the assumptions of the classical metapopulation model are relaxed a little, so that patches may differ in extinction probability according to size, and in recolonization probability according to location, a model can be produced that is sufficiently realistic to be parameterized. In at least some cases, such models also appear to have real conservation applications. Hanski (1994; 1997b) has developed this idea in some detail, and has named it the 'incidence function' approach.

There are many models in the literature, described as 'metapopulation models', that do not satisfy the above criteria fully (for example, Possingham *et al.*, 1992b; 1994; Lindenmayer and Lacy, 1995a; 1995b). To debate whether

these are 'real' metapopulation models is fairly sterile. However, to parameterize them, it is necessary to have information at the level of the individual patch on migration rates, birth and death rates, etc. This requires
approaches discussed earlier in this chapter, and in previous chapters. From
the perspective of parameterization, a metapopulation model is one that can be
parameterized by estimating patch-specific extinction and colonization rates.

Incidence function models

The incidence function approach developed by Hanski (1994) offers the promise of a metapopulation model realistic enough to be useful for quantitative
prediction of the extinction probability of a metapopulation, but simple enough
to be parameterized. I will therefore discuss the model and parameter estimation for the model in some detail.

   The model expands the classic Levins model by including information on
the spatial configuration of patches and patch sizes. Extinction of the study
species on a patch is allowed to be a function of patch area, and colonization
probability is allowed to be a function of patch isolation. Only presence or
absence of the study species is used in the analysis. If patch i is vacant in year t,
it has a probability C_i of being recolonized by year $t + 1$. If patch i is occupied in
year t, it has a probability E_i of being vacant by year $t + 1$.

   Given a large number of patches, differing in area and isolation, and
plausible functional forms relating extinction and colonization to these two
variables, it is possible, in principle, to estimate parameters for C_i and E_i directly. The logical approach would be some form of logistic regression (see
Chapter 2), and it has been used by Sjögren Gulve (1994) for this purpose.
The difficulty with this approach is that a substantial time series of extinctions
and colonizations is required (see Hanski, 1997b). Hanski cleverly circumvents
the problem by instead estimating the incidence J_i, or long-run probability
that patch i is occupied, as a function of patch size and isolation. Together with
some additional information, these estimates can then be used to estimate the
extinction and colonization rates, permitting the metapopulation process to
be simulated.

   Provided the system is in stochastic steady state (that is, the mean number
of occupied patches is neither increasing nor decreasing), J_i is related to the
extinction and colonization probabilities by

$$J_i = \frac{C_i}{C_i + E_i} \tag{7.8}$$

(Hanski, 1994). Given the steady-state assumption (which may well be a
problem, if the objective is to model a population that is in decline), incidence
can be modelled from a single snapshot of presence and absence on a large
number of patches. As is the case with direct estimation of extinction and

Table 7.2 Parameter and variable definitions for Hanski's incidence function metapopulation model

Parameter	Description	Comments
C_i	Probability of colonization of patch i in a single time interval	
E_i	Probability of extinction of the population on patch i in a single time interval	
J_i	The long-term probability of patch i being occupied	Defined as 'incidence' by Hanski
A_i	Area of ith patch	Obtained directly from data
x	Parameter describing how extinction scales with patch area	
e	Parameter describing the minimum patch size on which a population can persist	
M_i	Number of migrants arriving on patch i in a single time interval	
y	Half-saturation parameter of the sigmoid relationship between number of migrants and probability of successful colonization	Number of migrants per time interval that results in a probability of 0.5 of successful colonization
S_i	Isolation of patch i	Depends on distance to other patches, their size, and occupancy. Obtained from data, given assumption of steady state, and an independent estimate of the parameter α
β	Parameter relating isolation to number of migrants	
d_{ij}	Distance between patches i and j	Obtained directly from data
α	Parameter describing rate at which isolation increases with distance	
y'	Parameter combination linking probability of colonization to isolation	$y' = (y/\beta)^2$

colonization rates, it is necessary to assume plausible forms for the dependence of extinction and colonization on patch size and isolation. These are then substituted into eqn (7.8) to estimate the parameters themselves. Depending on the biology of the species concerned, a number of possible functional forms can be used.

The following description of the estimation procedure is based on Hanski (1997b), unless explicity stated otherwise. Table 7.2 summarizes the parameter and variable definitions.

Equation (7.8) should first be modified slightly by the addition of a 'rescue effect'. Colonists will arrive at a patch when it is occupied as well as when it is vacant. A patch with a high colonization rate should have a lowered extinction rate, as colonists will sometimes prevent extinction by arriving at a patch in a year when the patch population would otherwise have become extinct, as well as by recolonizing a patch once extinct. The incidence is thus

$$J_i = \frac{C_i}{C_i + E_i(1 - C_i)} = \frac{C_i}{C_i + E_i - E_i C_i}. \tag{7.9}$$

A plausible form for the extinction probability E_i is

$$E_i = \min\left(1, \frac{e}{A_i^x}\right), \tag{7.10}$$

here A_i is the area of the ith patch, and x is a parameter describing how extinction probability scales with patch area. The parameter e essentially determines the minimum patch size on which the species has any chance of surviving: if

$$A_i \leq e^{1/x}, \tag{7.11}$$

extinction is certain in one time period.

Colonization is harder to model plausibly. It will depend in some way on the mean number of immigrants M_i arriving at the ith patch per unit of time, which will in turn depend on both the distances to and sizes of nearby patches. A possible relationship between M_i and C_i is

$$C_i = \frac{M_i^2}{y^2 + M_i^2}. \tag{7.12}$$

This generates an S-shaped curve, so that small numbers of migrants have a very small chance of colonizing a patch, but large numbers of migrants are almost certain to do so. The parameter y tunes the shape of this curve. Other functional forms may be appropriate in some contexts.

The next problem is an appropriate form for M_i. Hanski suggests

$$M_i = \beta S_i, \tag{7.13}$$

where β is a parameter to be estimated, and S_i is a measure of the isolation of patch i. In Hanski's model,

$$S_i = \sum_{j \neq i} p_j A_j \exp(-\alpha d_{ij}). \tag{7.14}$$

Here, p_j equals 0 for empty patches, and 1 for occupied patches, d_{ij} is the distance from the ith to jth patches, and α is a parameter to be fitted. The model is thus assuming that the number of migrants a patch contributes to another is proportional to the product of the donor patch area and a negative exponential function of the distance between patches (following the dispersal model of eqn (7.1)). At a cost of an additional parameter, colonization can alternatively be made to scale as a power of A_j.

To make this workable, it needs to be further assumed that M_i is constant at steady state, so that the observed p_i distribution can be used to estimated the model parameters. In reality, even at stochastic steady state, the p_i set would change as individual patches winked on and off (thus causing the number of migrants to each patch to change).

Note, that by combining eqns (7.12) and (7.13), eqn (7.12) can be written as

$$C_i = \frac{S_i^2}{(y/\beta)^2 + S_i^2}$$ (7.15)

and thus only the combination $y' = (y/\beta)^2$ needs to be estimated.

Having made these assumptions, eqn (7.9) becomes

$$J_i = \left[1 + \frac{ey'}{S_i^2 A_i^x}\right].$$ (7.16)

Equation (7.16) may be written as

$$J_i = [1 + \exp[\ln(ey') - 2\ln S_i - x\ln A_i]^{-1}$$ (7.17)

and, furthermore,

$$\ln\left(\frac{J_i}{1 - J_i}\right) = -\ln(ey') + 2\ln S_i + x\ln A_i.$$ (7.18)

This is now a standard logistic regression, meaning that given a series of observed J_i, A_i and S_i, standard packages can be used to estimate the parameter x, and the parameter combination $-\ln(ey')$, using $2\ln S_i$ as an offset (see Crawley, 1993). Two problems remain. First, S_i does not consist solely of observable quantities: it includes the unknown parameter α. Second, e and y' cannot be disentangled, but an estimate of e itself is needed in eqn (7.10) if the metapopulation model is to be iterated.

Hanski (1994) suggests that α should be estimated from independent movement data. For example, α could be estimated as the decay parameter of eqn (7.1), using an experiment like that described by Hill et al. (1996). Note that it is not necessary to estimate the number of migrants dispersing to particular distances from a patch of area A, only the way in which the number of dispersers declines with distance from the source population. Alternatively, an iterative nonlinear method could be used to estimate this parameter, but that would considerably complicate things. Hanski (1994) asserts that the results of an incidence function simulation are not strongly dependent on the value of α that is used.

Hanski (1997b) suggests that e should be estimated from the size of the threshold patch size, beyond which extinction is certain. Using eqn (7.11), this will give an independent estimate of e. The area of the smallest occupied patch, A_0, could be used in the absence of other data. Following eqn (7.11),

$$e = A_0^x.$$ (7.19)

An example of an application of this model is shown in Box 7.1.

Box 7.1 Using the incidence function approach to predict the consequences of habitat destruction in an endangered butterfly

This example is based on Wahlberg *et al.* (1996). The false fritillary butterfly (*Melitaea diamina*) is endangered throughout most of its range in Europe. It requires habitat patches consisting of moist meadows where its larval host plant is present. Wahlberg *et al.* (1996) surveyed its only well-known metapopulation in Finland, recording the distribution and areas of habitat patches shown in (a) below. Solid circles represent occupied patches and open circles are unoccupied patches of suitable habitat (94 patches in total, 35 occupied). Circle diameters are proportional to patch areas (ranging from 0.005 to 4 ha).

As this is an endangered species occupying a relatively small number of patches, it is difficult to use data for the species itself to estimate parameters for the incidence function model. Hanski (1994) estimated parameters for the related butterfly, *Melitaea cinxia*, in 1600 habitat patches in Finland. These were used to attempt to predict the dynamics of the endangered species. The parameters used were:

x	ey'	e	α	$b*$	A_0
0.952	0.158	0.010	1.0	0.5	79 m^2

*In this example, a minor variant of eqn (7.12) was used, in which the effect of patch area on colonization was proportional to A_i^b.

With these parameters, the predicted incidence of patch occupancy is as shown below in (b), where the depth of shading is proportional to the

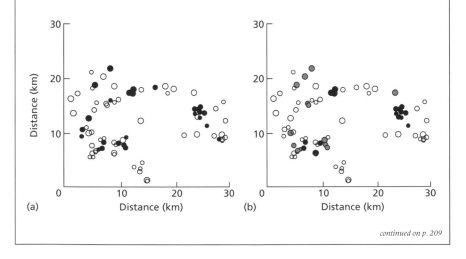

(a)   Distance (km)   (b)   Distance (km)

continued on p. 209

Box 7.1 *contd*

probability of the patch being occupied. The close correspondence to the observed occupancy is obvious.

The model was then iterated to investigate the consequences of various patterns of habitat destruction. In (a) 47 patches (representing 55% of the total habitat area) were destroyed. These are shown as crosses. Note that the destroyed patches are around the periphery of the patch network. The consequences are shown as the results of 10 replicate simulations over 400 time intervals. This pattern of destruction is not predicted to lead to extinction of the entire metapopulation. In contrast, (b) shows the consequences of destroying just 15 patches (44% of the total area) that are in the centre of the patch network. It is clear that this pattern of destruction is predicted to lead to extinction of the metapopulation over a short time horizon.

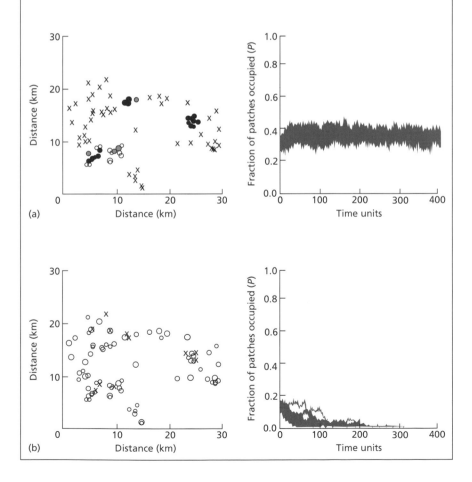

Diffusion models

Diffusion models are probably the most abstract form of spatial model. Space is considered to be continuous, rather than a series of patches or cells. The approach dates at least back to Skellam (1951). A recent, brief review is provided by Hastings (1996), and a fuller treatment can be found in Murray (1989).

In two dimensions, the simplest possible diffusion model is

$$\frac{\partial n}{\partial t} = rn + \frac{1}{2}s\left\{\frac{\partial^2 n}{\partial x_1^2} + \frac{\partial^2 n}{\partial x_2^2}\right\} \tag{7.20}$$

(van den Bosch *et al.*, 1992). Here, n is the population density at time t at a position in a two-dimensional plane defined by the coordinate pair (x_1, x_2), r is the intrinsic rate of growth and s is the diffusion constant. The model assumes that animals move randomly throughout life, at a constant rate that determines the diffusion coefficient s. In this simplest version of the diffusion model, the rate of spread s is the same in all directions, although this assumption can be relaxed fairly easily. Neither demographic parameters nor movement parameters depend on age or local population density.

The principal result from this simple model is that the square root of area occupied is predicted to increase as a linear function of time, with a rate coefficient

$$C = \sqrt{2rs}. \tag{7.21}$$

Surprisingly, this highly unrealistic model often fits observed data on area occupied as a function of time since release quite well, particularly after the initial stages (see Fig. 7.7). The most important general point to be gained from eqn (7.21) is that the rate of spread of an invading population is a function not only of the dispersal rate of individuals, but also of the intrinsic growth rate of the population.

Potentially, eqn (7.21) could be used directly to estimate the diffusion coefficient, given an observed rate of spread, and an estimate of the intrinsic rate of growth r (see Chapter 5). In reality, however, this approach is unlikely to be useful in many cases. The practical problem is usually one of predicting the rate of spread of an invasive species or genotype, and if the rate of spread is required to estimate the diffusion coefficient, this is begging the question. Equation (7.20) is a highly abstract model. In common with many of the abstract models discussed in this book, its parameters are compound entities that do not correspond closely to directly observable ecological quantities. It is difficult to estimate s directly from data, even if one is prepared to accept that the assumptions the model makes can reasonably be justified.

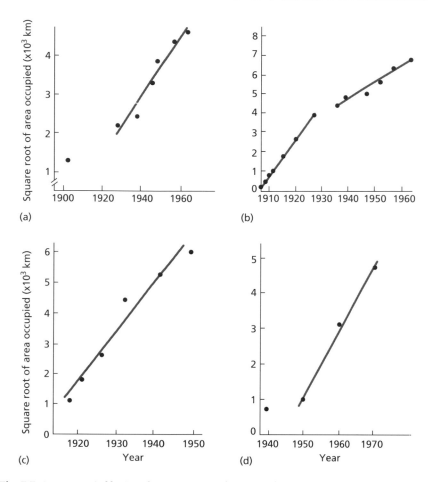

Fig. 7.7 Area occupied by invading species as a function of time since introduction: (a) The collared dove (*Streptopelia decaocto*) introduced into Europe from Asia in about 1900. (b) The muskrat (*Ondatra zibethicus*), escaped from a fur farm near Prague in 1905. A large-scale trapping program commenced in about 1930. Two separate regression lines, up to, and after, 1930 are therefore shown. (c) The starling (*Sturnus vulgaris*), released in New York City in 1890/91. (d) The cattle egret (*Bubulus ibis ibis*), invaded the Americas in about 1930. From van den Bosch *et al.* (1992).

Van den Bosch *et al.* (1992) present a method to predict the rate of range expansion of a population, given demographic data and dispersal data for individuals. The method is based on an integral equation, details of which can be found in the above paper. In general terms, the model assumes that the population is age-structured, with constant age-specific fecundity and mortality rates. It derives a relationship between the number of births at time *t* at a particular point in space and the number of births in the past at all other possible positions. Dispersal is assumed to be independent for each individual, to have no particular preferred direction, and not to depend on either local

Table 7.3 Parameters for the van den Bosch *et al.* (1992) models of range expansion. In this table, *i* represents the age class; M_i represents age-specific fecundity; L_i represents age-specific survivorship (the proportion of individuals surviving from the first census to census *i*); and l_i represents the midpoint of the age interval *i*. For dispersal data, it is assumed that *n* observations are available of individual dispersal distances d_j. These are assumed to be records of distances moved between birth and adulthood for animals that disperse only as juveniles. For animals that disperse at a constant rate throughout life, the data available are likely to be distances moved between successive census intervals. This complicates the estimation of dispersal parameters (see van den Bosch *et al.* (1992) for details)

Parameter	Parameter definition	Estimation formula
R_0	Net reproductive rate	$\hat{R}_0 = \sum_i L_i M_i$
μ	Mean age of reproduction	$\hat{\mu} = \dfrac{1}{R_0} \sum_i l_i L_i M_i$
υ	Variance in age of reproduction	$\hat{\upsilon} = \dfrac{1}{\hat{R}_0} \sum_i l_i^2 L_i M_i - \hat{\mu}^2$
σ_r^2	Variance of spatial distribution of recaptured animals	$\hat{\sigma}_r^2 = \dfrac{\sum_j d_j^2}{2n}$
γ_r	Kurtosis of spatial distribution of recaptured animals	$\hat{\gamma}_r = \dfrac{1}{\hat{\sigma}^4} \dfrac{3}{8} \dfrac{1}{n} \sum_j d_j^4 - 3$

population density or the current position in space. These are quite restrictive assumptions.

Van den Bosch *et al.* (1992) develop various approximations that predict the rate of range expansion. In the simplest possible case, individual animals could be assumed to move randomly throughout life, with dispersal distance following a normal distribution. If this is the case, they show that eqn (7.21) represents the rate of expansion of the population, with *r* being the usual intrinsic rate of growth, and *s* being the rate at which the variance of dispersal distance increases with age. Given a large set of data giving distance to which individuals had dispersed from their natal position, as a function of age, it would be possible to estimate the variance of the dispersal distance as a function of age. If dispersal were indeed continuous, then the relationship between that variance and age would be a straight line passing through the origin, and the slope of that line would estimate the parameter *s* in eqn (7.21). Unfortunately, such data are rarely available, and if they were, it is unlikely that there would be a simple linear relationship between the variance of dispersal distance and age. Few animals move at random at a constant rate throughout life. More reasonable assumptions are required in practice.

Table 7.3 defines parameters for species demography and dispersal that can be used to estimate the rate of population expansion. Using these definitions, van den Bosch *et al.* (1992) show that, for relatively slowly reproducing species ($R_0 < 1.5$),

$$C \approx \frac{\sigma_r}{\mu} \sqrt{2 \ln R_0}. \tag{7.22}$$

Comparing eqn (7.22) with eqn (7.21), it is possible to identify the approximations

$$r \approx \frac{\ln R_0}{\mu} \tag{7.23}$$

and

$$s = \frac{\sigma_r^2}{\mu}. \tag{7.24}$$

For more rapidly reproducing species ($1.5 < R_0 < 7$), van den Bosch *et al.* suggest the following approximation:

$$C \approx \frac{\sigma}{\mu} \sqrt{2 \ln R_0} \left\{ 1 + \left[\left(\frac{v}{\mu} \right)^2 - \frac{1}{12} \gamma \right] \ln R_0 \right\}. \tag{7.25}$$

This should be a reasonable approximation provided $v/\mu < 0.06$, and if dispersal occurs only in juveniles. If adults move as well, a more complicated approximation is needed. For details, see van den Bosch *et al.* (1992).

The relationship between these predicted rates of spread and observed rates of spread is shown in Fig. 7.8 for some vertebrate species. The correspondence between observed and predicted rates is encouraging. It is not possible to determine whether the discrepancies are a result of failures in the model assumptions or inaccuracies in the input data. Van den Bosch *et al.* (1992) used literature values for dispersal distances. As has been discussed earlier in this chapter, such data are likely to contain a variety of biases and inaccuracies. As the equations rely on the second and fourth moments of the distribution of dispersal distances, any inaccuracies are likely to be magnified. It is interesting to note that the more elaborate eqn (7.25) appears, for these examples at least, to provide somewhat more accurate predictions than the far simpler eqn (7.22), but the difference is not dramatic.

In most of the examples shown in Fig. 7.8, the actual rate of spread is underestimated by the simple model. A possible explanation is that rare, long-distance dispersal events increase the rate of spread. Hengeveld (1994) discusses this problem, together with some partial solutions. Rare events are always hard to detect, so it is unlikely that there are complete solutions. Another limitation of the model is that density dependence is not included.

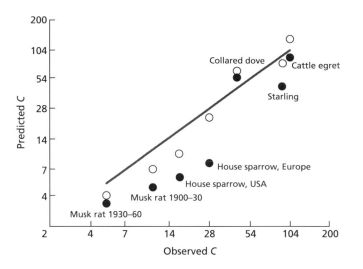

Fig. 7.8 The observed rate of spread of some vertebrates compared with their predicted rate of spread. Data points represented by black circles used expected rates of spread calculated from eqn (7.22), whereas open circles show expected rates calculated from eqn (7.25). If predictions were perfect, all points would fall on the 45° line. From data in van den Bosch *et al.* (1992).

Population growth will inevitably decline as local density increases, although it may be that direct density dependence is not of crucial importance when considering the initial stages of population spread. To some extent, an increased propensity to disperse at high densities may compensate for lower population growth. Alternatively, there may be inverse density dependence (an Allee effect) at very low densities: the rare long-distance dispersers may have difficulty finding mates. This may have a significant impact on the rate of spread of an invader (Kot *et al.*, 1996; Veit & Lewis, 1996). Some approaches that may be helpful for 'fat-tailed' distributions (that is, those in which there are some very long-distance dispersers) are outlined by Kot *et al.* (1996).

Summary and recommendations

1   An unbiased representation of the frequency distribution of dispersal distances is difficult to obtain. The distribution is almost always highly skewed, with a few individuals moving very long distances. Some methods, such as mark–recapture studies in a finite study area, will underestimate long-distance dispersal. Others, such as band recovery, tend to overestimate it. Using radiotelemetry to follow individuals is expensive, but is the best way to obtain unbiased data.

2   The probability distribution of dispersal distance of propagules from a parent organism is often estimated by placing collectors of a given area at different distances and directions from the parent. If you use this approach, it is important to recognize that a uniform probability distribution of dispersal distance leads to the number of propagules per unit area declining as the square of distance, and to correct accordingly.

3 Formal mark–recapture methods can be used to estimate migration rates between subpopulations or habitat patches, but it is difficult to differentiate between survival and migration rates. Recapture rates must be reasonably high for these approaches to be useful.

4 Molecular genetic data are useful for determining whether there is gene flow between populations, and can shed light on the spatial pattern of gene flow. However, a small amount of migration is sufficient to prevent genetic differentiation. Methods based on statistics such as F_{ST} are most useful over an evolutionary, rather than an ecological, time scale. They are relatively poor at estimating the amount of migration in ecological time. The recent development of 'assignment tests' holds promise for measuring ecologically relevant migration rates.

5 Many spatially divided populations are described as 'metapopulations'. For parameter estimation, a metapopulation is a system that can adequately be characterized by describing patches as 'occupied' or 'vacant', without using information about the population size or structure on each patch. The parameters that then need to be estimated are patch-specific probabilities of extinction and colonization.

6 The incidence function approach has potential to be realistic enough to be useful, but simple enough to be parameterized. How widely applicable it is remains to be seen.

7 Diffusion models for invading species are very abstract, but their key prediction that the square root of area occupied should be a linear function of time since invasion is often surprisingly accurate. To use these models predictively, it is necessary to be able to estimate the diffusion coefficient from demographic data. This is not easy. Table 7.3 and eqns (7.22) to (7.25) provide some suggestions on how to proceed, but these require an accurate description of the probability distribution of individual dispersal distances, up to the fourth moment (kurtosis).

CHAPTER 8

Competition

Introduction

Interspecific competition has been the subject of almost as much acrimonious debate in ecology as has density dependence (Connor & Simberloff, 1986; Diamond, 1986; Connell, 1983; Roughgarden, 1983; Simberloff, 1983). The basic principle that two coexisting species will have a negative impact on each other, if they share a limiting resource, is not in dispute. However, inferring the strength, or even the existence, of competitive interactions using data from undisturbed field populations is difficult, and experiments designed to measure its impact are often trenchantly criticized (Underwood, 1986).

The simplest model of interspecific competition is the Lotka–Volterra model. It models two species N_1 and N_2, each independently following logistic growth with intrinsic growth rates r_1 and r_2 and carrying capacities K_1 and K_2. The presence of each species also has an impact on the carrying capacity of the other via competition coefficients α_{12} and α_{21}. The equations can be expressed in various forms, including:

$$\frac{dN_1}{dt} = r_1 N_1 \left(1 - \frac{N_1 + \alpha_{12} N_2}{K_1} \right) \equiv f_1(N_1, N_2), \tag{8.1}$$

$$\frac{dN_2}{dt} = r_2 N_2 \left(1 - \frac{N_2 + \alpha_{21} N_1}{K_2} \right) \equiv f_2(N_1, N_2). \tag{8.2}$$

From these, it can be seen that the competition coefficient α_{12} measures the effect of species 2 on the growth rate of species 1 relative to the impact of species 1 on its own growth rate. Whereas one individual of species 1 decreases the growth rate of its own species by a factor $1/K_1$, each individual of species 2 decreases the growth rate of species 1 by an amount α_{12}/K_1. This is a very general and abstract representation of interspecific competition, which can be extended to any number of species in a fairly obvious fashion.

It is easy to become confused about the meaning of subscripted parameters. The first subscript is the *target* of the competition, and the second the *agent*. Thus α_{ij} measures the effect of species j on species i. The distinction between α_{ij} and α_{ji} is important, as empirical evidence suggests strongly that competition is often very asymmetric (Lawton & Hassell, 1981; Connell, 1983).

The main use of this general model is to answer general questions, and most basic ecology textbooks provide a discussion of the competitive exclusion principle and conditions for coexistence of competitors based on this simple model (e.g. Krebs, 1985; Begon *et al.*, 1990). Box 8.1 reviews the use of phase-plane analysis to study the behaviour of the model.

Box 8.1 Phase-plane analysis of the Lotka–Volterra competition equations

Phase-plane analysis is a powerful graphical method for exploring the qualitative behaviour of two-species differential equation models. A phase plane is simply a graph with the population size of one species on the horizontal axis, and the population size of the other on the vertical axis. Any point on the graph then represents a pair of species abundances. An arrow at that point can be used to show the trajectory of the system, which is the direction in which the model predicts the densities will change. Lines along which the rate of change of one of the species is constant can also be drawn. These are called isoclines. The most useful isoclines are the zero net growth isoclines (sometimes simply called zero isoclines). The zero net growth isocline of species 1 is the line along which the growth rate of species 1 is zero. Where the zero isoclines for both species cross, there is an *equilibrium*, a point where the system will remain, if it is already there. Trajectories drawn around the isoclines can be used to investigate the stability of the equilibrium. The zero isoclines for N_1 and N_2 can be found by setting $dN_1/dt = 0$ and $dN_2/dt = 0$ and then solving each equation to give N_2 in terms of N_1.

For example, in the simple Lotka–Volterra eqns (8.1) and (8.2), the zero isocline for N_1 is the solution of

$$1 - \frac{N_1 + \alpha_{12}N_2}{K_1} = 0 \Rightarrow N_2 = \frac{K_1}{\alpha_{12}} - \frac{N_1}{\alpha_{12}},$$

and the zero isocline for N_2 is the solution of

$$1 - \frac{N_2 + \alpha_{21}N_1}{K_2} = 0 \Rightarrow N_2 = K_2 - \alpha_{21}N_1.$$

Both of these zero isoclines are straight lines. The zero isocline for N_1 divides the phase space into two regions. Close to the origin, N_1 is increasing, but beyond the line, it decreases. Similarly, the N_2 isocline divides the phase space into two regions. Simple vector addition ('nose to tail') can be used to predict the overall trajectory of the system at any point.

continued on p. 218

Box 8.1 *contd*

There are four qualitatively different outcomes from this pair of isoclines, as shown in the phase plots below. In each of these, the N_1 isocline is shown with a solid line, and the N_2 isocline is dashed.

$$\frac{K_1}{\alpha_{12}} > K_2 \text{ and } \frac{K_2}{\alpha_{21}} > K_1. \tag{i}$$

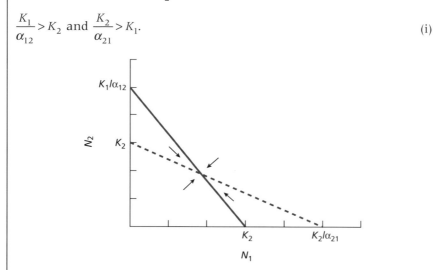

In this case, there is an equilibrium with both species present, and the trajectories show that it is stable. The two species can coexist. Careful inspection of the conditions shows that one additional member of each species has a bigger effect on its own species' growth rate than it does on the growth rate of the other species: intraspecific competition is stronger than interspecific competition.

$$\frac{K_1}{\alpha_{12}} < K_2 \text{ and } \frac{K_2}{\alpha_{21}} < K_1. \tag{ii}$$

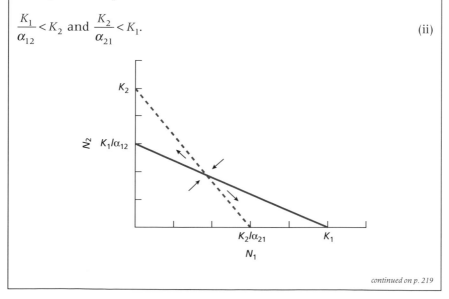

continued on p. 219

Box 8.1 *contd*

Now the zero isoclines still intersect, but the trajectories show that the equilibrium is unstable. One species will replace the other, but which one wins depends on the initial conditions. Interspecific competition is stronger than intraspecific competition.

$$\frac{K_1}{\alpha_{12}} > K_2 \text{ and } \frac{K_2}{\alpha_{21}} < K_1. \tag{iii}$$

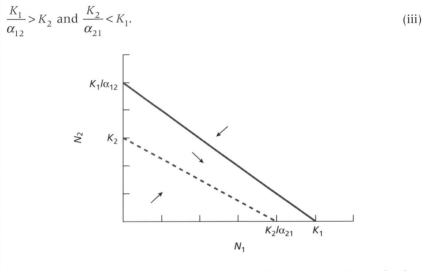

In this case, the isoclines do not intersect, and the N_1 zero isocline is further from the origin than is the N_2 zero isocline. Species 1 will always outcompete species 2.

$$\frac{K_1}{\alpha_{12}} < K_2 \text{ and } \frac{K_2}{\alpha_{21}} > K_1. \tag{iv}$$

This is the reverse of (iii). Species 2 will always outcompete species 1.

Several authors have attempted to understand and model particular competitive interactions through parameterizing the model. Such approaches are most likely to be useful when modelling competition between microbes or protozoa, rather than large, long-lived organisms. More commonly, the model is used as a caricature of a competitive interaction. An attempt is made to estimate a competition coefficient, and this is used to infer something about competition between the two species, without the full model actually being used to predict the time-course of an interaction between the two species.

Defining interaction coefficients

There are at least four ways that interaction coefficients between species can be defined (Laska & Wootton, 1998). These are:

1 Per capita direct effects that one single individual of one species has on one member of another species. If eqns (8.1) and (8.2) were, in fact, an accurate description of a competitive interaction, then the per capita effect of species 2 on species 1 would be $\alpha_{12}r_1/K_1$. Frequently, this will be scaled relative to the density-dependent effect that the target species has on itself, yielding simply the coefficient α_{12}.

2 Elements in a Jacobian matrix, evaluated at equilibrium. This approach was originally advocated by May (1974b). The Jacobian matrix is a matrix of first partial derivatives of the functions describing the interaction, with respect to each of the population variables in turn. For example, for the model described by eqns (8.1) and (8.2), the matrix would be:

$$\begin{pmatrix} \dfrac{\partial f_1(N_1,N_2)}{\partial N_1} & \dfrac{\partial f_1(N_1,N_2)}{\partial N_2} \\[2mm] \dfrac{\partial f_2(N_1,N_2)}{\partial N_1} & \dfrac{\partial f_2(N_1,N_2)}{\partial N_2} \end{pmatrix}, \tag{8.3}$$

evaluated at the equilibrium point (N_1^*, N_2^*). With a little algebra, the effect of species 2 on species 1 would therefore be:

$$\frac{\partial f_1(N_1^*,N_2^*)}{\partial N_2} = \frac{-\alpha_{12}r_1(K_1 - \alpha_{12}K_2)}{K_1(1 - \alpha_{12}\alpha_{21})}, \tag{8.4}$$

provided that an equilibrium with both species present exists.

Equation (8.4) appears a little messy. However, if it is scaled relative to the equilibrium population density of the target species, which is

$$N_1^* = \frac{K_1 - K_2\alpha_{12}}{1 - \alpha_{21}\alpha_{12}}, \tag{8.5}$$

it becomes simply

$$\frac{\partial f_1}{\partial N_2} = \frac{-\alpha_{12}r_1}{K_1}.$$

This definition is well suited to theoretical investigations, as it provides a single quantity to measure the interaction strength, even if the effect of one species on the other is nonlinear. It is rather difficult to use empirically, as it describes the effect on species 1 of an infinitesimally small change in species 2, and it requires that the system should be in equilibrium before manipulation.

3 The inverted negative Jacobian matrix. If the interaction in question is embedded in a multispecies community, the actual effect of reducing the density of species j on species i may be quite different from the direct effect. For example, suppose three species, A, B and C compete with each other. If B competes with A, then one would expect that reducing the density of B would allow A to increase. But suppose that the effect of B on A is small, compared with the competitive effect that B has on C, and suppose further that C competes strongly with A. Then reducing B's density would allow C to increase. C would then compete with A, producing an overall detrimental effect on A. The effect of reducing B would then be the opposite of the beneficial effect expected from the direct competitive interaction. The elements in the Jacobian matrix (expression (8.3) above), inverted and multiplied by -1, measure the overall (direct and indirect) effect of a constant small perturbation in any of the species in the community upon each of the others.

4   The removal matrix. An obvious way to investigate a competitive interaction experimentally is to remove the putative agent of the competition, and then record the change in abundance of the target. Laska and Wootton (1998) suggest measuring the strength of the interaction simply as the change in abundance of the target species when the competitor is removed. Paine (1992) suggests scaling the change relative to the abundance of the target in the community without the agent present, and then expressing the effect per individual agent. The interaction strength I_{ij}, representing the effect on target species i of removing the agent species j, would then be:

$$I_{ij} = \frac{N_{ij+} - N_{ij-}}{N_{ij-}N_j} \tag{8.6}$$

where N_{ij-} is the abundance of the target i after the agent j has been removed, N_{ij+} is the abundance of the target i with j present, and N_j is the abundance of the agent before removal. (Note that in Paine (1992), 'controls' are treatments from which the agent has been removed, and 'treatments' have the agent still present. This is rather confusing!) For the simple Lotka–Volterra model of eqns (8.1) and (8.2),

$$I_{12} = -\frac{\alpha_{12}}{K_1}. \tag{8.7}$$

Provided two species only are interacting, and the effect of competition is linear, all four of the above definitions thus are equivalent, subject to minor considerations of scaling. All real competitive interactions, however, are embedded in larger food webs. Indirect interactions may have a substantial influence on measured competition coefficients. In several of the few cases that have been looked at systematically (see below), isoclines have also been found to be nonlinear.

Estimating competition coefficients from resource usage

The most intuitively obvious way to estimate competition coefficients without direct experimentation is to use the extent of overlap in resource use. This idea was proposed by McArthur and Levins (1968). They suggested

$$\alpha_{ij} = \frac{\sum_h p_{ih} p_{jh}}{\sum_h p_{ih}^2},$$
(8.8)

where there are h potential resources, and p_{ih} is the relative resource utilization of resource h by species i. The relative resource utilization p_{ih} is the fraction of all its resources that species i obtains from resource h. If resource utilization patterns of the two species are identical, then α_{ij} will equal 1, and it will be less than 1 as the patterns of usage differ. There are substantial problems in both defining and estimating the p_{ih} terms, but even if this is accomplished, the problem remains that eqn (8.8) is an index of resource utilization overlap. To use an index of resource overlap as an index of competition, it would be necessary to make some adjustment for the total amount of resource use of individuals from each species. Furthermore, resource overlap is neither a necessary nor a sufficient condition for competition (Holt, 1987). If a particular resource is not limiting, overlap in usage will not result in competition. Interference competition may also occur without resource overlap. Schoener (1974b) has a detailed discussion of this approach and its variants, and a more recent discussion can be found in Krebs (1989). Whilst the indices thus calculated may be useful as measures of resource overlap, they cannot be considered as measures of competition coefficients that might have value in predicting the outcome of competitive interactions in the field.

Experimental approaches

If they can be carried out at appropriate temporal and spatial scales, there is no doubt that experimental manipulations of the density of potential competitors are a more satisfactory way to study competition than is observation. The most obvious way to investigate competition experimentally is to remove the putative agent of competition entirely from some replicated experimental units, leaving matched intact replicates as controls. Some time later, the population size or density of the putative target of competition is then compared between treatments and controls. This basic approach has been taken in a number of studies. Fairly recent examples include Paine (1992), Pfister (1995) and Fox and Luo (1996). Each of these authors considered this removal method to be a standard against which other approaches could be compared.

Provided only two species are interacting, provided the isoclines are linear, and provided the system can be assumed to have reached equilibrium after manipulation, the competition coefficient can be estimated from the observed data using eqn (8.6). Each of these provisos needs discussion. I have already briefly mentioned the problem of indirect interactions. Bender *et al.* (1984)

discuss it in detail. In general, there are two possible types of perturbation experiment. In 'press' experiments, the agent of competition is removed, or altered in abundance, and is maintained at this new level indefinitely. Once a new equilibrium is reached, the change in abundance of the target species or community is compared between treatment and control replicates. In 'pulse' experiments, a perturbation is applied at a single time, and the short-term response of the entire community, including both agent and targets, is recorded as the system returns to equilibrium. A pulse experiment would ideally involve a small perturbation, rather than a complete removal of the agent.

Press experiments measure the total effect of removal of the agent species from the community, and therefore include indirect, as well as direct effects. If you want to estimate the coefficients α_{ij} themselves, this poses a problem. In an n-species community, each would need to be perturbed, the response of every other species would have to be measured, and the resulting matrix of total effects manipulated to extract the interaction coefficients (Bender $et\ al.$, 1984). If you are interested in a particular pair of species, this is a difficulty more of theoretical than practical importance. The question is usually how one species affects another, in their particular community context, rather than one of resolving all the pathways of that interaction. In any event, Lotka–Volterra models subsume indirect interactions via species not included explicitly in the model into the interaction coefficients. If you want to describe all interactions in a community, then the problem of indirect interactions cannot be ignored, and is difficult to resolve with press experiments.

Pulse experiments record the short-term response of each species in a community to a single perturbation away from equilibrium. They are therefore attempting a direct measurement of the change in dN_i/dt in response to the perturbation. In principle, they should be able to estimate the components of the interaction equations, providing estimates of the direct interaction coefficients α_{ij} themselves (Bender $et\ al.$, 1984). In practice, there are major problems in applying the pulse approach to measuring competition in the field, and it has rarely been used (but see Seifert & Seifert 1976; 1979). Ideally, the perturbation should be infinitesimally small, so that the response of the system in the neighbourhood of equilibrium is determined. For practical purposes, however, the perturbation needs to be substantial, to produce a response big enough to be detected over random variation.

Bender $et\ al.$ (1984) suggest the following procedure to estimate interaction coefficients between a pair of species i and j, which may be components of a wider interaction community. This procedure should be able to deal with temporal variation in the intrinsic growth rates and carrying capacities of the two species, provided such variation is the same across replicates.

1   Divide the available replicates into matched triples.

2 Perturb the putative agent species j in one of each triple, perturb the putative subject of competition i in another, and leave the final member as an undisturbed control.

3 A short time later, measure N_i^C and N_j^C, the population size of the target and agent species in the control community. At the same time, also measure N_{ij}^T and N_{jj}^T, which are the population sizes of the target and agent species in the matched treatment community in which j has been perturbed. Also measure N_{ii}^T and N_{ji}^T in the matched treatment community in which i has been perturbed.

4 Simultaneously, also estimate the per capita rates of change of each population in the control and treatment communities. Call these r_i^C, r_j^C, r_{ii}^T, r_{ij}^T, r_{ji}^T and r_{jj}^T.

5 Estimate the interaction coefficients using

$$\alpha_{ij} = \frac{(r_{ij}^T - r_i^C)(N_i^C - N_{ii}^T)}{(r_{ii}^T - r_i^C)(N_j^C - N_{jj}^T)}. \tag{8.9}$$

This equation relies on estimating per capita rates of change, as well as actual population sizes. To measure a competition coefficient between a pair of species, it is necessary to manipulate *both* species, not just the putative agent of competition. This is necessary because the competition coefficient α_{ij} is defined as the impact that species j has on species i *relative to* the impact that species i has upon itself. Whatever its theoretical advantages, this method will be difficult to apply in practice, particularly to vertebrate populations. Rates of population growth are not easy to estimate (see Chapter 5), and will also inevitably fluctuate more in response to environmental noise and sampling error than will estimates of population size themselves.

How long should be left between the perturbation and measurement of the response? This is not a simple question, nor is the answer. Obviously, if it is left too long, the impact of the perturbation may have disappeared entirely. It is also possible to measure the impact too soon. The method is intended deliberately to measure direct, and not indirect effects. However, the most common form of competition is resource competition. This is often a form of indirect interaction, as the limiting resource or resources are often one or more species of organism. Whether the pulse perturbation method will detect resource competition depends on the characteristic time scale on which the resource dynamics operate. There are three possible situations: resource dynamics that are much faster than the exploiters; resource and exploiter dynamics on comparable time scales; and resource dynamics on a much slower time scale than the exploiters. If the resource responds more rapidly than the exploiter populations, then a pulse experiment with the response recorded after the resource has settled to a new equilibrium will measure resource competition coefficients correctly. If the resource responds slowly, then resource competition will not be detected at all. If time scales are comparable, then it is important to include the resource species in the perturbation experiment – see Bender *et al.* (1984) for a more detailed discussion.

Mapping competition isoclines

A few recent studies have attempted to determine the shape of competitive isoclines experimentally. Mapping the zero net growth isocline for species 2, subject to competition by species 1, requires experimentally holding the species 1 population size constant at several levels, and then determining, at each of these levels, the population size of species 2 at which there is no net growth in the population of species 2. This will never be easy to do directly. In principle, it would be possible to use an elaboration of the 'pulse' approach advocated by Bender *et al.* (1984). Another way of approaching the problem is to see if the competition coefficient α_{12}, which measures the effect of one individual of species 2 on one individual of species 1, is the same for all densities of each species. If it is, then the isocline is linear. If not, then α_{12} is the gradient of the isocline, and the form of the isocline itself can be obtained by integration with respect to N_2, provided α_{12} is a function of N_2 alone.

Abramsky *et al.* (1991; 1992; 1994) used an ingenious method to estimate the competition coefficients between two species of gerbil. They constructed replicated paired enclosures in the Negev Desert, Israel. Two gerbil species occupied the area: *Gerbillus allenbyi* (mean adult body mass about 26 g), and the larger *G. pyramidum* (mean adult body mass about 40 g). Their approach relies on the assumption that animals will distribute themselves between two different patches so that their average fitness is the same in each patch. This is the 'ideal free distribution' (Fretwell, 1972). If one further assumes that the per capita growth rate of a population is a measure of its average fitness, the animals should therefore distribute themselves so that the per capita growth rate is the same in each patch. If two patches differ only in that an agent of competition is present in differing densities, then the target of competition should distribute itself so that its density is higher in the patch where the competitor is less common than in the patch where the competitor is more common. The difference in the population density of the target species between the two patches, divided by the difference in density of the agent, is an estimate of the competition coefficient in that region of the state space.

To apply this method in practice, Abramsky and co-workers needed to devise a means of allowing one species of gerbil to move freely between adjacent enclosures, whilst constraining the other species to the enclosure into which they were placed. So that *G. pyramidum* could be prevented from moving between the paired enclosures, they used small gates through which only the smaller *G. allenbyi* could fit (Abramsky *et al.*, 1992). Designing a means of allowing the larger *G. pyramidum* to move through gates, whilst constraining the smaller species was more difficult (Abramsky *et al.*, 1994). They developed the ingenious device of weight-dependent latch mechanisms. The method requires that the quality of each enclosure must be identical, so that the animals should distribute themselves equally between the enclosures, if the competitor is not present. This is testable. It also requires that the animals should distribute themselves according to the ideal free distribution if a competitor is present in differing abundance between patches. This assumption cannot easily be tested, although Abramsky *et al.* (1991) use similar weight changes in animals in each pair of enclosures as evidence that fitness is similar in each pair.

The results Abramsky *et al.* obtained are shown in Fig. 8.1. If the animals do indeed distribute themselves so that their fitness is the same in adjacent enclosures, and stochasticity is neglected, then each of the lines shown in the

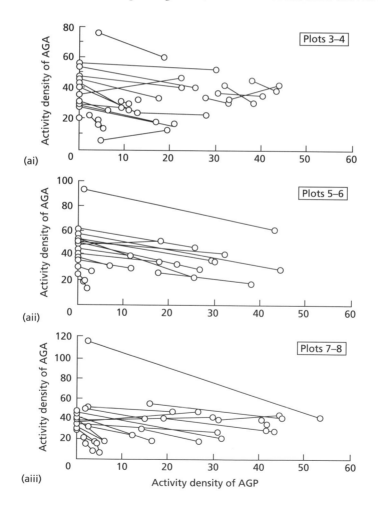

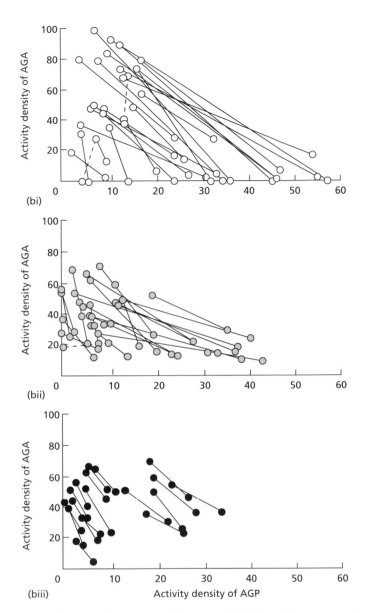

Fig. 8.1 Competition coefficients for gerbils. The axes in the figure show the population density of *Gerbillus allenbyi* (AGA, vertical axis) and *G. pyramidum* (AGP, horizontal axis), as measured by an activity index based on track counts on sandplots. Abramsky *et al.* felt that this method of estimating density was less disruptive to normal behaviour patterns than a trapping-based index. In each graph, lines connect data from simultaneous censuses on adjacent plots, separated by a 'semipermeable' fence. The gradient of the lines connecting these pairs was used to estimate the competition coefficient in the region of the state space in which the line falls. Dashed lines identify the few cases where the gradients of the lines were positive. (a) *G. pyramidum* as the agent of competition (with densities fixed in each plot), and *G. allenbyi* as the target (free to move between members of a pair). The individual panels are for each pair of enclosures. (b) *G. allenbyi* as the agent of competition (with densities fixed in each plot), and *G. pyramidum* as the target (free to move between members of a pair). In this case, each panel represents a series of experiments carried out in different pairs, but in the same month. From Abramsky *et al.* (1992, 1994).

figure should be chords (straight lines) connecting points on an isocline for the species that is free to move. (Recall that an isocline for a species is a line in state space along which the species' growth rate is constant.) All the lines will not be on the same isocline, however, and no line will necessarily be on the zero net growth isocline. However, each line will be an estimate of the gradient of an isocline in the region of state space sampled.

Abramsky *et al.* used regressions to predict the gradient of the isoclines as a function of the activity density of each species. They estimated activity density by the number of gerbil tracks left on sandplots. This index would estimate either actual density or total activity level with substantial error, potentially leading to substantial 'measurement error bias' (see Chapter 11). Their results are shown in Fig. 8.2. They then integrated the equations to obtain zero net growth isoclines for each species. To do this, it is necessary to specify a constant of integration, which determines where the isocline is located in state space. Zero net growth isoclines must pass through any equilibrium point, so forcing a zero isocline to pass through an observed equilibrium is an appropriate way to determine its constant of integration. Of course, this makes circular any attempt to 'test' the isoclines by seeing if they successfully predict the location of the observed equilibrium.

The zero isoclines presented by Abramsky *et al.* (1994) are shown in Fig. 8.3. Neither is linear, and the overall shape of each can be explained in terms of the habitat use and behaviour of the two species (see Abramsky *et al.*, 1994, for details). The isoclines predict that the leftward equilibrium where both species coexist should be only locally stable: if *G. pyramidum* activity density increases beyond a breakpoint of about 45, *G. pyramidum* should be able to exclude *G. allenbyi*. There is some anecdotal evidence consistent with this prediction.

This method offers the possibility of experimentally estimating the shape of isoclines from field data. However, it relies totally on the assumption that the animals will distribute themselves in accordance with the ideal free distribution. It also relies on the assumption that the quality of the environment on either side of the semipermeable membrane is identical. Abramsky *et al.* (1991) review briefly some factors that may distort the ideal free distribution. These include travel costs (in this context, costs of crossing the fence), exposure to predation, and 'despots', which are high-ranking individuals that exclude others from favourable habitats. Whilst the ideal free distribution has received much empirical support across a range of taxa, there are numerous examples in which it has been found to be an unsatisfactory predictor of animals' distribution (Kennedy & Gray, 1993; Guillemette & Himmelman, 1996; Cowlishaw, 1997; Kohlmann & Risenhoover, 1997).

Chase (1996) describes a different way of estimating the shape of competitive isoclines, using two competing grasshopper species. Potential competitors were placed together in screen cages on natural grassland. The objective was

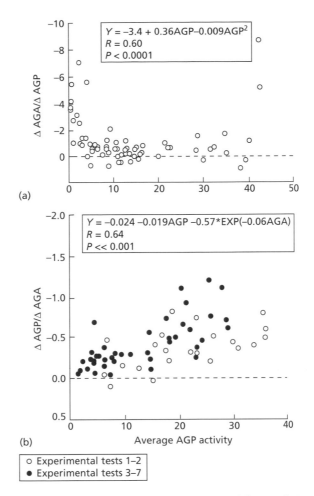

Fig. 8.2 The relationship between competition coefficients and the population density of agents of competition in gerbils. (a) The relationship between the estimated competition coefficient of *G. pyramidum* competing with *G. allenbyi* and average *G. pyramidum* activity density. The equation of the quadratic regression describing the relationship is shown on the plot. Note that competition is very strong at low and high *G. pyramidum* densities, but almost absent at intermediate densities. (b) The relationship between the estimated competition coefficient of *G. allenbyi* competing with *G. pyramidum* and average *G. allenbyi* density. The regression describing the relationship is shown on the plot. Note that, in this case, the regression depends on the density of the target of competition, as well as the agent. This means that the isoclines for *G. pyramidum* are not a family of parallel curves. From Abramsky *et al.* (1992, 1994).

to measure the zero isocline of one species (the 'target') as a function of the density of another (the 'neighbour' – 'agent' in my terminology). Several replicated densities of neighbours were used. These densities were maintained constant, by addition of cage-acclimatized additional grasshoppers, if

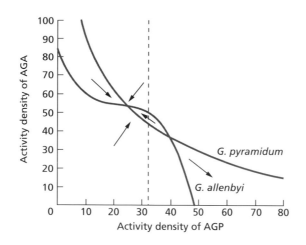

Fig. 8.3 Zero net growth isoclines for the competitive interaction between *G. allenbyi* and *G. pyramidum*. The isoclines are shown in a state space of activity density for both species. The *G. allenbyi* isocline is the inverse sigmoid curve, and the *G. pyramidum* curve is approximately hyperbolic. A stable equilibrium is indicated by the converging arrows, whereas the other intersection of the isoclines represents an unstable equilibrium. From Abramsky *et al.* (1994).

necessary. The targets were initially added at fairly high densities, which then declined towards a constant level over a period of a few days. The approximately constant level at which the target population ceased to decline was then taken to be the location of the zero isocline at that neighbour density (see Fig. 8.4(a)). The nonlinear isoclines thus obtained are shown in Fig. 8.4(b).

Whether these isoclines can be used to predict the outcome of competitive interactions between these two species in the field is doubtful. The competition investigated is only very short-term competition in a situation where emigration is prevented. No account is taken of either differing reproductive or dispersal ability. It is also difficult to see how such a method may be applied to species that are larger or disperse more widely or rapidly.

Estimating competition coefficients from census data

There has been a lengthy and continuing debate over whether competition coefficients can usefully be estimated from field data without some sort of manipulation (Crowell & Pimm, 1976; Pimm, 1985; Rosenzweig *et al.*, 1985; Schoener, 1985; Pfister, 1995; Fox & Luo, 1996). Two main approaches have been suggested. The static approach uses data from a single snapshot recording of abundance of the two putative competitors at a number of sites, whereas the dynamic approach uses information on changes through time in the population size of the putative competitors.

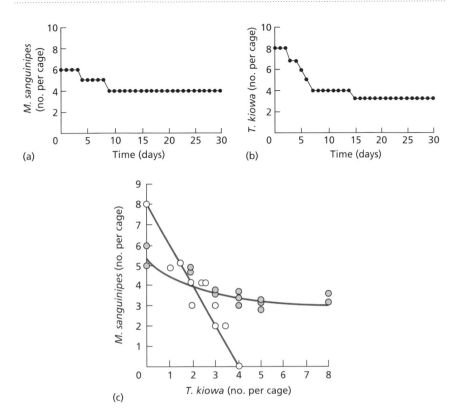

Fig. 8.4 Derivation of nonlinear isoclines for competing grasshoppers. (a,b) Sample data from cages in which one grasshopper species (the agent) was held at a constant density in cages (by replacement of animals that died), whilst the other (the target) was allowed to decline without replacement. (a) shows *Melanoplus sanguinipes* as the target, and *Trachyrachys kiowa* as the agent. In (b) the target and agent are reversed. (c) The isoclines obtained by a series of experiments similar to those shown above. The open circles represent data points where *T. kiowa* was the target, with the straight line passing through them being the estimated zero isocline for *T. kiowa*. The shaded circles represent data points where *M. sanguinipes* was the target. The nonlinear curve passing through them is the estimated zero isocline for *M. sanguinipes*. The curve was assumed to be nonlinear a priori, based on a model of interspecific competition developed by Schoener (1974a) for species with exclusive, as well as shared, resources. From Chase (1996).

Static approaches

In principle, these methods are based on the reasonable idea that, if two species compete, there should be a negative correlation between the population densities at given sites, once confounding effects of habitat quality are removed. More formally, the zero-growth isocline for species N_1 in a Lotka–Volterra model will be given by an equation of the following form:

$$N_1 = K_1 - \alpha_{12}N_2. \tag{8.10}$$

Thus, if a series of sites can be found over which K_2 varies, but α_{12} and K_1 are constant, a regression of N_1 versus N_2 should have a slope of α_{12}, provided the system is near equilibrium. There are a lot of provisos in this argument, sufficient that this very simple approach cannot be expected to work in practice.

The approach that has been used in practice was dubbed the 'Pimm–Schoener' method by Rosenzweig *et al.* (1985). The idea is to obtain data on the population density of the putative competitors at a number of sites and times, together with a range of variables describing the habitat from each of these observations. The objective is then to remove the confounding effects of habitat heterogeneity from the data statistically, so that the underlying relationship between the two species of interest is revealed. Using the population density of the 'target' species (or usually its logarithm) as a dependent variable, habitat variables are entered in a stepwise regression. The significance level for entry into the regression model would normally be set rather more loosely than the conventional $p = 0.05$, as environmental variation needs to be removed if it is possibly there. Following this, the density of the putative agent of competition is allowed to enter the regression, and the competition coefficient is then its regression parameter. The potential for measurement error bias in this approach is high.

Rosenzweig *et al.* (1985) used a number of variants of this method in a study of possible competition between desert rodents in Israel. The six variants were grouped into two subsets. In the first, principal components analysis was used to reduce the dimensionality of the habitat variables. Three subvariants were: (i) using a stepwise regression first, and then entering the census variables; (ii) using stepwise regression for habitat variables for both competitors, and then using regression on the residuals thus derived; and (iii) using a free regression, where census and habitat variables were entered together, with the order determined by the stepwise procedure itself. The second set used raw habitat descriptor variables instead of principal components, and then applied the three regression variants described above.

Their conclusion was that there was unacceptable inconsistency between methods, depending on the details of the method, although Pimm (1985) finds the consistency in the same analysis to be acceptable. There are certainly some statistical biases in using this approach. For example, the greater the ratio of the variance in the population size of the target of competition to the variance in population size of the agent, the larger is the estimated competition coefficient. This is simply because greater variation in density supplies more 'support' for a regression line. Without any extrinsic measure of competition, however, it is difficult to compare the various methods with any validity. Several studies have since compared the static regression approach with competition coefficients estimated from manipulations.

Abramsky *et al.* (1986) used two species of bumble-bee to compare the regression model for detecting competition with direct manipulation of populations. Using a number of meadows that were separated by unsuitable habitat, the population size of each species was compared between control meadows with both species present and experimental meadows from which one or other of the species had been removed. The conclusion was that, whereas the manipulations detected competition, the regression model, based on only the unmanipulated meadows, could not. This conclusion is not really surprising. A complete removal is bound to be more powerful at detecting competition than is observation of the naturally existing range of densities.

Pfister (1995) conducted an elegant study of competition between fish in tidepools. She compared competition coefficients estimated by the Pimm–Schoener method with coefficients estimated by experimental manipulations, concluding that whereas the experimental manipulations provided strong evidence for competition, the regression analysis provided weak and inconsistent evidence. She suggested that part of the problem may have been that the carrying capacities for each pair of species were strongly correlated. This is what would be expected in tidepools, where volume is probably a dominant component of the carrying capacity. The regression method functions essentially by correcting for the carrying capacity of the target of competition, before examining the effect of the agent. If carrying capacities are strongly correlated, once this is done, there is very little signal left. Pfister (1995) also used a novel census-based approach, which used changes in population density through time, rather than simply population density, as the analysis variable. This approach appeared to work much more satisfactorily, and is described in the following section.

Fox and Luo (1996) have recently revisited the static regression approach, in a study comparing regression-based competition coefficients with estimates based on experimental removals. They examined competition between two Australian native rodents, *Rattus lutreolus* and *Pseudomys gracilicaudatus*, at four study sites in wet heathland. They suggest that the simple device of standardizing population densities before analysis (that is, subtracting the mean density from each observation, and then dividing by the standard deviation) will correct for the statistical artefact caused by differing variances. Experiments removing *R. lutreolus* were carried out at all four sites, to estimate the competition coefficient of *R. lutreolus* on *P. gracilicaudatus*. There was good agreement between these estimates and those based on regression using standardized data, but not with estimates based on raw densities.

Dynamic methods

Dynamic methods are not based on experimental manipulations of population sizes, but use the way in which densities of the competing species change through time to estimate competition coefficients. Recent advances in statistical methods for the analysis of nonlinear time series have made this approach possible, and we can expect further developments and applications of this approach in the future. At present, few applications have appeared in the ecological literature.

Pfister (1995) based her method on a discrete-time version of the Lotka–Volterra model:

$$\ln\left[\frac{N_1(t+1)}{N_1(t)}\right] = r[K_1 - N_1(t) - \alpha_{12}N_2(t)]/K_1. \tag{8.11}$$

Here, all parameters and variables are as defined in eqn (8.1), except that $N_1(t+1)$ and $N_1(t)$ represent the population size of N_1 in successive time intervals. In principle, given a time series of N_1 and N_2, it is possible to fit this equation, estimating the parameter combinations r/K_1, $1/K_1$ and α_{12}/K_1 using methods described by Dennis and Taper (1994) (see Chapter 6). Pfister (1995) estimated the parameters of eqn (8.11) using a time series over five censuses and five unmanipulated tidepools for the two most common fish species (*Oligocottus maculosus* and *Clinocottus globiceps*). The statistical method was a straightforward multiple regression with $\ln[N_1(t+1)/N_1(t)]$ as the response variable, and $N_1(t)$ and $N_2(t)$ as the predictor variables. The only difficulty was that, because of the autoregressive structure of eqn (8.11), the usual standard errors, F tests, etc. could not be used. Pfister used a parametric bootstrap to assess the significance of the results (see Chapter 6 for details). She used all five replicate tidepools in a single analysis, yielding 20 data points. The results are shown in Table 8.1. They suggest that *O. maculosus* biomass has a significant negative effect on the change in *C. globiceps* biomass, but that there is no evidence of a reciprocal effect, nor of any effects on density. The only significant interspecific competitive interaction detected by removal experiments was that the growth of *C. globiceps* individuals was greater in pools from which *O. maculosus* had been removed, supporting the results of the dynamic regression. No significant interspecific interactions were detected by any of several variants of the static regression method applied to the same data.

Pascual and Kareiva (1996) used a more elaborate version of the dynamic approach to estimate parameters of a Lotka–Volterra model applied to Gause's classic *Paramecium* data. Rather than using a simple difference-equation version of the Lotka–Volterra equations, they retained the original differential equation form, using the Runge–Kutta method (a standard numerical integration method) to integrate the equations numerically between the times of observation. This complicates the numerical calculations considerably, as standard statistical packages can no longer be used. In principle, however, the process is the same as Pfister's approach.

Table 8.1 Results of dynamic regression models for tidepool fish. The table shows the results of dynamic regression models using eqn (8.11) on data from five replicate tidepools over four time intervals. The variable N_1 represents either the numbers or the biomass of the fish *Oligocottus maculosus*, and N_2 represents either the numbers or biomass of the fish *Clinocottus globiceps*. Asterisks represent significance based on conventional multiple regression tests (*$p < 0.05$; **$p < 0.01$; ***$p < 0.001$). As is explained in the text, the autoregressive nature of the data means that these significance levels may be unreliable. The numbers in brackets are significance levels obtained from a bootstrap based on 1000 simulations (see text for details). Notice that the bootstrap results are much less significant than those based on multiple regression, which shows that conventional regression results may be quite misleading. The table also shows the overall R^2 from the multiple regression. This cannot be interpreted in a hypothesis-testing framework, but does show that the model based on biomass explained more variation than the one based on numbers. The overall interpretation of these results is that there is strong evidence of intraspecific competition in both species, particularly when biomass is used as the variable, but that the only interspecific effect detectable is a negative impact of *O. maculosus* biomass on the change in *C. globiceps* biomass. Modified from Pfister (1995)

	Independent variables		
Dependent variable	$N_1(t)$	$N_2(t)$	R^2
Variables expressed as numbers per tidepool:			
$\ln(N_1(t + 1)/N_1(t))$	−0.718* (0.35)	−0.153	0.261
$\ln(N_2(t + 1)/N_2(t))$	−0.061	−0.364** (0.028)	0.338
Variables expressed as mass (g):			
$\ln(N_1(t + 1)/N_1(t))$	−0.909*** (0.025)	0.070	0.624
$\ln(N_2(t + 1)/N_2(t))$	−0.425** (0.041)	−0.285*** (0.020)	0.597

The nature of the errors in the data is a critical factor that determines the appropriate method of parameter estimation. As I discuss in Chapter 2, ecological time-series data will include two forms of error. Process error is uncertainty about the actual value of $N_1(t + 1)$, given known values of $N_1(t)$ and $N_2(t)$. It is generated by such things as environmental variability and demographic stochasticity. Observation error occurs because the recorded value of $N_1(t + 1)$ is not the actual value of $N_1(t + 1)$, due to sampling error. Gause's data could be expected to include both sources of error, but parameter estimation with both acting simultaneously is very difficult, especially if the relative size of the errors is not known a priori. Pascual and Kareiva (1996) took the approach suggested by Hilborn and Walters (1992) of running the estimation procedure assuming first, only process error, and second, only observation error. They then compared the results.

Assuming only process error, they proceeded as follows. For given estimates of the parameters and starting populations $N_1(t)$ and $N_2(t)$, they predicted the values of $N_1(t+1)$ and $N_2(t+1)$ by integrating eqns (8.1) and (8.2). When they had done this over all the time intervals in the data for one set of parameter estimates, they repeated the process with another set of parameters. Finally, they selected the set of parameter estimates that minimized the discrepancy between the observed and predicted population changes. This 'one-step-ahead' prediction method is necessary for process error because the error accumulates: the starting point of the system for the next time step depends on the error in the previous step. However, it is assumed that the actual starting points for each step are known exactly.

If all the error is observation error, the fitting process is more similar to a conventional regression. For given estimates of the starting conditions and a set of parameter estimates, Pascual and Kareiva used the Runge–Kutta method to integrate an entire pair of trajectories. They then measured the discrepancy between the observed and predicted data points, and chose the set of parameters and starting points that minimized this discrepancy. This method is appropriate because the observation error model assumes that the error is in the eye of the beholder. The underlying process continues deterministically, given a set of starting values and parameter estimates, and an observation error at one time has no impact whatsoever on the future trajectory.

An important issue that they needed to resolve was how to measure the 'discrepancy' between observed and predicted values. The most appropriate method is maximum likelihood (see Chapter 2). This requires some assumption to be made about the probability distribution of the errors. Pascual and Kareiva assumed that it followed a lognormal distribution, both for process and for observation errors.

Table 8.2 shows the parameter estimates resulting from assuming either pure observation or pure process errors, and Fig. 8.5 shows the best fit of an observation error model. There is a reasonable degree of consistency in these results: both methods suggest that competition is asymmetrical, with the effect of *P. aurelia* on *P. caudatum* being much greater than the reverse effect. However, the best point estimates assuming only process error predict that *P. aurelia* should always win, whereas the best estimates assuming only observation error predict that the outcome should depend on the initial conditions (see Box 8.1).

Dynamic methods offer a powerful means of estimating interaction coefficients from census data. Their most important drawback, aside from the computational complexity, is that the model must be specified correctly. For Gause's experiments, the simple Lotka–Volterra model without time delays, age structure, etc. can probably be justified. In longer-lived species with substantial maturation times and complex social behaviour, living in seasonal environments, the basic Lotka–Volterra model is unlikely to be a good

Table 8.2 Maximum likelihood estimates for the Lotka–Volterra model, fitted to Gause's *Paramecium* data. The table shows maximum likelihood parameter estimates (MLEs), together with 95% confidence intervals. The parameters were fitted assuming either only observation error (columns 2 and 3) or only process error (columns 4 and 5). Subscripts on the parameters identify the species: *Paramecium aurelia* (species 1) and *Paramecium caudatum* (species 2). From Pascual and Kareiva (1996) (note, however, that the species identifications were reversed in this paper)

Parameter	Observation error fit		Process error fit	
	MLE	95% CI	MLE	95% CI
r_1	0.82	0.79–0.86	0.85	0.68–1.02
K_1	5380	4820–6000	4920	4180–5900
α_{12}	2.83	1.95–3.82	1.79	0.60–3.14
r_2	0.63	0.60–0.66	0.64	0.48–0.79
K_2	2000	1770–2280	1890	1520–2430
α_{21}	0.46	0.36–0.56	0.40	0.25–0.61
σ^2	0.11		0.10	

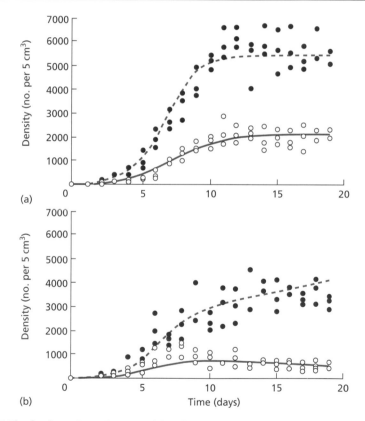

Fig. 8.5 The fit of a Lotka–Volterra model with observation error to Gause's *Paramecium* data. *Paramecium aurelia* numbers are shown with solid circles, and *P. caudatum* numbers are shown as open circles. (a) shows the case of the two species grown in separate cultures, whereas (b) shows the case when they were in mixed culture. The lines show the results predicted using an observation error model. From Pascual and Kareiva (1996) (note, however, that the species identifications were reversed in this paper).

description of the underlying dynamics. Fitting an inappropriate model, albeit by a sophisticated method, produces meaningless results. Pascual and Kareiva (1996) point out, quite correctly, that their general approach can be applied using any dynamic model that can be written down explicitly. Nevertheless, identifying an appropriate model is an essential precursor to any fitting procedure. A criterion such as Akaike's information criterion (see Chapter 2) may help in model selection, but is no substitute for grounding models in biological reality.

Mechanistic models of competition

As discussed in Chapter 1, it is absolutely essential to be clear on the purpose of any modelling exercise before proceeding. Parameter estimation for a model is rarely a goal in itself: it is usually only worthwhile if the parameterized model is useful either for prediction or understanding of a particular sort of ecological interaction. The most likely use of a parameterized model of competition is to predict the outcome of a competitive interaction. Here, as Tilman (1990) has pointed out, many of the above methods become rather circular. The only entirely satisfactory way to estimate competition coefficients directly is via experiments in which the outcome of competitive interactions is observed. Thus, we need the answer to the question our model is intended to address before we can proceed with the model. Furthermore, Tilman (1990) points out that the experiments would rapidly become unworkable if questions about competitive interactions in even fairly small communities are addressed. If there are y species, $(y^2 + y)/2$ pairwise series of experiments would be required.

The Lotka–Volterra model and its variants discussed above make no attempt to model explicitly the processes involved in competition. The Lotka–Volterra model itself subsumes the entire process of competition into the competition coefficients. Schoener (1974a) developed a model based on the proposition that there are both shared and exclusive resources, but the resources had no dynamics. An alternative approach, championed by Tilman (1982; 1987; 1990), is to recognize explicitly that resource competition involves interactions between the two or more potential competitors and their shared resources.

Tilman has applied his method primarily to plants and unicellular organisms, in which the situation is rather simpler than it is for animals, because the number of resources available to be shared is far smaller, but the approach is worth discussing in a book primarily directed towards animal ecology, as it offers the prospect of usefully predictive models. The principal limitation of the approach is that it is quite explicitly a means of looking at resource

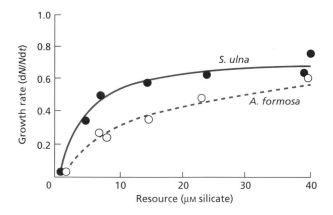

Fig. 8.6 Population growth of diatoms as a function of silicate concentration. These curves show the dependence of the per capita reproductive rate of the diatoms *Synedra ulna* and *Asterionella formosa* on silicate concentration at 24°C. From Tilman (1982, p. 50).

competition, and will not assist with understanding or predicting the outcome of interference competition.

The simplest version of the model is as follows. Suppose the population size of competitor i is represented N_i, and that there is a single resource R. Then

$$\frac{dN_i}{N_i dt} = f_i(R) - m_i \tag{8.12}$$

and

$$\frac{dR}{dt} = y(R) - \sum_i Q_i N_i f_i(R). \tag{8.13}$$

Here, $f_i(R)$ is the per capita growth rate of the population of competitor i, as a function of the resource level R; m_i is the per capita death rate of i; and $y(R)$ is the replenishment rate of the resource. The final term in eqn (8.13) describes the resource consumption by all the competitors. Tilman assumes that it is the product of the total growth rate of each competitor, multiplied by a competitor-specific proportionality constant Q_i.

In simple cases, the functions $f_i(R)$ and $y(R)$ can be measured experimentally. For example, Fig. 8.6 shows the growth rate of two species of diatom as a function of silicate concentration. It may also be possible to measure $y(R)$ and Q_i experimentally.

The main prediction of eqns (8.12) and (8.13) is that, following the competitive exclusion principle, one species will displace the others. The one that will do so is the competitor that reaches equilibrium with the lowest concentration

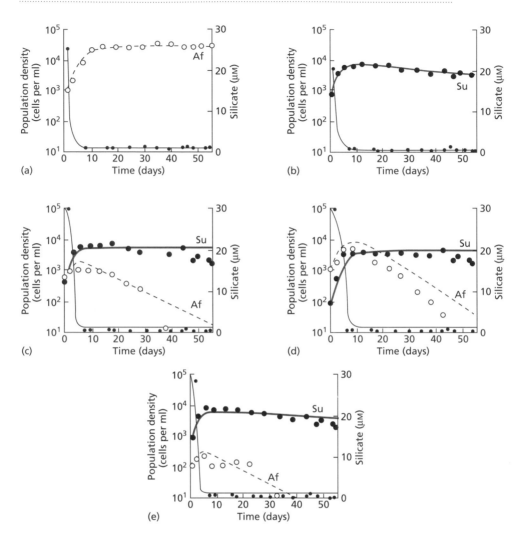

Fig. 8.7 The outcome of competition experiments between diatoms. (a) The population density of the diatom *Asterionella formosa* (Af) growing by itself, and the concentration of silicate, the limiting nutrient, in the same flask (small dots) at 24°C. (b) The population density of the diatom *Synedra ulna* (Su), also growing by itself and the concentration of silicate (small dots) at 24°C. (c–e) Competition between the two diatoms, with three initial starting densities at 24°C. In each case *Synedra* outcompetes *Asterionella*. From Tilman (1982, p. 52).

of the resource, R^*, when grown in monoculture. Some experimental results obtained by Tilman supporting this prediction are shown in Fig. 8.7.

In principle, this idea can readily be generalized to a number of resources (Tilman, 1982). Suppose that there are n competitors, that the population size of competitor i is represented by N_i, and that there are k resources R_j. Then

$$\frac{dN_i}{N_i dt} = f_i(R_1, R_2, \ldots, R_k) - m_i \tag{8.14}$$

and

$$\frac{dR_j}{dt} = y_j(R_j) - \sum_i N_i f_i(R_1, R_2, \ldots, R_k) q_{ij}(R_1, R_2, \ldots, R_k). \tag{8.15}$$

Here f_i is a function describing the dependence of the rate of growth of species i on the availability of the resources R_j, and q_{ij} is the generalization of Q_i, describing how the consumption rate of each resource depends on the growth rate of each competitor.

The outcome of an interaction described by eqns (8.14) and (8.15) depends on the form of the functions $f_i(R_1, \ldots, R_k)$, which describe how the growth rate of each species depends on the level of each resource. Tilman (1982) developed a powerful graphical technique to predict the outcome. For each competitor, a zero net growth isocline can be constructed as a function of resource concentration. Figure 8.8 shows some forms that these isoclines might take if there are two resources. Tilman experimentally obtained such isoclines for several species of diatoms, and showed that he could successfully predict the outcome of competition in 'chemostat' experiments, in which resources were supplied at a constant rate.

This general approach offers the prospect of genuinely predictive models of resource competition. The challenge that remains is to apply the models to multicellular animals in heterogeneous environments, rather than to unicellular organisms or plants.

Summary and recommendations

1 Resource overlap indices are exactly that and no more: they cannot be used to infer either the existence of or extent of competition.

2 If a pair of interacting species is embedded in a larger community (and this is almost always the case), indirect effects may mean that the overall effect on one species of perturbing the population size of another may be quite different from the effect expected from the direct interaction coefficient.

3 Manipulative experiments offer the best prospect of estimating the strength of a competitive interaction. As with any ecological experiment, adequate replication and appropriate controls are essential. It is equally important that the experimental setup should be reasonably representative of the natural situation, and that the spatial and temporal scales of the experiment should be appropriate.

4 Removal experiments are the most straightforward way to estimate competition coefficients. In such experiments, the population size of a target

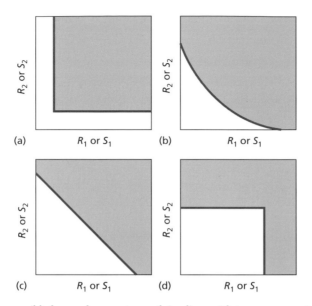

Fig. 8.8 Some possible forms of zero net growth isoclines with two resources. In each of these graphs, the concentration of two resources is shown on the two axes. The zero net growth isocline for a consumer is shown as a solid line. In the shaded region the consumer's population can increase, whereas in the unshaded region insufficient resources are available for growth. (a) Essential resources. Each resource is needed for growth, so minimum levels of resources 1 *and* 2 must both be available to the consumer. (b) Complementary resources. Neither resource is absolutely essential, but the growth rate is higher if both are consumed together than would be predicted by simply adding the growth rates expected from consuming either resource alone. The zero net growth isocline is therefore bowed in towards the origin, if both resources are available. (c) Substitutable resources. Each resource can replace the other, so the growth rate is determined by the resource concentrations added together. (d) Switching resources. The consumer can use either resource, so that minimum levels of resource 1 *or* resource 2 must be available to the consumer. From Tilman (1982, p. 62).

species is compared between experimental units that are intact, and units where the putative agent of competition has been removed. The interaction coefficients estimated by removal experiments include indirect as well as direct effects. They characterize the competitive interaction completely only if the isoclines are linear.

5   Pulse experiments are theoretically the best way to measure direct interaction coefficients in complex communities. In such experiments, individual species are perturbed one at a time, and the response of the population growth rate of all species to the perturbation is recorded. They are, however, logistically difficult in practice, and have rarely been used.

6   In the few cases that have been looked at systematically, competitive isoclines have been found to be nonlinear. This means that the interaction cannot adequately be characterized by a single competition coefficient.

7 Purely observational methods may give an indication of whether competition is occurring, particularly if time-series data are used, but are unlikely to provide sufficient information to develop a predictive model of a competitive interaction.

8 Models based on explicit resource dynamics offer the best prospect of producing predictive models of competitive interactions. Their potential, however, has yet to be fully realized.

CHAPTER 9

Predator–prey, host–parasitoid and plant–herbivore models

Introduction

Ecological interactions between exploiters and victims have the common feature that one species is consuming biomass from another, to the cost of the latter. In predator–prey and host–parasitoid interactions, the exploiter kills the victim as an inevitable part of the interaction. Host–parasite, host–pathogen and herbivore–plant interactions, in contrast, do not involve the inevitable death of the victim. Some damage will occur, although it may be so hard to detect that its existence may even be questioned, and death may occur in some cases.

In this chapter, I deal with predator–prey, host–parasitoid and plant–herbivore interactions, which share many common features. To model these interactions adequately, it is necessary to include two aspects of the exploiter population's response to changes in the victim population. First, the numerical response describes the way in which the exploiter population size responds to changes in the victim population size. A weak numerical response means that the exploiter population size responds slowly and/or little to changes in the victim population. Conversely, a strong numerical response means that the exploiter population responds rapidly, or a great deal, to changes in the population size of the victim. The functional response describes the way in which the victim consumption rate per exploiter changes as the victim population changes. It is determined by the way in which the exploiter searches for victims, and also by how the exploiter deals with or handles victims once they have been located.

By definition, parasites and pathogens live in close association with one host individual per life-history stage. This means that the 'numerical response' of the parasite is very closely linked to the host dynamics. Conventionally, models of host–parasite or host–pathogen interactions also pay less attention to the searching behaviour of the exploiter. These two considerations make the population dynamics and the parameter estimation problems associated with such interactions qualitatively different from those of predator–prey, parasitoid–host or plant–herbivore interactions. I have therefore dealt with them in a separate chapter. As will be seen in Chapter 10, modern host–parasite and host–pathogen models are giving increasing attention to complexities of how infection passes from one individual to another.

Basic structure of predator–prey models

The most elementary predator–prey model is the Lotka–Volterra model:

$$\frac{dN}{dt} = aN - \alpha NP,$$ (9.1)

$$\frac{dP}{dt} = -bP + \beta NP.$$ (9.2)

Here, N is the population size of the prey and P is that of the predator. The prey are assumed to have an intrinsic growth rate a per unit of time, whereas the predator population decays at a rate b if there are no prey present. The death rate of each prey individual from predation is assumed to be αP, and the per capita growth rate of the predator population is assumed to be linearly dependent on prey density via the proportionality constant β. This model has the virtue of extreme simplicity, with only four parameters. The model is neutrally stable, meaning that it oscillates indefinitely with an amplitude determined solely by the starting conditions. Largely because of this property, described as 'pathological' by May (1976), the model is of little practical use in describing real populations, and has never been parameterized successfully for any particular interaction.

The following is the simplest plausible predator–prey model:

$$\frac{dN}{dt} = r_N N(1 - N/K) - Pf(N,P),$$ (9.3)

$$\frac{dP}{dt} = Pg(N,P).$$ (9.4)

Here, r_N is the intrinsic rate of increase of the prey population, $f(N,P)$ is the predator functional response, K is the carrying capacity of the prey population, and $g(N,P)$ is the numerical response of predator population.

Functional responses

Functional responses are generally classified as belonging to one of three types (Holling, 1959), as shown in Fig. 9.1. A type I functional response is a straight line, with prey consumption per predator increasing linearly with prey density. If predator and prey encounter each other at random (the 'blundering idiot' searching strategy) a type I response will occur. One parameter only is necessary to describe the response. Type I functional responses may occur in filter feeders, and are often assumed in host–parasite models. A type II functional response is far more common in host–parasitoid or predator–prey

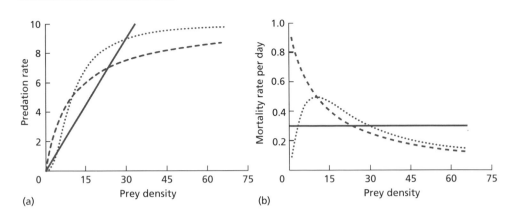

Fig. 9.1 Functional responses. (a) The prey consumption rate per predator, per unit of time, as a function of prey density N. Each of these was generated from Real's equation (no. 5 in Table 9.1). Solid line, type I response, $\alpha = 0.3$, $\gamma = 0$, $c = 1.0$; dashed line, type II response, $\alpha = 1.0$, $\gamma = 0.1$, $c = 1.0$; dotted line, type 3 response, $\alpha = 0.1$, $\gamma = 0.01$ $c = 2.0$. (b) The prey mortality per prey individual ϕ, as a function of prey density N, arising from each of the above type I, II and III functional responses, assuming that the predator population remains constant. Predation mortality is density-independent if the response is type I, inversely density-dependent at all prey densities if the response is type II, and directly density-dependent at low prey densities if the functional response is type III.

interactions. At low prey densities, the predation rate per predator increases approximately linearly with prey density, being limited by the ability of the predator to find prey. At high prey densities, however, the predation rate levels off, because either the time taken to consume the prey (the handling time) becomes the limiting factor, or because the predator becomes satiated. Such a response will require a minimum of two parameters to describe it adequately: one for the searching efficiency and one for the asymptote. Finally, in a type III functional response, the predation rate initially increases faster than linearly, before levelling off in a manner like that of a type II response. The initial increase in the predation rate can be attributed to predators switching on to the prey item as it becomes more available, learning how to search for it, or responding in particular ways to patchiness. Although a sigmoid curve can be described with two parameters, a general representation of a type III response will require the estimation of three parameters. Table 9.1 shows some explicit functional forms that have been used for functional responses. Occasionally, a type IV or 'domed' functional response is defined, in which the predation rate declines at high prey densities (Crawley, 1992). Such a response might occur if the prey species uses group defence mechanisms against predators.

A further possibility is that the predation rate per predator may increase more slowly than linearly, but may never reach an asymptote like a type II response. An example is a 'square root' functional response, which Sabelis

Table 9.1 Explicit forms for functional responses. In each of these, the prey or host population size is represented by N, and the equation gives the predation or parasitism rate per individual exploiter. In each of these, the parameters must be ≥ 0

Number	Type	Equation	Comments	Sources
1	I	αN		
2	II	$\dfrac{\alpha N}{1+\frac{\alpha}{\beta}N}$	Initial gradient is α, tends to β at high prey densities	Holling (1959)
3	II	$\beta\left(1-e^{-\frac{\alpha}{\beta}N}\right)$	As above. Less easy to manipulate algebraically than previous version	Ivlev (1961)
4	III	$\dfrac{kN^2}{N^2+D^2}$	Tends to k at high prey densities. D is the prey density at which the predation rate is $k/2$	Holling (1959)
5	I/II/III	$\dfrac{\alpha N^c}{1+\gamma N^c}$	Tends to α/γ at high prey densities. If $c=1$, equivalent to response 2; if $c=2$, equivalent to response 4; if $c=1$ and $\gamma=0$, equivalent to response 1	Modified from Real (1977)

(1992) proposed may occur if the predation rate is limited by gut fullness rather than handling time, and if the predator retains some propensity to feed whenever its gut is not entirely full.

For discussion of the implications of these responses for population dynamics, see Hassell (1978), Crawley (1992) and May (1976).

Numerical response of predators

Compared to functional responses, rather less attention has been given to generating explicit functional forms for numerical responses. Two main approaches have been commonly used. One, following from the standard Lotka–Volterra model, is to assume that the predator's rate of increase is proportional to the amount of food ingested, and hence the functional response, and that they have a constant death rate. Thus,

$$g(N,P) = -b + cf(N,P) \tag{9.5}$$

where b is the death rate of the predators, f is the functional response, and c can be thought of as a 'conversion rate' of prey numbers consumed per unit time into predator numbers. A second approach is to assume that the predators follow logistic population growth, with the carrying capacity determined by the size of the prey population. Hence:

$$g(N,P) = r_P\left(1 - \frac{P}{\kappa N}\right) \tag{9.6}$$

where r_p is the intrinsic rate of increase of the predator population, and κ is a ratio for converting prey numbers into the number of predators that they can support.

The model based on substituting eqn (9.6) into eqn (9.3) and eqn (9.4) is sometimes known as the Rosenzweig–MacArthur (Rosenzweig & MacArthur, 1963) model, and has been used quite extensively for building parameterized models (e.g. Hanski & Korpimäki, 1995).

Total responses

Some authors, particularly those studying mammals, work with the death rate from predation per prey individual, as a function of prey density (e.g. Messier, 1994; Pech et al., 1995). This is called the total response. Using a total response means that the predator dynamics are not considered explicitly. If predator dynamics occur on a much faster time scale than those of the prey (an unlikely scenario, unless the numerical response occurs through highly mobile predators aggregating on prey patches), then the shape of the total response is a product of the functional and numerical response. Alternatively, if the numerical response is slow or absent, the total response has the same shape as the functional response.

Ratio-dependent predation

A recent, and controversial, suggestion has been that predator–prey interactions are best understood in terms of the ratio of predators to prey, rather than in terms of the prey numbers themselves. The basic argument (see Berryman, 1992) is that it is unreasonable to assume that the predation rate per predator is a function only of prey density, as is assumed by the functional responses in Table 9.1. Rather, at high predator densities, there should be fewer prey per predator, and the attack rate should go down. Further, it is argued that the Holling functional response measures the attack rate of predators on prey on a very short behavioural time scale, but that eqns (9.3) and (9.4) represent population behaviour on the longer population dynamic time scale. The consequence of these assumptions, according to the proponents of the ratio-dependent view, is that the predator isocline is vertical (see Fig. 9.2), meaning that predator populations will decrease below a certain threshold prey density, and increase above it, irrespective of the nature of the intrinsic density dependence in the prey species. In practice, this leads to the 'paradox of enrichment', in which increasing the food supply of the prey does not lead to any increase whatsoever in the prey population, but merely to an increase in the predator population. Further, it means that it is very difficult to generate a model in which the prey are regulated to a very low level by the predators,

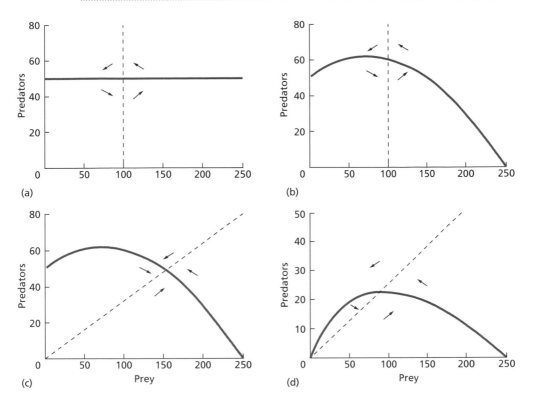

Fig. 9.2 Graphical analysis of some common predator–prey models. In each of the figures, the zero net growth isocline for the prey N is shown as a solid line, and the zero net growth isocline for predators P is shown dashed. The arrows show the trajectory of the system around the equilibrium point, where the isoclines intersect. In all cases, the trajectories show a tendency to cycle. It can be shown that the equilibrium point is stable if the predator isocline intersects a descending part of the prey isocline, but is unstable if the intersection is on an ascending part of the isocline (see, for example, McCallum, 1987). These isoclines are very easy to construct and explore on a spreadsheet. (a) The standard Lotka–Volterra model (eqns (9.1) and (9.2)). Parameters are $a = 1.0$, $\alpha = 0.02$, $b = 0.05$, $\beta = 0.0005$. (b) Logistic prey density dependence (eqn (9.3)), a Holling type II functional response, and a numerical response of the form of eqn (9.5). Parameters are: $r_N = 1.0$, $K = 250$, $\alpha = 0.02$, $\beta = 2.0$, $b = 0.2$, $c = 0.2$. The position of the vertical predator isocline does not depend on either K or r_N. This means that increasing the prey carrying capacity or growth rate does not affect the equilibrium prey population size. (c) Logistic prey density dependence (eqn (9.3)), a Holling type II functional response, and a numerical response of the form of eqn (9.6). Parameters are: $r_N = 1.0$, $K = 250$, $\alpha = 0.02$, $\beta = 2.0$, $\kappa = 0.3$ (The position of the isoclines does not depend on r_p.) (d) Logistic prey density dependence (eqn (9.3)), a ratio-dependent type II functional response (eqn (9.7)), and a numerical response of the form of eqn (9.5). Parameters are: $r_N = 1.0$, $K = 250$, $m = 4.0$, $w = 2.4$, $b = 0.5$, $c = 0.2$.

although this does in fact appear to be the outcome of some successful biological control programs.

The simplest ratio-dependent functional response can be obtained by substituting the ratio N/P for N in the Holling type II functional response (Berryman, 1992):

$$f(N/P) = m(N/P)/(w + N/P) = mN/(wP + N),\qquad(9.7)$$

where m and w are constants. A disadvantage of eqn (9.7) is that it does not rest on a straightforward biological argument in the same way as the Holling type II response does.

Arditi and Saïah (1992) suggested a more flexible predator-dependent functional response:

$$f(N,P) = \frac{\alpha NP^{-m}}{1 + (\alpha/\beta)NP^{-m}},\qquad(9.8)$$

where α and β have the same meaning as in the usual type II functional response in Table 9.1, and m is a parameter measuring the extent of dependence on predator population size, ranging from $m = 0$, corresponding to the usual prey-dependent model, to $m = 1$, corresponding to a ratio-dependent model.

The arguments in favour of ratio dependence have been trenchantly rejected by some authors, particularly Abrams (1994) – but see also the responses by Berryman *et al.* (1995) and Akçakaya *et al.* (1995). The problem with the vertical predator isocline only occurs if the assumption is made that the numerical response of the predator is some constant times the functional response. There is no reason why this should necessarily be the case, as there is certainly a difference in time scale between functional and numerical responses. A numerical response of the form of eqn (9.6) does not result in a vertical isocline, although Berryman (1992) argues that, as it involves the ratio P/N, it is, in fact, a ratio-dependent model. The time scale problem, as applied to the prey dynamics, is spurious, as Abrams points out. A differential equation model is concerned with instantaneous rates, and the functional response enters into the prey equation as a component of the instantaneous death rate of the prey. The translation to the longer population dynamic time scale is accomplished by integration of the equations. The incompatibility of the behavioural and population dynamic time scales occurs in the predator dynamics, and then only if the numerical response is assumed to be proportional to the functional response. Finally, it is certainly the case that high predator densities may influence the searching efficiency of individual predators. Prey depletion is taken care of by the prey equation, but mutual interference, or alternatively group hunting, may cause the searching efficiency of predators to be a function of their density. Another criticism of ratio-dependent models is that they assume that the amount of prey available per predator increases to infinity as predator density declines to zero (Gleeson, 1994).

Ultimately, whether ratio dependence is a more meaningful base assumption from which to build predator–prey models than prey dependence needs to be decided on the basis of usefulness: what is the relative performance of the two approaches in practice? Models of the form of eqns (9.3) and (9.4) are very abstract indeed. In particular, they omit spatial heterogeneity. Appeals to the logic of model construction in this very simple model will not necessarily ensure that the optimal model is obtained to approximate the behaviour of more complex systems. The issues of spatial heterogeneity and spatial scale are dealt with later in the chapter.

Whether ratio-dependent models are more often appropriate in practice than prey-dependent models remains an open question. Proponents of each approach can cite examples in which their favoured approach has worked better than the alternative. My advice is to stick to prey-dependent functional responses, unless there is strong empirical evidence to the contrary, because the parameters have a simpler mechanistic interpretation than do their ratio-dependent counterparts. However, as predator reproduction and prey consumption operate on very different time scales, a numerical response of the form of eqn (9.6) should generally be used in preference to one that links explicitly to the functional response such as eqn (9.5).

Host–parasitoid models

Parasitoids are insects that lay their eggs in the immature developmental stages of other insects. Successful development of the parasitoids then inevitably kills the host into which they have been laid. More attention has probably been directed at the modelling of host–parasitoid interactions than at any other form of interspecific interaction. There are several reasons for this interest. First, host–parasitoid interactions are extremely common – there are about 68 000 described species of parasitoid, but there may be up to 2 million species in total (Godfray, 1994). Second, parasitoids are often significant natural enemies of pest insect species, and are often investigated as agents of biological control. Finally, the biological properties of host–parasitoid interactions make them relatively straightforward to model. Generations of host and parasitoid usually run in step, there is often a one-to-one replacement of the developing host by the parasitoid, and, in temperate regions at least, there is often one single, non-overlapping generation of parasitoid and host each year, lending the system to a simple difference equation (see Chapter 1) modelling approach.

A major objective in modelling host–parasitoid interactions is to develop models to predict the likely success of biological control (May & Hassell,

1988; Godfray & Waage, 1991), although the extent to which many models have performed this task successfully is debated (Murdoch *et al.*, 1984; 1985; Murdoch & Briggs 1996).

Hassell (1978) is the classic reference on host–parasitoid models. For a recent review see Mills and Getz (1996). The simplest possible host–parasitoid model is the Nicholson–Bailey model:

$$N_{t+1} = \lambda N_t \exp(-aP_t), \tag{9.9}$$
$$P_{t+1} = qN_t[1 - \exp(-aP_t)]. \tag{9.10}$$

Here N_t is the number of hosts in generation t, and P_t is the number of parasitoids. The number of hosts in the subsequent generation is the overall reproductive rate λ multiplied by the current population size and the proportion of hosts that escape parasitism, $\exp(-aP_t)$. The number of parasitoids in the next generation is simply equal to the number of hosts that do not escape times q, the number of female parasitoids emerging per parasitized host. The exponential term for the proportion of hosts attacked is obtained from two straightforward assumptions: first, that the number of encounters that parasitoids have with hosts is proportional to the densities of the two multiplied together; and second, that these encounters are randomly distributed amongst hosts. Together, these assumptions mean that the probability of a host escaping from parasitism is the zero term of a Poisson distribution with mean aP_t. The model is thus assuming a type I functional response, with the parameter a measuring the search efficiency of the parasitoids.

As does the Lotka–Volterra model of predation (to which it is closely related) the Nicholson–Bailey model generates unrealistic dynamic behaviour (it is unstable, leading to diverging oscillations) and has rarely been fitted to a real system. An exception is the laboratory interaction between a greenhouse whitefly and a chalcid wasp described by Burnett (1958), which appears in a number of textbooks (for example, Hassell, 1978; Begon *et al.*, 1996b).

More realistic host–parasitoid models are discussed in detail by May (1978), Hassell (1978), Hassell and Godfray (1992) and Mills and Getz (1996). They may involve substituting a type II or type III functional response for the aP_t term in eqn (9.9) (using one of the functional forms in Table 9.1); adding density dependence to the prey population (using a functional form from Table 6.1); nonrandom parasitoid searching; or including age or spatial structure in the host population.

Aggregation and heterogeneity in predator–prey and host–parasitoid models

It is very well established that heterogeneity in parasitoid or predator attack rates has major consequences for the dynamics of the corresponding interactions

– see Murdoch and Oaten (1975), or Mills and Getz (1996) for a recent review. Heterogeneity can come from a wide variety of sources. The simplest idea is that of spatial refuges: areas of the environment in which the prey are protected from the predators. Without going into the details of the models, the existence of such refuges stops the predators reducing the prey to low levels, which would in turn lead to very low predator populations, and hence exaggerated coupled cycles in abundance. A similar 'refuge effect' may also result from the existence of a prey age class that is invulnerable to attack (Murdoch *et al.*, 1987).

In general, spatial variation in the attack rate of parasitoids, without explicit refuges, also leads to stability, whether or not the attack rate of the parasitoids depends on host density (Pacala *et al.*, 1990; Hassell *et al.*, 1991b; Pacala & Hassell, 1991; Godfray & Pacala, 1992). Using a model based on differential equations, Murdoch and Stewart-Oaten (1989) suggest that aggregation will not necessarily lead to stability if generations are overlapping, as is typical for predator–prey interactions. The situation is further complicated if functional responses are of type II or III rather than type I (Ives, 1992).

Given the importance of aggregation to the qualitative dynamics of host–parasitoid models in particular, it is unlikely that a parameterized model that does not take aggregation into account will describe an interaction adequately. From the point of view of parameter estimation, however, a major problem is that many of the models that have been developed are very abstract. The parameters they contain are compound entities that cannot easily be reduced to items that can reliably be estimated in the field or laboratory.

Pacala *et al.* (1990) have shown that, for a wide variety of host–parasitoid models, the interaction is stable provided the coefficient of variation squared (σ^2/μ^2) of the parasitoid attack rate in the vicinity of individual hosts exceeds 1. This '$CV^2 > 1$ rule' suggests that considerable progress is possible in predicting the outcome of a host–parasitoid interaction, provided that σ^2/μ^2 can be estimated from (s^2/y^{-2}). They modified eqns (9.9) and (9.10) by replacing the proportion of hosts that escape parasitism, $\exp(-aP_t)$, by a term $\exp(-ap)$, where a is the searching efficiency of the parasitoids within a patch and p is the local (within-patch) density of searching parasitoids. The quantity that needs to be estimated is the coefficient of variation of p. The estimation process is unfortunately not entirely straightforward, as data on the searching behaviour of individual parasitoids are rarely obtainable. Pacala and Hassell (1991) provide a maximum likelihood estimation procedure to obtain the CV^2 and related data from the frequency distribution of the proportion of parasitized hosts per patch, and host density per patch. These are data that are often relatively easy to obtain. Jones *et al.* (1993) describe an application of the approach in practice. These papers should be consulted for further details.

A second approach to handling spatial heterogeneity is to build a model that considers the distribution in space of victims and exploiters explicitly, and models interactions on a local scale. This approach is demanding both of data and computing power, but examples are beginning to appear in the ecological literature (e.g. DeAngelis *et al.*, 1998).

Experimental approaches

There are many examples of experiments measuring functional responses in experimental arenas, using different densities of hosts or prey, and measuring the number of individuals attacked over a certain time interval (Hassell, 1978; Fan & Petitt, 1994). In fitting functional responses there are various statistical considerations that have caused some debate, particularly in the entomological literature. One common problem is that such experiments are typically run over time periods long enough that the number of prey individuals consumed by the end of the experiment is a substantial proportion of those initially present, and prey are not replaced during the experiment. If this is the case, the functional forms in Table 9.1 cannot be used to describe the data directly. Instead, the functional response must be integrated over the course of the experiment. The number of prey still present at time t, $N(t)$, will be a solution of the equation

$$\frac{dN}{dt} = -f(N,P),$$
(9.11)

where $f(N,P)$ is the appropriate functional response from Table 9.1. For most functional responses, the solution to eqn (9.11) does not have a simple closed form, substantially complicating the estimation process. For example, the integrated form of the Holling disc equation (no. 2 in Table 9.1) can be written as

$$N_e = N_0[1 - \exp(\tfrac{\alpha}{\beta}N_e - \alpha T)],$$
(9.12)

where N_e is the number of prey eaten during the course of the experiment, N_0 is the number of prey present initially, and T is the duration of the experiment. Equation (9.12) is an implicit equation, meaning that it cannot be rearranged so that N_e appears on the left-hand side only. This equation is known as the random predator equation (Royama, 1971; Rogers, 1972). The problems in estimating its parameters from data have been studied in some detail by Juliano and Williams (1987). In particular, they compared a regression approach using a linearization of eqn (9.12) developed by Rogers (1972),

$$Y = \tfrac{\alpha}{\beta}N_e - \alpha T,$$
(9.13)

where $Y = \ln((N_0 - N_e)/N_0)$, with a nonlinear least squares approach first developed by Cock (1977). The linearization approach is easier in practice, but it has theoretical problems, in that N_e is present on both sides of the equation. This will cause standard regression approaches to be unreliable. Using simulated data where the parameters were known, Juliano and Williams (1987) also found that it returned very biased estimates for the parameters.

The nonlinear approach requires numerical solutions of eqn (9.12), with values of the parameters α and β being selected that minimize the residual sum of squares. This process can be accomplished fairly easily using the nonlinear regression procedures built into most modern statistics packages. In general, Juliano and Williams (1987) found that the nonlinear approach performed well on their simulated data.

More recently, Fan and Petitt (1994) used simulation to compare nonlinear least squares parameter estimation for the simple Holling disc equation with linearized versions. Their conclusion, in contrast to Juliano and Williams (1987), was that the linearization

$$\frac{N_0}{N_e} = \frac{1}{\alpha} + N_0/\beta \qquad (9.14)$$

performed better than a nonlinear fit, particularly if errors were multiplicative. As Williams and Juliano (1996) point out, this conclusion is valid only if the effects of prey depletion can be neglected. Presumably, nonlinear fitting after transformation (probably logarithmic) to achieve a normal error distribution would perform better that either of the above alternatives (see Chapter 2). Equation (9.14) is probably an adequate and simple approach if highly precise answers are not required, and prey depletion can be neglected.

More fundamental than the technical details of curve fitting, however, is the question whether a functional response measured in the laboratory bears any resemblance to the functional response between the same pair of species in the field. There are two basic issues here, which are closely related.

First, there is the question of scale. Unless the system being parameterized is a laboratory system, or conceivably a small greenhouse system, the objective will be to apply parameter estimates obtained in small arenas to a much larger 'natural' environment. A dimensional analysis of any functional response can provide some information on the appropriate way to apply parameter estimates obtained on a small scale to larger arenas. For example, consider the Holling type II equation, with prey measured as a density per unit area:

$$f(D) = \frac{\alpha D}{1 + \frac{\alpha}{\beta}D}. \qquad (9.15)$$

D is the density of prey per unit area. $f(D)$ has units of prey consumed per unit area per unit time. α and β thus both have units of time^{-1}. This suggests, to a first approximation, that it should be possible to use estimates of both parameters measured in small arenas directly in larger-scale situations, provided prey are measured in consistent units of density per unit area. Similar arguments would also apply to predators foraging in three dimensions, except that density would need to be measured per unit volume. However, the foraging behaviour of a predator or parasitoid in a 1 m^2 enclosure, surrounded by barriers, is likely to be quite different from that of a predator foraging within 1 m^2 of an infinite plane with the same prey density. Simple application of parameter estimates from the small scale to the large scale is likely to be unreliable, even in the improbable event that the prey distribution can be considered to be homogeneous.

Second, there is the question of habitat and environmental complexity. It is almost inevitable that artificial experimental arenas will be less spatially heterogeneous than the natural environment, and that they will have fewer species of alternative prey or competing exploiters present. It is very well established that the shape of a functional response depends on the spatial arrangement of the prey (Murdoch & Oaten, 1975; Murdoch & Stewart-Oaten, 1989; Godfray & Pacala, 1992; Mills & Getz, 1996). Type III functional responses frequently occur as a result of predators switching from alternative prey, as the species under consideration becomes more common. Hence, the abundance and spatial arrangement of alternative prey may have a major influence on the qualitative shape of a functional response.

The net result of the above discussion is that it is most unlikely that a functional response measured in a small, artificial arena will be able to provide even qualitative insight into the dynamics of the interaction between the same predator and prey species in a natural environment. It is therefore essential that models be parameterized using field data.

Using field data

Functional responses

Measuring the functional response itself in the field requires recording the predation rate, per predator, as a function of prey density. In principle, this could be done using either a prey-centred approach or a predator-centred approach.

A prey-centred approach involves recording, for a sample of prey individuals, the death rate from predation, by the predator in question, over a range of prey densities. In addition, the population size of the predator in question must also be known or estimated. Otherwise, it is a total response, rather than a functional response, which is being estimated. Ideally, one would also want the prey density to be manipulated by the observer, rather than natural spatial or temporal variation being used. This would eliminate the possibility of confounding variables affecting both the predation rate and prey density. Furthermore, in this ideal experiment, one would want a range of predator densities to be explored, to examine the possibility of complications due to varying predator density.

A predator-centred approach involves recording the prey consumption rate of a sample of the predators in question, for a variety of prey densities. As this method records the predation rate per predator directly, it estimates a functional and not a total response. Ideally, as in the prey-centred approach, one would prefer the prey and predator densities to be manipulated, rather than merely observed.

We do not, of course, live in an ideal world, so most real attempts to measure functional responses fall short of these ideals. The best strategy will often be determined by particular idiosyncrasies of the biology of predator and prey in the particular environment being studied, or by logistic considerations. I will therefore review briefly some case studies in which functional responses have been measured, in the hope that readers may find an approach that can be modified for their situation, rather than attempting to make general recommendations.

Functional responses of predators

Boutin (1995) reviewed attempts to measure functional responses of predators preying on small mammals with fluctuating populations. These studies looked at prey remains, either in scats, stomach contents or raptor pellets. For example, Erlinge et al. (1983) attempted to determine the functional response of a variety of mammalian and avian predators feeding on small rodents in Sweden. The information came from the proportions of various prey species identified in the scats or pellets of the predators. Together with data on the biomass of the prey, the food requirements per day of each predator, and the abundance of the predators, the number of prey individuals consumed was determined. For technical information on appropriate correction factors for diet analysis, see Simmons et al. (1991) (raptors) or Reynolds and Aebischer (1991) (red fox).

Boutin (1995) points out that such methods are likely to be unreliable, particularly at high and low prey densities. At high densities, surplus killing, caching and partial prey consumption are likely to lead to an underestimate of the actual number of prey consumed, and at low densities, if food is in short supply, fewer scats or pellets may be produced, leading to a decrease in prey consumption not fully being reflected in the reduction of the proportion of remains per scat or pellet. Furthermore, the overall food requirements of the predators usually are obtained from a diversity of sources (as was the case in the study of Erlinge et al., 1983), some of doubtful accuracy, such as laboratory feeding trials or allometric calculations (Box 9.1 summarizes some allometric relationships for calculating food consumption rate).

Box 9.1 Allometric relationships for ingestion rates

Peters (1983), in his book on allometric relationships in ecology, describes several relationships that may be helpful in estimating maximal prey consumption rates, as a function of predator body size. As was noted in Chapter 5, the apparently excellent relationships predicted by allometric equations need to be treated with a degree of caution, as they are based on double log-transformed data, usually over several orders of magnitude of body size. This means that apparently minor scatter around the regression line represents quite considerable error in the prediction of a consumption rate for a particular species.

For an endothermic tetrapod, the following equation gives the ingestion rate I, measured in units of watts (joules per second) as a function of body mass W, in kilograms:

$$I = 10.7 \, W^{0.703}.$$

The data on which this relationship is based are shown below. Somewhat surprisingly, there is no evidence of a difference in ingestion rate between herbivores and carnivores.

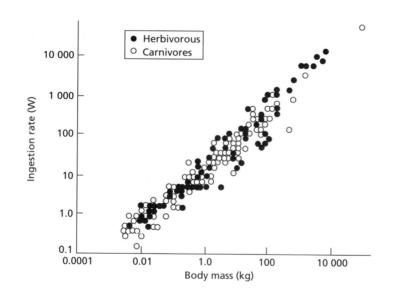

As one would expect, the ingestion rate for poikilotherms is much lower than that of homoiotherms, although it scales in a similar way with body size. For carnivorous poikilotherms, the ingestion rate is

$$I = 0.78 \, W^{0.82}.$$

continued on p. 259

Box 9.1 *contd*

Relationships for nontetrapod consumers or poikilothermic herbivores do not seem to be as firmly established.

To use either of the above relationships in practice, it is necessary to translate the ingestion rate in watts to a rate of biomass per unit time, or number of individuals per unit of time. For animal prey, Peters (1983) suggests that 1 W = 0.012 kg fresh mass per day. For herbivores, given the very different energetic content of different sorts of vegetable matter, calorimetry would be needed.

Using further allometric relationships between predator and prey size, Peters gives the following relationships between numbers of prey killed per day Kr and predator body mass W (kg) for large prey eaters (e.g. raptors or carnivores) and small prey eaters (e.g. insectivores): for large prey-eating homoiotherms, $Kr = 3.00 \ W^{-0.47}$; small prey eating homoiotherms, $Kr = 137 \ W^{-0.49}$; large prey eating poikilotherms, $Kr = 0.22 \ W^{-0.34}$; small prey eating poikilotherms, $Kr = 10.1 \ W^{-0.36}$. As the average mass of prey items can usually be estimated fairly easily, better accuracy can be obtained by using the relationship given above between biomass consumed per day and watts, together with the relationships between ingestion rate and body mass.

In some cases, it may be possible to do better than this. Sinclair *et al.* (1990) estimated the functional responses of raptors feeding on house mice by examining pellets and counting the number of right mandibles, or by halving the number of upper or lower incisors. As these raptors had been found to produce very close to one pellet per day, the counts of bone fragments could be translated into prey consumed per day, per predator, without needing to estimate prey mass or predator food requirements. Unfortunately, the resulting functional responses (see Fig. 9.3(a)) contain so much variation that it is hard to determine even if they are of type I, II or III.

As an alternative to examining scats or pellets, several workers have attempted to determine prey consumption rates directly from stomach contents. The problem is to convert field data on the average number of prey items, or amount of food, per stomach into a prey consumption rate per unit of time. This requires a model of the stomach evacuation rate, together with some assumptions about the feeding rate of the predators in question. For example, Pech *et al.* (1992) estimated the functional response of foxes preying on rabbits in Australia, using an approach based on the proportion of fox stomachs containing rabbits, the average amount of rabbit per stomach, an estimate of the stomach emptying time, and an assumption that the feeding rate was uniform over the nocturnal foraging period (Fig. 9.3(b)).

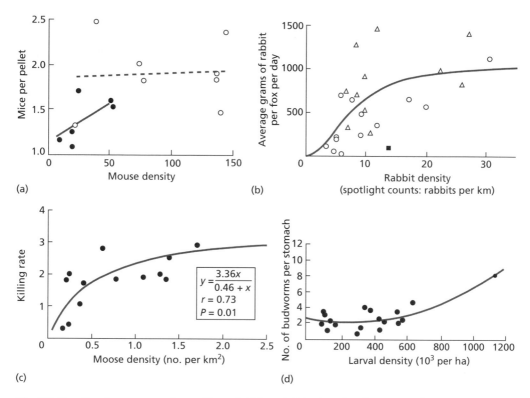

(a)

(b)

(c)

(d)

Fig. 9.3 Functional responses estimated from field data. (a) The functional response of diurnal raptors preying on mice in Australia (Sinclair *et al.*, 1990). The response is based on the number of mouse skeletal elements per pellet (see text), and the prey density is estimated from mark–recapture. Data points are based on sampling through time. The solid dots and solid line represent the functional response in a non-outbreak phase in the mouse population, and the open circles and dashed line represent the functional response in the outbreak phase. (b) The functional response of foxes preying on rabbits in semi-arid Australia (Pech *et al.*, 1992). The response is based on the weight of rabbit in fox stomachs sampled by shooting, and the prey density is based on spotlight counts from vehicle transects. Data are based on samples through time: triangles are from the spring–summer period, and circles are from the autumn–winter period. The solid square is an anomalous point. (c) The functional response of wolves preying on moose in North America (Messier, 1994). The killing rate is the number of moose killed per wolf per 100 days. Data are based on samples through both time and space. The solid line is a type II functional response, with parameters shown in the body of the figure. (d) The functional response of birds (species pooled) preying on spruce budworms (Crawford & Jennings, 1989). Data are based on replication through space. The fitted curve is a quadratic, and the concave form is described as evidence for a type III functional response. The large influence of the outlying data point at a larval density of 1.1×10^6 ha^{-1} means that any tests of significance of the quadratic term should be interpreted with caution.

Crawford and Jennings (1989) estimated the functional response of forest birds preying on spruce budworm (*Choristoneura fumiferana*). They used natural spatial variation in budworm density, and used shot samples to collect songbirds in two-hour time blocks from dawn onwards. The number of individual budworms in each bird gut was determined by counting pairs of left and right mandibles. Given previous work, which indicated that 35% of budworm remains could be identified 2.5 h after feeding, the total number of budworms consumed was determined by dividing the observed count by 0.35. This correction is not entirely satisfactory, but will not influence the shape of the functional response. The results indicated that the gradient of the functional response increased with prey density (Fig. 9.3(d)), consistent with the lower portion of a type III functional response.

The problems of estimating prey consumption rates from stomach contents have been explored in some detail in the fisheries literature. For most fish (Elliott & Persson, 1978) and most mammals and birds (Warner, 1981), the rate of stomach emptying in the absence of feeding is exponential. Thus, the change in $S(t)$, the amount of food in the stomach of an animal at time t, can be described by the following equation:

$$\frac{dS}{dt} = \alpha(t) - cS, \tag{9.16}$$

where $\alpha(t)$ is the rate of ingestion of food ($\alpha(t)$ may be a function of time), and c is the exponential rate of stomach emptying. This basic equation can often be solved to predict daily food consumption rates from observed data on stomach contents.

The simplest possible case is one in which predators forage for a period τ each day, and the rate of prey consumption during this foraging period is a constant α. This would mean that the daily food intake or ration R was simply.

$$R = \alpha\tau. \tag{9.17}$$

If the rate of passage through the stomach is reasonably quick, as it is for most mammalian or avian predators (Warner, 1981), then the stomach can be assumed to be empty at the start of foraging. Given these assumptions, the stomach fullness at time t after the onset of foraging, of an average predator, can be found by solving eqn (9.16), with an initial condition of $S(0) = 0$, yielding

$$S(t) = \frac{\alpha}{c}(1 - e^{-ct}). \tag{9.18}$$

If the time from the start of the foraging period when each predator was sampled had been recorded, eqn (9.18) could be fitted directly to a set of stomach fullnesses and sampling times, and both α and c and hence R could be estimated.

If the only data available are records of stomach contents themselves, then an independent estimate of the evacuation rate c is needed, and some assumption needs to made about the intensity, as a function of time, with which the predators are sampled. The simplest assumption is that sampling intensity is constant throughout the foraging period, in which case it will have a uniform distribution with probability density $1/\tau$. Then, the average stomach fullness $\bar{S}$ is the expected value of $S(t)$:

$$\bar{S} = \int_0^\tau \frac{1}{\tau}(1 - e^{-ct})\frac{\alpha}{c}dt \tag{9.19}$$

or

$$\bar{S} = \frac{\alpha(c\tau + e^{-c\tau} - 1)}{c\tau}. \tag{9.20}$$

Finally, eqn (9.17) can be substituted into eqn (9.20), and the result rearranged to give the average daily ration $\bar{R}$ as a function of $\bar{S}$:

$$\bar{R} = \frac{c^2\tau^2\bar{S}}{c\tau + e^{-c\tau} - 1}. \tag{9.21}$$

Estimates for c for a wide range of mammals and birds are given by Warner (1981). (Note that, if the evacuation follows an exponential function, c is the inverse of the mean retention time.) The fisheries literature provides modifications of this basic approach for cases where the stomach cannot be assumed to be empty at the start of feeding (Sainsbury, 1986), or for cases where satiation within a feeding period limits the ingestion rate.

If stomach contents are being collected with the objective of determining a total prey consumption rate, it is important to record the time at which individual samples were collected, relative to the foraging period of the predator. By using eqn (9.16) directly, time-dependent data may allow for the consumption rate to be determined without an independent estimate of the evacuation rate, and may enable some of the restrictive assumptions that were necessary to obtain eqn (9.21) to be relaxed, or at least tested.

These methods were derived, and have been tested on, fish eating small prey items (usually copepods). They may not apply well to situations where the prey are sufficiently big that a single successful prey capture event satiates the predator for most or all of the remainder of the foraging period.

For medium-sized vertebrate prey, Boutin (1995) suggests that a prey-centred approach, in which radio-collared prey are followed and their cause of death determined, may give more reliable results than methods based on predator diet composition. The radiotelemetry approach has been applied, with considerable success, in a long-term study of the snowshoe hare (Trostel *et al.*, 1987; Krebs *et al.*, 1995). The obvious limitation of using radiotelemetry to estimate functional response parameters is its cost, both in maintaining enough radio-collared prey to obtain worthwhile results, and in sufficient personnel to follow up dead animals rapidly enough to determine the cause of death.

Some of the best vertebrate functional response data have been obtained from long-term studies of wolves (*Canis lupus*) preying on moose (*Alces alces*) in North America (Messier and Crête, 1985; Petersen and Page, 1988; Messier, 1994). Both predators and prey are large, predation is relatively infrequent (less than 1.1 moose per 100 wolf days) and wolves remain near a carcass for a considerable time (at least 8 days for adult moose; Messier and Crête, 1985). It is therefore possible to use a predator-centred approach, in which individual wolf packs are intensively tracked and all successful predation events for an individual pack are recorded. Some typical results are shown in Fig. 9.3(c).

Distinguishing between type II and type III functional responses

Theory suggests that the qualitative effects of predation depend critically on whether the functional response is of type II or type III. A type II functional response is destabilizing, as it has an inversely density-dependent effect at all prey densities, whereas a type III functional response is stabilizing over a range of prey densities, as it is directly density-dependent at low prey densities. Distinguishing between these two types of response is therefore important. Unfortunately, it is not straightforward.

One apparently attractive method that does not work is to compare the fit of a cubic regression with that of a quadratic regression. It is possible to test a null hypothesis that a response is type I versus an alternative that it is non-linear by testing whether a term which is quadratic in prey density improves the fit of a regression of predation rate per predator versus prey density. This approach is valid because the type I functional response is indeed linear. A type II functional response may, in some cases, be adequately fitted with a quadratic regression, and cubic regressions will often be an adequate fit to curve with an inflection point (which a type III response has). However, none of the suggested functional forms suggested for a type II functional response in Table 9.1 is, in fact, a quadratic. Cubic and even higher-order terms will, in general, improve the fit of a polynomial to a convex curve that reaches an asymptote as does a type II functional response. A cubic term improving the fit over a quadratic is quite consistent with a type II functional response, and conversely, the failure of a cubic term to improve the fit over a quadratic is not evidence that the relationship is no more complex than a type II response.

When nonlinear relationships are being compared, the most appropriate way to determine whether a more complex model fits a given set of data better than a simpler one is to choose models so that they are nested, meaning that the simpler model can be obtained from the more complex one by setting some parameters to particular values. Of the forms for functional responses in Table 9.1, the model of Real (1977),

$$f(N) = \frac{\alpha N^c}{1 + \gamma N^c},$$

(9.22)

with three parameters, α, γ and c, is the most suitable for this purpose. If γ is set to 0, and c to 1, eqn (9.22) yields a type I response. If c is set to 1, the response is type I or II, and if all three parameters are allowed to vary, the response is I, II or III, with a type III response occurring if $c > 1$.

Whether using a type III response rather than a type II response is justified can then be determined by the improvement in fit when c is allowed to vary rather than being fixed at 1. The model should be fitted by maximum likelihood, which requires some assumption to be made about the form of the error distribution. Probably the most reasonable assumption, in the absence of specific information to the contrary, is that the error is lognormal. This would be the case if the coefficient of variation in the observed prey consumption rate was constant over a range of prey densities. Nonlinear least squares analysis on log-transformed data would then generate maximum likelihood estimates of the parameters, and an F ratio could be used to determine which functional response form best fitted the observed data (Box 9.2).

Box 9.2 Fitting functional responses to data

Foxes consuming rabbits (Pech *et al.*, 1992)
The following data are rabbit number (km^{-1}, N) and daily prey consumption (g) per fox, f. See text for a discussion of how these were collected.

N	f	N	f
3.58	113.64	9.58	922.90
4.58	53.29	9.60	524.80
5.08	191.09	10.61	267.43
5.08	215.61	11.58	1477.40
5.75	704.80	11.91	360.55
5.80+	20.58+	13.84*	107.47*
6.54	738.59	17.07	666.03
6.88	332.92	19.93	579.88
7.47	655.40	22.21	996.86
8.15	1277.83	25.88	834.96
8.38	703.83	26.95	1424.42
9.11	244.75	30.54	1135.87
9.19	486.65		

continued on p. 265

Box 9.2 *contd*

The asterisked data point is identified as an outlier by Pech *et al.* (1992), and is omitted from the analysis that follows. The data were log-transformed, and the logged form of eqn (9.22),

$$\ln(f(N)) = \ln \alpha + c \ln N - \ln(1 + \gamma N^c),$$

was fitted to the data using nonlinear least squares. After inspection of the results, the data point identified by + was also evidently an outlier, and is not included in the analysis.

The table below shows parameter estimates; residual sums of squares (*SSE*); error degrees of freedom; error mean squares (*MSE*); partial sums of squares compared to the full type III model (*SSP*); and the *F* ratio comparing the full model to the model in question,

$$F = \frac{SSP/(f - p)}{SSE_f/(n - f)},$$

where *f* is the number of parameters in the full model, *p* is the number of parameters in the partial model, and *n* is the number of data points.

Type	α	γ	c	SSE	df	MS	SSP	F	Prob.
I	49.78	0	1	8.874	22	0.403	1.789	2.525	0.10
II	58.43	0.014 9	1	8.656	21	0.412	1.571	4.435	0.05
III	0.659	0.000 811	3.791	7.085	20	0.354			

It can be seen that the type II model is very little better than the type I model, and that the type III model is a marginally significant improvement over the type II model.

The resulting functional responses are shown below, with $f(N)$ on both logarithmic and linear scales. In each case, the type I response is shown as a dotted line, the type II as a solid line, and the type III as a dashed line.

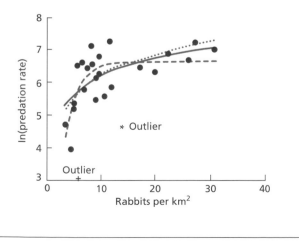

continued on p. 266

Box 9.2 *contd*

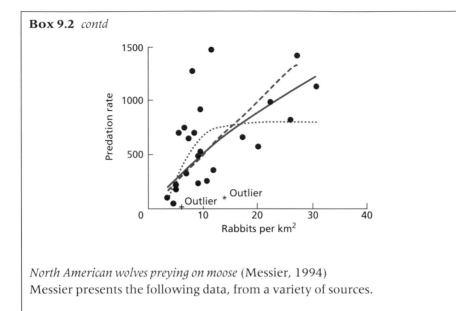

North American wolves preying on moose (Messier, 1994)
Messier presents the following data, from a variety of sources.

Moose density (km^{-2})	Killing rate (100 d^{-1})
0.17	0.37
0.23	0.47
0.23	1.9
0.26	2.04
0.33	2.78
0.37	1.12
0.42	1.74
0.8	1.85
1.1	1.88
1.14	2.44
1.2	1.96
1.37	1.8
1.73	2.81
2.49	3.75

Using the fitting procedure explained above, the following results are obtained:

Type	α	γ	c	SSE	df	MS	SSP	F	Prob.
I	2.741	0	1	5.658 075	13	0.435 2	2.750	11.35	0.005
II	6.163	1.763	1	2.907 577	12	0.242 3	0.773	3.98	0.07
III	5663.8	2597.8	5.352	2.134 865	11	0.194 1			

continued on p. 267

Box 9.2 *contd*

In this case, a type II functional response is strongly supported over a type I ($P = 0.005$), and there is some suggestion that fitting a type III response may be justified ($P = 0.07$). The resulting functional responses are shown below, with $f(N)$ on both logarithmic and linear scales. In each case, the type I response is shown as a dotted line, the type II as a solid line, and the type III as a dashed line.

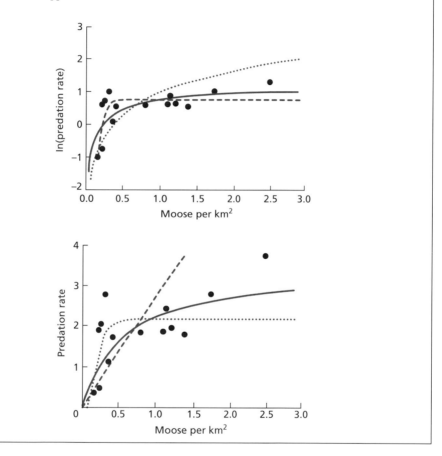

If the raw data are numbers of prey individuals consumed over a fixed time interval, then it may be that a model using Poisson errors is appropriate (see Chapter 2). In most cases, however, it is likely that the data will be overdispersed, with a variance greater than that assumed by the Poisson distribution. Wilson and Grenfell (1997) provide some ideas on how to proceed in this situation, which involves quasi-likelihood methods. These are beyond the scope of this book, but discussions can be found in McCullagh and Nelder (1989) or Crawley (1993).

Predator numerical responses in the field

Almost all published studies of predator numerical response have attempted to estimate the strength of the response by simply regressing predator abundance against prey abundance. Observations have been based on existing spatial variation in prey abundance (e.g. Crawford & Jennings, 1989), temporal variation in prey density (Newsome *et al.*, 1983; Rohner, 1995) or sometimes a combination of both (Messier, 1994). Such correlative approaches are not ideal, as a third factor (e.g. habitat quality) could affect both variables. The time delays that are a typical feature of predator–prey interactions are also not taken into account by a simple regression of predator abundance versus prey abundance, although lagged responses can easily be included if the variation in prey density is through time (e.g. Rohner, 1995). It is generally better to link prey abundance to predator reproduction (e.g. Shine & Madsen, 1997) or survival, rather than to predator abundance. If this is done, the mechanistic link is clearer, although not necessarily unequivocally demonstrated.

Plant–herbivore models

In their basic structure, the most commonly used plant–herbivore models do not differ fundamentally from Lotka–Volterra type predator–prey models (Crawley, 1983). However, as herbivores often (although not invariably) consume biomass from plants rather than killing them outright, the variable describing the plant population is usually biomass, rather than numbers of individuals. A detailed discussion of plant population models themselves, and the best way to parameterize them, is beyond the scope of this book. Crawley (1983; 1997) or Harper (1977) are good places to start for information in this area.

For herbivores themselves, the parameterization problems are the same as those for predators. It is necessary to quantify the functional response, which is the way in which the grazing off-take depends on the amount of forage present, and the numerical response, which is the way in which the population growth rate of the herbivore depends on the amount of forage. In some unusual cases, such as herbivores feeding on planktonic algae, or seed predation, the methods needed to parameterize these responses are identical to those for predator–prey models. Normally, however, the immobile nature of plants and the likelihood of partial consumption mean that the solutions to parameterization problems are quite different from those needed for predator–prey or host–parasitoid models.

Models for vertebrate herbivores

Caughley and Gunn (1993) provide a framework for modelling the population dynamics of mammalian herbivores in variable environments. There are four basic components:

1   The plant response function, which is the growth rate of plant biomass as a function of standing biomass and relevant environmental variables. In deserts, the particular focus of the article, the fundamental environmental variable driving the plant response is rainfall. The plant response is analogous to the $r_N N(1 - N/K)$ component in eqn (9.3), with the environmental variable driving variability in r_N.

2   The herbivore's functional response, which is typically of type II.

3   The herbivore's numerical response, which is the rate of increase r as a function of forage biomass.

4   Possible density dependence in the herbivore population that is not directly resource related (the 'intrinsic response').

This basic structure is broadly the same as that of eqns (9.3) and (9.4), with the addition of the climatic forcing, and the expectation that the model will be run with short discrete time steps, rather than as a set of continuous differential equations.

Figure 9.4 shows the first three of these components for a red kangaroo (*Macropus rufus*) model parameterized using data from Kinchega National Park, New South Wales.

The plant response model is based on data from Robertson (1987), which should be consulted for a detailed description of the methods. In brief, the data are based on 0.25 m² caged plots, in which pasture growth over periods of three months was recorded as a function of initial pasture biomass and rainfall over the period.

The functional response is based on data from Short (1987), and is based on a 'graze-down' experiment, in which kangaroos were penned in small enclosures. The food intake per animal per day was then recorded through time as a function of the decreasing vegetation cover, until the pens were essentially devegetated. As Short himself discusses, this is not the ideal way to obtain a functional response of a free-ranging grazer in a heterogeneous environment. Spalinger and Hobbs (1992) suggest a series of mechanistic models of mammalian herbivore functional responses that may be useful alternative ways of parameterizing herbivore functional responses.

Finally, the numerical response was based on analysis by Bayliss (1987), modified by Caughley (1987). Over a four-year period, finite rates of increase for three-month segments were converted to exponential annual rates of increase r, and related to pasture biomass. The strongest relationship was observed with no lag between vegetation and rate of increase. The data were then fitted to an Ivlev (1961) curve of the form

$$r = -a + c(1 - e^{-dV}),$$
(9.23)

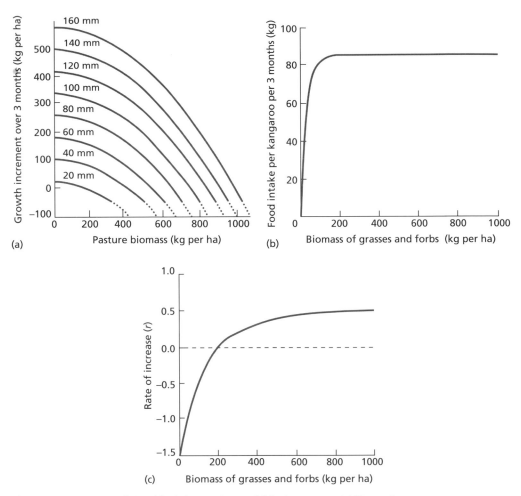

Fig. 9.4 Components of Caughley's interactive model for kangaroos. (a) Vegetation response model. This shows the increment, over a three-month period, in biomass of grasses and forbs at Menindee, New South Wales, Australia, as a function of rainfall over that period and initial biomass. One would expect the instantaneous growth increment to have an approximately parabolic shape, as small starting biomass would lead to a small growth increment. The three-month time period over which growth is measured is relatively long for these fast-growing plants, and so only the descending arm of the growth curve is detected. (b) Herbivore functional response. The fitted food intake in kilogram dry weight per kangaroo over three months, as a function of availability of forage. (c) Herbivore numerical response. From Caughley and Gunn (1993).

where V is the vegetation biomass, and a, b and d are constants. In principle, eqn (9.23) could be fitted to observed data by nonlinear least squares or maximum likelihood, provided the problems of non-independence of successive error terms was ignored. However, Bayliss (1987) used an *ad hoc* procedure in which the asymptote was fitted first by inspection. Caughley (1987) further adjusted the estimates in order to make them more biologically 'sensible'. Such fitting procedures would offend statistical purists, but if the objective is to produce a well-functioning mechanistic model, rather than to estimate parameters as an end in themselves, a certain amount of adhocery can be defended.

The outcome of a sample run of this model is shown in Fig. 9.5. Perhaps the most interesting result is the way in which the dynamics of the plant–herbivore interaction transform the essentially uncorrelated white noise of the rainfall into long runs of population increase and decrease in the kangaroo population. Repeated runs of a model of this type can be used to generate frequency distributions of animal numbers, in a similar way to population viability analyses (e.g. Burgman et al., 1993). However, the correlated noise introduced by the dynamics in this model is likely to produce quite different (and probably more reliable) predictions than the simple white noise incorporated into most population viability analysis models.

It should be apparent, however, that to parameterize this model for this particular species and location, albeit in a not entirely satisfactory way, has required an extensive research effort.

Approximate methods when few data are available

In most cases where it might be helpful to apply a predator–prey model, we cannot conduct extensive experiments, nor are there long runs of field data at various predator and prey densities. A variety of ad hoc approaches has therefore been applied. Quite obviously, these approaches are not as satisfactory as a controlled experiment or the analysis of extensive data. However, a model based on approximate estimates is often better than no model at all, provided its limitations are recognized. By their very nature, it is difficult to generalize about ad hoc models. I will therefore review a particular example.

The interaction between voles and their predators in northern Europe has attracted a great deal of attention over many years. Vole population dynamics, particularly at high latitudes, are characterized by pronounced oscillations. One of the main hypotheses put forward to explain these cycles is that predators drive them. The following model is the most recent in a series of relatively simple predator–prey models (Hanski et al., 1991; 1993a; Hanski & Korpimäki, 1995; Turchin & Hanski, 1997) that have been parameterized to investigate the hypothesis that the cycles are driven by predators.

Two sources of predation are included. Generalist predators have a range of possible prey species. Their dynamics are therefore decoupled from the vole population dynamics, and are not explicitly included in the model. The vole death rate from generalist predators is determined by their functional response. Generalists switch on to voles when voles become sufficiently common, and will switch to alternative prey when voles are rare. They should

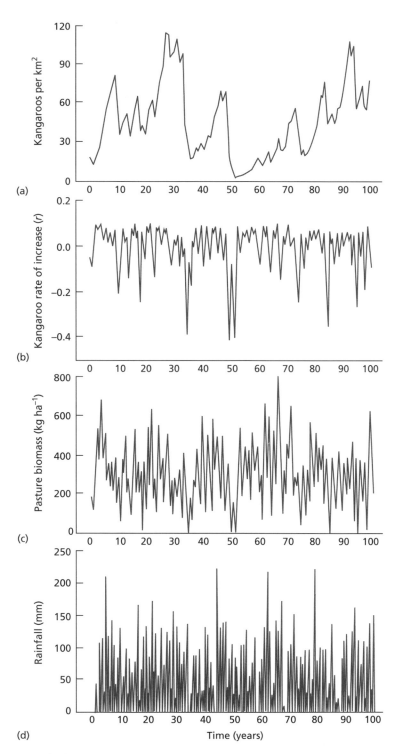

Fig. 9.5 Results from a single run of Caughley's interactive model. (a) Kangaroos per square kilometre. (b) Kangaroo rate of increase. (c) Pasture biomass. (d) Rainfall, as generated by random draws from historic records.

therefore have a type III functional response. Mustelids, mainly weasels, are the main specialist predators of voles in northern Europe. As they do not have many alternative prey, their dynamics are coupled to those of the voles, and are included in the model. They would be expected to have a type II functional response.

The structure of the model is closely based on eqns (9.3) and (9.6):

$$\frac{dN}{dt} = rN\left(1 - \frac{N}{K}\right) - \frac{GN^2}{N^2 + H^2} - \frac{CNP}{N + D},\tag{9.24}$$

$$\frac{dP}{dt} = sP\left(1 - Q\frac{P}{N}\right).\tag{9.25}$$

Here, N is the population density of the voles, and P is the population density of the specialist predators. The first term in eqn (9.24) is a simple logistic term, with two parameters to be estimated: the intrinsic rate of growth, r, and the carrying capacity, K. The second term represents mortality from the generalist predators. Its two parameters are G, the asymptotic rate of generalist predation, when voles are very common, and H, which is the half-saturation constant (the vole density at which mortality from generalist predation drops to $G/2$). The final term is mortality from specialist predators. This is a type II response depending on P, the population density of the specialists, and has two parameters: the maximum consumption rate, C, and the half-saturation constant, D. The predator equation (eqn (9.25)) has a simple logistic structure, with a predator intrinsic rate of increase, s, and a parameter, Q, that represents the ratio of prey to predators at equilibrium.

Equations (9.24) and (9.25) do not include any environmental variation, whether in predictable seasonal fluctuations, or in random environmental fluctuation. Turchin and Hanski (1997) assumed that all the seasonal fluctuation occurred in the growth rates r and s, but did not influence the density-dependent parts of eqns (9.24) and (9.25). They further assumed that seasonality could be represented as a simple sine wave with an amplitude e and a period of one year. Having done this, they scaled N relative to its carrying capacity K, and P relative to K/Q. The actual equations used in the study were then:

$$\frac{dn}{dt} = r(1 - e\sin 2\pi t)n - rn^2 - \frac{gn^2}{n^2 + h^2} - \frac{anp}{n + d}\tag{9.26}$$

and

$$\frac{dp}{dt} = s(1 - e\sin 2\pi t)p - s\frac{p^2}{n}.\tag{9.27}$$

Here, the new scaled variables are $n = N/K$ and $p = KP/Q$, and the scaled parameter combinations are $g = G/K$, $h = H/K$, $d = D/K$ and $a = C/Q$. I discuss scaling

Table 9.2 Parameters of the vole–mustelid model. See the text for explanation of how these estimated values were chosen. Where more than one value is given in the table, Turchin and Hanski (1997) investigated results for each of the listed values. Asterisked parameters are those used in the scaled predator–prey equations

Parameter	Definition	Value(s)
r^*	Intrinsic rate of increase of voles	6 y^{-1} (5–7)
s^*	Intrinsic rate of increase of specialist predators	1.25 y^{-1} (1–1.5)
K	Carrying capacity for voles	150 ha^{-1}
G	Asymptote of generalist predator functional response	70° − 10 L (L is latitude)
H	Generalist predator half-saturation constant	13.8 ha^{-1}
C	Asymptote of specialist predator functional response	600 y^{-1}
D	Specialist predator half-saturation constant	6 ha^{-1}
Q	Equilibrium prey–predator ratio	20
e^*	Strength of seasonality	1
σ^*	Standard deviation of environmental noise	0.12
d^*	D/K	0.03, 0.04, 0.06
a^*	C/Q	7.5, 10, 15, 20, 30
g^*	G/K	
h^*	H/K	

in Chapter 2. In this model, the scaling reduces by two the number of parameters that need to be examined in any sensitivity analysis, and shows that it is the ratios G/K, etc. that determine the dynamic behaviour of the model. Unless it is possible to estimate such ratios directly, however, all the original parameters still need to be estimated to relate the model to the real world.

Environmental fluctuations were included by assuming that each parameter changed in each year i, randomly and independently of other parameters, such that the parameter p was replaced by

$$p_i = p(1 + N(0,\sigma^2)) \tag{9.28}$$

where $N(0,\sigma^2)$ is a random number drawn from a normal distribution with mean 0 and standard deviation σ. Equation (9.28) is one of the least satisfactory parts of this model. In general, one would expect that environmental variation (such as an unusually warm winter) would cause several parameters to co-vary, and there is also no reason to expect that the additive variance in parameters should be constant. The model is deliberately highly simplified, to keep the number of parameters that need to be estimated down, as far as is possible. Nevertheless, eight parameters must be estimated. These are summarized in Table 9.2.

I discuss methods for estimating aggregated demographic parameters such as intrinsic growth rates and carrying capacities in Chapters 5 and 6. The intrinsic rate of increase of voles, r, required for this model is a yearly average: the peak rate of increase in summer would be $r(1 + e)$, and the minimum rate of increase would be $r(1 - e)$ (see eqn (9.26)). Turchin and Hanski (1997) used an estimated $r = 6.0$ y^{-1}. This was based on a detailed study (Turchin & Ostfeld, 1997) of another vole species (*Microtus pennsylvanicus*) in another habitat entirely (Millbrook, NY, USA). In that study, r was estimated from the intercept of a regression of the realized rate of increase (see eqn (5.5)) versus population density. Turchin and Hanski estimated the intrinsic rate of increase of the specialist predators, s, in an even more approximate fashion. Weasels produce one generation per year, in a maximum of two litters, with a mean litter size of five. Assuming a 1:1 sex ratio, and no mortality whatsoever, the annual rate of increase would be $\ln(10/2) = 1.6$. Turchin and Hanski adjusted this downwards to allow for density-independent mortality, yielding $s = 1.25$. In their study of *M. pennsylvanicus* in New York State, Turchin and Ostfeld used a regression-based method to estimate the seasonality parameter e at 0.53. On the basis that the boreal environments they were modelling were more extreme, Turchin and Hanski used $e = 1.0$ in their study.

As I discuss in Chapter 6, the carrying capacity K for voles should be the equilibrium density that would occur in the absence of predation. An additional complication here is that the structure of eqn (9.26) means that the carrying capacity in summer should be $2K$ if $e = 1$. Turchin and Hanski estimate that the peak carrying capacity in summer is 300 ha^{-1}, and hence $K = 150$. The prey–predator ratio Q, which determines the carrying capacity of the specialist predators, proved very difficult to estimate adequately from the data available. Turchin and Hanski suggest that Q should be in the range of 20–40, based on an argument that a weasel needs about one vole per day to maintain itself, and that about 10 female voles will produce an average of one offspring per day. Given that the sex ratio is 1:1, 20 voles in total will then have a productivity of one vole per day.

The maximum consumption rate of specialist predators was derived from calculations of food requirements of weasels per gram of body weight per day, with an *ad hoc* adjustment for surplus killing and higher predation on juvenile voles. Turchin and Hanski describe the estimate of $C = 600$ y^{-1} as conservative. The half-saturation constant D was based on the observation that there is a critical minimum vole density N_{crit}, below which weasels cannot breed. Estimates from the literature put this at about 10 ha^{-1}. At this prey density, the consumption rate of voles by an individual weasel would be C_{crit}, which was estimated (see above) at about 1 vole per day $= 365$ y^{-1}. Using the third term in eqn (9.24),

$$C_{crit} = \frac{CN_{crit}}{N_{crit} + D} \Rightarrow D = N_{crit}\left(\frac{C - C_{crit}}{C_{crit}}\right) \tag{9.29}$$

or $D \approx 6$ voles ha^{-1}.

Turchin and Hanski estimated the parameters for the generalist predators using literature data on avian predators feeding on voles in several locations in

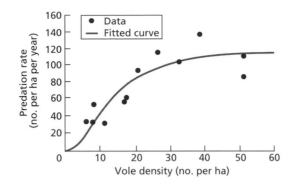

Fig. 9.6 A type III functional response for avian predators feeding on voles. Data are from Revinge, southern Sweden (56°N), and the predation rate is shown as a function of the vole density two months previously. From Turchin and Hanski (1997).

Scandinavia. Avian predators appear to aggregate in areas of high vole density, with a lag of about two months, so the predation rate was analysed as a function of vole density two months previously. Figure 9.6 shows a type III functional response fitted by nonlinear least squares data from Revinge in southern Sweden. The estimated parameters are $G = 71$ voles ha^{-1} and $H = 13.8$ voles ha^{-1}. Furthermore, assuming that G should be proportional to generalist predator densities, Turchin and Hanski (1997) used literature values on predator densities to estimate that $G = 70° - 10L$, where L is the latitude in degrees.

The noise parameter σ was estimated using a time series on vole fluctuations at Revinge. The between-year standard deviation in log-transformed density was 0.3 at this site, and the model was run with different values of σ, until an estimate was found that duplicated this observed standard deviation.

Some of the results of the model are shown in Fig 9.7. The agreement between the observations and predictions is impressive.

I have dealt with this example at some length because it illustrates several important points about parameter estimation and ecological modelling in general. The first, and most obvious, point is that few of the parameters are based on a solid statistical footing. Most of the estimates rely on a fair amount of *ad hoc* reasoning (but hopefully not too much *post hoc* reasoning!). This system is relatively simple and, compared with most predator–prey systems, has been

Fig. 9.7 (*opposite*) Dynamics predicted from a vole–predator model, compared with actual data. (a, b) Compares sample trajectories for the model of Turchin and Hanski (1997) with varying amounts of generalist predation (a) with observed trajectories at locations with similar estimated levels of generalist predation (b). (c) Shows the relationship between latitude and the predicted amplitude of population fluctuations (solid line, dashed lines are 90% confidence interval) and compared with observed data (open circles). From Turchin and Hanski (1997).

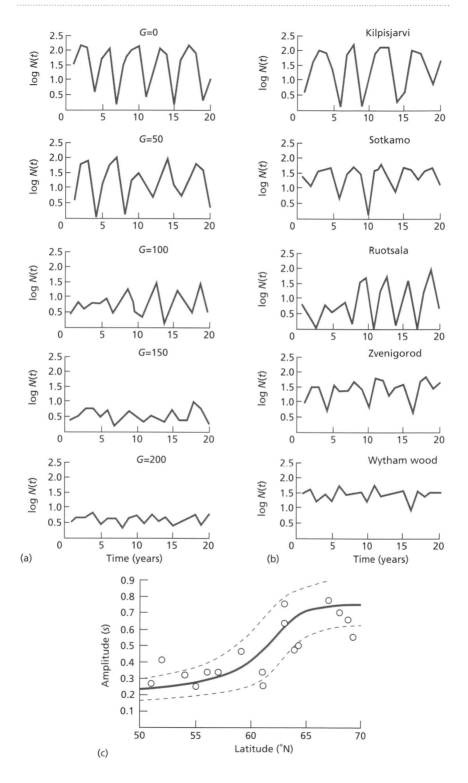

(a) Time (years)

(b) Time (years)

(c)

studied very intensively over a long period. If it cannot be done for this system, there is little hope of producing solid, statistically valid estimates of all the parameters that are necessary to characterize other predator–prey systems. One response to this rather depressing conclusion is that it is a waste of time even to attempt construction of mechanistic models of predator–prey inter-actions. However, it is abundantly clear that the exercise has been very pro-ductive indeed, both from the results of this model, some of which are shown in Fig. 9.7, and from the published antecedents of the model (Hanski *et al.*, 1993a; Hanski & Korpimäki, 1995). The model has lead to important insights into the role of predation in population dynamics in general and the likelihood of complex dynamical behaviour in ecological models. The example also shows some of the tactics that can be employed in attempting to derive para-meter estimates, such as using information from related species, and scaling to reduce the number of independent parameters that must be used in sensitivity analysis. Finally, there is the important point, particularly when formal estima-tion methods are not used, that parameter estimates should not be derived from the same time series with which the model results will be compared. If you do so, your logic becomes circular.

Analysis of time series

Relatively few attempts have been made to fit predator–prey models to time series directly. As I have just said, this is an exercise that should always be approached with caution. Hanski and Korpimäki (1995) used the properties of observed time series to estimate some parameters, as did Akçakaya (1992) in a similar parameterization of a hare–lynx model. However, neither used the time series in a systematic or statistically rigorous fashion to estimate all parameters simultaneously. An exception is the work of Carpenter *et al.* (1994), who used statistical techniques to estimate parameters for an inter-action between phytoplankton and cladoceran herbivores in two freshwater lakes.

A major problem in fitting models to time series is that data usually include process error, which is random variation in the actual numbers of organisms present, and observation error, which occurs because the data are estimates only of those numbers. This problem is referred to briefly in Chapters 2 and 5, and is discussed in some detail by Hilborn and Mangel (1997). Provided only one of process and observation error is present at one time, it is possible to estimate both the parameters of a time-series model and the variance of the error. However, if both sources of error are present simultaneously, things are less straightforward, but possible, as the following example shows.

Carpenter *et al.* (1994) fitted a discretized version of the standard Lotka–Volterra model (eqns (9.3) and (9.4)), with both process and experimental error. Defining the number of prey individuals at time t as N_t and the number of predators as P_t, the model was:

$$N_{t+1} = \{rN_t - sN_t^2 - f[N_t, P_t]\}\exp[v_{N_t}], \tag{9.30}$$

$$P_{t+1} = \{af[N_t, P_t] - mP_t\}\exp[v_{P_t}], \tag{9.31}$$

$$N_{obs,t} = N_t \exp[\omega_{N_t}], \tag{9.32}$$

$$P_{obs,t} = P_t \exp[\omega_{P_t}]. \tag{9.33}$$

Here, r is the rate of increase of the prey population; s is self-limitation of the prey; f is the functional response of the predator, which may have one or two parameters, depending on whether the response is of type I or II; a is the conversion efficiency of prey into predators, and m is the death rate of the predators. The v terms are the process errors, and the ω terms are the observation errors. The terms v and ω themselves are assumed to have normal distributions, with the exponents meaning that the error in N and P follows a lognormal distribution. Equations (9.32) and (9.33) represent the observation error. The quantities that are actually observed are $N_{obs,t}$ and $P_{obs,t}$, not N_t and P_t themselves.

Given that the observation errors are obtained independently, this model has five or six parameters (depending on whether the functional response is type I or II) to be fitted from the data. A further two parameters for the variances of the process error also must be estimated from the data. A full description of the fitting process is beyond the scope of this book, but is given in Carpenter *et al.* (1994). In principle, the process is as follows. Given a particular set of parameter estimates, the predicted values of P_t and N_t can be calculated at each time step, using the previously observed values of P and N, and eqns (9.30) and (9.31). A loss function, which measures the discrepancy between these expected values and the observed values, is then defined, and a numerical process is used to select the parameters which minimize this function. If there is no observation error, the most appropriate loss function is the negative log-likelihood, and the fitting process is one of maximum likelihood estimation (see Chapter 2). With observation error, the situation is a little more complicated, but the parameter estimates generated do not depend markedly on the particular loss function that is used (Fig. 9.8). An additional complication in this case is that there is some evidence of serial correlation in the data sets used by Carpenter *et al.* (1994), meaning that autoregressive terms requiring a further two parameters need to be added. Given the number of parameters being fitted, a long time series is clearly necessary to have any prospect of obtaining worthwhile results.

Carpenter *et al.* (1994) tested their fitting procedure using simulated data with a known underlying model structure and known errors. They also applied the method to weekly records of edible phytoplankton biomass and

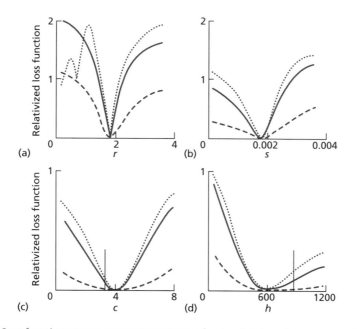

Fig. 9.8 Loss functions versus parameter estimates for simulated predator–prey models. Three loss functions (ways of measuring the discrepancy between predictions and observed data) are shown versus parameter values for 100 data points simulated from eqns (9.30)–(9.33). Using a given loss function, the parameter estimate would be the parameter value at which the loss function was minimized. Dotted lines in each figure represent a weighted least squares loss function, dashed lines represent a lognormal likelihood loss function, and the solid lines are a normal likelihood loss function. (a) r, intrinsic rate of increase of prey population; actual value = 1.8; (b) s, resource limitation on prey population; actual value = 0.0018, (c) c, maximum consumption rate; actual value = 5; (d) h, functional response half saturation constant; actual value = 600. Ordinary least squares estimates are indicated by vertical lines. The key point is that all loss functions produce similar results. From Carpenter *et al.* (1994).

herbivorous zooplankton biomass, which were obtained from two North American lakes over a period of seven years. One lake (Paul) was not manipulated in any way. The zooplankton biomass in the other lake (Tuesday) was perturbed, first by the removal of planktivorous fish, and then for a second time by their reintroduction two years later.

Results from the simulated data with realistic observation errors showed that the fitting procedure was frequently unable to identify the correct model (i.e. the nature of the functional response, and whether there was density dependence). When perturbations were added to the simulated data, the ability of the procedure to discriminate between models was substantially improved. The fit of the model to the observed data is impressive (Fig. 9.9), but it should be noted that these are one-step-ahead predictions, meaning that the observed data at each time step are used to generate the predicted populations for the next time step. The results with the simulated data caution that it

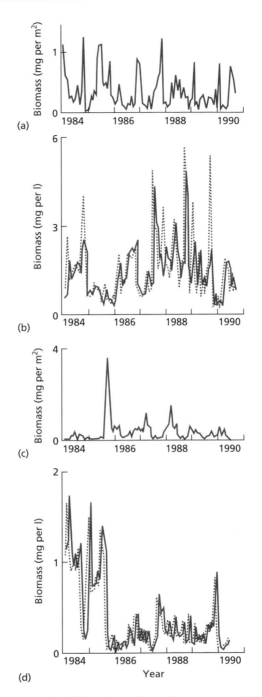

Fig. 9.9 Observed series and model predictions for predator–prey models of plankton in lakes. (a) Biomass of cladoceran herbivores (mg m^{-2}) in Lake Paul. (b) Biomass of edible algae (mg l^{-1}) in Lake Paul, observed (dotted line) and predicted (solid line). (c) Biomass of cladoceran herbivores (mg m^{-2}) in Lake Tuesday. (d) Biomass of edible algae (mg l^{-1}) in Lake Tuesday, observed (dotted line) and predicted (solid line). From Carpenter *et al.* (1994).

should not be assumed that the correct model has necessarily been identified by the fitting procedure.

In summary, this is a mathematically sophisticated approach, which appears to reproduce observed data well. There are two critical questions that need to be considered to assess its usefulness. First, does the fitting procedure help to identify the underlying processes correctly? This would be the objective if the model was being built to understand the nature of the predation process. The results Carpenter *et al.* (1994) obtained with simulated data suggest that the modelling procedure may not do this correctly. Second, is the observed fit better than could be obtained using a simpler nonlinear time-series method, similar to those discussed in Chapter 5 or by Tong (1990)? If the objective was simply to predict plankton biomass, and given that the process may not be identified correctly anyway, it might be better to use a simpler model with fewer parameters.

Stenseth *et al.* (1997), for example, fitted autoregressive models to the well-known time series of snowshoe hares and lynx fur returns. To analyse the data, they took the natural logarithm of population size, and then scaled the result to have zero mean and unit variance. This transformed variable was called X_t. They found that the model

$$X_t = \alpha_1 X_{t-1} + \alpha_2 X_{t-2} + \alpha_3 X_{t-3} + \varepsilon_t \qquad (9.34)$$

was adequate to describe both lynx and hare populations, although there was evidence of some nonlinearities. (For details on how to fit such autoregressive models, see Chapter 6.) They found that the lynx series could be fitted adequately with two terms, with $\alpha_1 > 0$, and $\alpha_2 < 0$, whereas the hare series required $\alpha_1 < 0$, $\alpha_2 \approx 0$ and $\alpha_3 < 0$. The essentially two-dimensional lynx dynamics are consistent with the lynx being regulated primarily by the trophic level below (i.e. the hares), whereas the three-dimensional hare dynamics suggest that the hares are regulated both from below (by vegetation), and from above. Stenseth *et al.* (1997) show how the observed lag relationships can be derived from explicit predator–prey–vegetation equations. Rather than fitting food-web equations directly, they have used autoregressive time-series approaches, and have then interpreted the results in the light of food-web equations.

Summary and recommendations

1 Exploiter–victim models entail the description of two aspects of the interaction between exploiter and victim: the numerical response of the exploiter, which is the way its population size responds to changes in the victim population, and the functional response of the exploiter, which is the way in which the exploitation rate per exploiter depends on victim density.

2 There is some debate over whether functional and numerical models should depend primarily on the number of victims ('prey-dependent models') or primarily on the ratio of victims to exploiters ('ratio-dependent models'). Both sorts of model are worth exploring for a particular system, but there is as yet no convincing evidence that prey-dependent models need to be replaced in many cases by ratio-dependent models. However, numerical responses that are simply a 'conversion factor' times the functional response may often not perform as well as numerical responses assuming that the predator carrying capacity is proportional to prey density.

3 Realistic functional responses are usually type II or type III. Distinguishing between these from data is rarely easy, although the predicted dynamic behaviour resulting from each is quite different. The best approach is to fit Real's model (eqn (9.22)), using nonlinear regression, and then to use extra sum of squares analysis (see Box 9.2) to select the model most appropriate for your data.

4 Spatial heterogeneity is of central importance in determining the behaviour of exploiter–victim interactions. For host–parasitoid interactions, the '$CV^2 > 1$ rule' is a relatively simple way to include the effect of heterogeneity. An alternative approach is to build models that include spatial relationships between victims and exploiters directly, but such models are demanding, both of data and computing resources.

5 Laboratory or microcosm experiments are of limited value in estimating parameters that are intended to apply at a larger spatial scale, or in a more heterogeneous environment.

6 Direct estimation of parameters from time series of predator and prey populations will rarely be practical or advisable.

7 The maximum prey consumption rate per unit of time (the asymptote of the functional response) is usually relatively easy to estimate approximately. Either energetic requirements or direct observation can be used. Surplus killing may, however, sometimes be a substantial complication.

8 The parameter or parameters describing searching efficiency, which describe the victim consumption rate when victims are rare, are far more difficult to measure from field data than is the maximum prey consumption rate.

9 In principle, there are two possible approaches to estimating parameters for functional responses. An exploiter-centred approach records the victim consumption rate of exploiters, over a range of victim densities. A victim-centred approach records the loss rate from the victim population to the exploiter in question, as a function of victim density. Which method is preferable depends primarily on practical considerations.

10 *Ad hoc* approximate methods are not ideal, but will frequently be the only way that useful progress can be made.

CHAPTER 10

Host–pathogen and host–parasite models

Introduction

Host–parasite and host–pathogen interactions are similar to predator–prey or host–parasitoid interactions, because they are interactions where one species uses the other as a resource. The extremely close association between exploiter and victim, which is characteristic of host–pathogen and host–parasite interactions, means that the structure of models of these interactions is rather different from those of predator–prey interactions. The key processes that must be understood and parameterized are transmission and the effect that the parasite or pathogen has upon the host.

For obvious reasons, models of host–parasite and host–pathogen interactions are most highly developed for organisms that cause disease in humans (see, for example, Anderson & May, 1991). Covering the problems of parameterizing models of disease in humans is beyond the scope of this book. The data available in human disease epidemiology are typically of much better quality than those available for wildlife disease (for example, records of measles infection are available covering 70 years, on a monthly basis, in most developed countries; nothing approaching this quality is available for any free-ranging animal). Methods for human disease are therefore able to be much more demanding of data, and may correspondingly be more complex than methods that can usefully be applied to animal populations. Nevertheless, many approaches currently used in wildlife epidemiology were first developed for human disease, and no doubt this cross-fertilization will continue in the future. Of course, many human diseases involve other animals as vectors, intermediate hosts or primary hosts. This has always provided a substantial impetus for studying the interactions of parasites and pathogens with animals.

Why parameterize host–parasite and host–pathogen models?

Parameterized models of host–parasite interactions in wildlife may be needed for a variety of reasons. If the parasite can be transferred to humans or domestic stock, the objective may be to evaluate various strategies of disease control, such as vaccination or culling. For example, strategies for control of rabies in Europe were modelled by Anderson *et al.* (1981). Barlow (1991) used a model to evaluate methods for reducing the level of tuberculosis in New Zealand dairy cattle by controlling the disease in its reservoir hosts, the introduced brushtail possum (*Trichosurus vulpecula*). Pathogens, or less commonly, parasites (McCallum & Singleton, 1989), may be considered as control agents for problem species. Models can be used to evaluate the likelihood of successful control (e.g. Barlow, 1997) or the optimal timing of release of the agent (e.g. McCallum, 1993). The objective of modelling may also be to determine the possible impact of infection on a given species, particularly if the species is of value as game (see, for example, Hudson & Dobson, 1989), or if the species is endangered (McCallum & Dobson, 1995).

The models that are required for such problems can span the range from strategic to tactical, as is discussed in the introductory chapter. A particularly important consideration in parameter estimation is whether the model is intended to be applied to the same system in which the parameter has been estimated. For example, Woolhouse and Chandiwana (1992) estimated the rate of infection of snails with *Schistosoma mansoni* in Zimbabwe, with the objective of using the estimated parameter to model the dynamics of the disease in that same area. If the objective of modelling is to predict the possible impact of an exotic disease, or a novel biocontrol agent, then it is clearly impossible to estimate the parameter in the environment in which the model is to be applied. The problems of extrapolation from laboratory studies to the field, or from one field situation to another, are considerable, particularly for the key process of transmission.

Basic structure of macroparasite and microparasite models

For modelling purposes, it is convenient to divide parasites into microparasites, which usually can reproduce rapidly within their hosts, and macroparasites, which usually must leave their host at some point in a complete life cycle. A virus or bacterium is a typical microparasite, and a helminth is a typical macroparasite. The simplest models of microparasitic infections consider hosts as 'infected', without attempting to differentiate between degrees of infection. As the microparasite can reproduce rapidly within its host, the level of infection depends primarily on the level of host response, rather than on the number of infective stages the host has encountered. In contrast, the impact of a macroparasitic infection on its host depends quite critically on the number of parasites the host is carrying, which depends in turn very strongly on the number of infective stages encountered. It is thus essential, even in the simplest models, to keep some track on the distribution of parasites between hosts. These generalizations are oversimplifications: recent research has emphasized the importance of the infective dose on the severity of disease in microparasites and the importance of host response to the level of infection in macroparasitic infections (Anderson & May, 1991). Nevertheless, they provide a point at which to start.

Models of microparasitic infections

Most recent models are based on the framework outlined by Anderson and May (1979) and May and Anderson (1979). This is, in turn, based on earlier models (Ross, 1916; 1917; Bailey, 1975), dating back to Bernoulli (1760). For ecologists, the major advance in Anderson and May's models is that the host population is dynamic and influenced by the disease, so that the ability of a disease to regulate its host population can be examined.

The basic model structure involves dividing the host population into three categories: susceptibles, infecteds and resistants. These models are often therefore called *SIR* models. Anderson and May use the variables X, Y and Z to identify these categories, respectively. Hosts in each category are assumed to have a *per capita* birth rate of a per unit of time, and to die at a disease-independent rate b. The death rate of infected hosts is augmented by a factor α. Infection is assumed to occur by direct contact between infected and naïve hosts by a process of simple binary collision. This means that hosts leave the susceptible class and enter the infected class at a rate βXY. Infected hosts recover to enter the resistant class at a rate γ per unit of time. Finally, resistance is lost at a rate v per unit of time.

These assumptions lead to the following equations:

$$\frac{dX}{dt} = a(X + Y + Z) - bX - \beta XY + vZ, \tag{10.1}$$

$$\frac{dY}{dt} = \beta XY - (b + \alpha)Y - \gamma Y, \tag{10.2}$$

$$\frac{dZ}{dt} = \gamma Y - vZ. \tag{10.3}$$

The full details of the behaviour of this model can be found in Anderson and May (1979) and May and Anderson (1979), and some of the many elaborations of the model are discussed in Anderson and May (1991). Two fundamental and related quantities can be derived from the model. These are the basic reproductive number of the disease, R_0, and the minimum population size for disease introduction, N_T. Estimation of these quantities, either directly or as compounds of other parameters, is often essential to predict the behaviour of a disease in a population.

The basic reproductive number is the number of secondary cases each primary infection gives rise to, when first introduced into a naïve population. If $R_0 > 1$, the disease will increase in prevalence, but if $R_0 < 1$, the disease will die out. In almost all disease models, R_0 depends on the population density, and N_T is the critical host population size at which $R_0 = 1$.

These two quantities can be obtained from eqn (10.2) in the limiting case when $Y \approx 0$ and total population size is thus X. Hence,

$$R_0 = \frac{\beta N}{b + \alpha + \gamma} \tag{10.4}$$

and

$$N_T = \frac{b + \alpha + \gamma}{\beta}. \tag{10.5}$$

Microparasite models are frequently much more complicated. Transmission may require vectors, there may be long-lived infective stages, and infection rates may depend on host age and behaviour. There is also often an exposed but not infectious stage. Infection may affect the reproductive rate as well as, or instead of, the death rate. Immunity may be lifelong or incomplete. Whatever the complications, the simple model illustrates that parameterizing a microparasite model involves estimating parameters for three main processes: transmission; the impact of the disease on the host; and recovery and immunity. In addition, estimation of R_0 and N_T will usually be of crucial importance.

Models of macroparasitic infections

The models of Anderson and May (1978) and May and Anderson (1978) form the basis of most recent host–macroparasite models. The system is represented by three variables: host population size, $H(t)$; the population of parasites on hosts, $P(t)$; and the free-living infective stages, $W(t)$. Hosts have a disease-free birth rate a and a disease-free death rate b. Their death rate is also increased by parasitic infection. This will depend in some way on the level of, as well as just the presence of, parasitic infection. The simplest form such dependence could take is that the death rate of a host infected with i parasites could increase linearly with parasite burden at a rate αi. Parasites are assumed to infect hosts in a straightforward binary collision process between hosts and infective stages, at a rate $\beta W H$. Once on hosts, parasites are lost from the system through a constant death rate γ, and are also lost if the host itself dies, through causes both related to infection (at the rate αi) and unrelated to infection (at the rate b). Parasites on hosts produce infective stages at a rate λ per unit of time. Finally, infective stages themselves are lost from the system either through their death, at a rate μ, or by infection of hosts.

All the above are easy to incorporate into a differential equation model, with the exception of the loss rate of parasites caused by parasite-induced death. Given that this occurs at a rate αi on a host with i parasites, the total loss over the whole system will occur at the rate $\alpha \Sigma i^2 p(i)$, where $p(i)$ is the probability that a host has i parasites.

Because this term involves the square of the parasite burden, it depends on the variance of the parasite distribution within the host population, not just its mean. (Recall that $\Sigma i^2 p(i) = \text{var}(i) + (\bar{i})^2$, where $\text{var}(i)$ is the variance of the parasite distribution amongst hosts, and $\bar{i}$ is its mean.) This is an essential point about macroparasitic infections. Their dynamics depend crucially on how parasites are distributed within the host population. In general, this distribution can be expected to be determined by the dynamics of the host–parasite interaction, but Anderson and May make the simplifying assumption that it is a negative binomial distribution with a fixed parameter k. The smaller the value of k, the more aggregated is the distribution.

These assumptions lead to the following differential equations:

$$\frac{dH}{dt} = (a - b)H - \alpha P, \tag{10.6}$$

$$\frac{dP}{dt} = \beta WH - (b + \gamma + \alpha)P - \frac{\alpha(k+1)}{k}\frac{P^2}{H}, \tag{10.7}$$

$$\frac{dW}{dt} = \lambda P - (\mu + \beta H). \tag{10.8}$$

In many cases, the lifespan of the infective stages is much shorter than that of the hosts or parasites on hosts. In such cases, W can be assumed to be at an equilibrium, W^*, with whatever are the current values of P and H, leading to

$$W^* = \frac{\lambda P}{\mu + \beta H}. \tag{10.9}$$

This can be substituted into eqn (10.7), leading to:

$$\frac{dH}{dt} = (a - b)H - \alpha P, \tag{10.10}$$

$$\frac{dP}{dt} = \frac{\lambda PH}{\left(\dfrac{\mu}{\beta} + H\right)} - (b + \gamma + \alpha)P - \frac{\alpha(k+1)}{k}\frac{P^2}{H}. \tag{10.11}$$

As is the case with microparasitic infections, R_0 is a crucial quantity in the understanding of parasite dynamics, although it has a slightly different meaning. In the macroparasitic case, R_0 is the number of second-generation parasites each parasite replaces itself by, when first introduced into a host population. From eqn (10.11),

$$R_0 = \frac{\lambda H}{\left(\dfrac{\mu}{\beta} + H\right)(b + \gamma + \alpha)}. \tag{10.12}$$

There is also a threshold host population for parasite introduction:

$$H_T = \frac{\left(\dfrac{\mu}{\beta}\right)}{\left(\dfrac{\lambda}{[b+\alpha+\gamma]} - 1\right)}. \tag{10.13}$$

This makes it clear that parasite reproduction must exceed parasite death (the first term in the denominator). Otherwise there is no host density at which the parasite will persist, but once this occurs, the higher the transmission rate relative to the death rate of the infective stages, the lower the host density at which the infection can persist.

This model is a vast oversimplification of the dynamics of most macro-parasite infections. Many macroparasites have complex life cycles with one or more intermediate hosts. This can be accommodated by coupling two or more sets of these basic equations together (Roberts *et al.*, 1995). Infection rates may depend on the age of the host, or infection may occur as a process more complex than binary collision. Parasites may affect host fecundity as well as mortality, and the death rate may not be a simple linear function of burden. Host death may not lead to parasite death. The model also does not include any effects of density dependence within the individual host, either on pro-duction of infective stages or on the death rate of parasites within hosts.

Nevertheless, this very simple model identifies clearly the key processes that must be quantified to understand a host–macroparasite infection. They are the infection process, the production of transmission stages, the effect of parasites on hosts, and the loss rate of parasites on hosts. The reproductive number R_0 and the threshold host population H_T are also of crucial import-ance. The important distinction between macroparasite and microparasite infections is that, for macroparasites, it is vital to consider the effect of the para-site burden of a host, not simply whether or not it is infected. At a population level, it is essential to quantify the nature of the parasite distribution between hosts, and not simply to measure the mean burden.

The transmission process

The rate at which disease is transmitted between hosts, or the rate at which infective stages infect hosts, is obviously of central importance in determining whether and how fast a disease or parasite will spread in a population (see eqns (10.4) and (10.12)). It is, however, a particularly difficult process to quantify in the field. Unfortunately, it is also very difficult to apply the results of experiments on laboratory or captive populations to the field. Transmission involves at least two distinct steps. First, there must be contact between the susceptible host and the infective stage, infected host or vector. Second, following contact, an infection must develop, at least to the point that it is detectable, in the susceptible host. Most methods of estimating transmission rates confound these two processes.

Force of infection

In general terms, the quantity that needs to be measured is the force of infection, which is the rate at which susceptible individuals acquire infection. In principle, the force of infection can be estimated directly by measuring the rate at which susceptible individuals acquire infection over a short period. The usual approach is to introduce 'sentinel' or uninfected tracer animals; these are removed after a fixed period and their parasite burden (for macroparasites) or the prevalence of infection in the sentinels (for microparasites) determined. This has been done in a number of studies (Scott, 1987; Quinnell, 1992; Scott & Tanguay, 1994), although none of these used an entirely unrestrained host population.

In practice, estimating the force of infection by using sentinels in a natural situation is not a straightforward task. First, it is necessary to assume that the density of infected hosts or parasite infective stages remains constant over the period of investigation. Second, if sentinels are animals that have been reared in captivity, so that they can be guaranteed to be parasite-free, it is most unlikely that, once released, they will behave in the same way as established members of the wild population. The extent of their exposure to infection may therefore be quite atypical of the population in general. If they are members of the wild population that have been selected as sentinels because they are uninfected, then they clearly are not a random sample of the whole population with respect to susceptibility or exposure to infection. The force of infection will therefore probably be underestimated. Finally, if sentinels are wild animals that have been treated to remove infection, there is the possibility of an immune response influencing the results.

A similar approach to using sentinels is to use the relationship between age of hosts and prevalence of infection (microparasites) or parasite burden (macroparasites) to estimate force of infection. Here, newborns are essentially the uninfected tracers. This approach has been used with considerable success in the study of human infections (Anderson & May, 1985; Grenfell & Anderson, 1985; Bundy et al., 1987). In the simplest case, the force of infection is assumed to be constant through time and with age. Then, in the case of microparasitic infection,

$$\frac{dx}{da} = -\Lambda a, \tag{10.14}$$

where x is the proportion of susceptible individuals of age a and Λ is the force of infection. Equivalently,

$$\frac{dF}{da} = \Lambda(1 - F), \tag{10.15}$$

where F is the proportion of a cohort that has experienced infection by age a (Grenfell & Anderson, 1985). This equation means that Λ is simply the reciprocal of the average age at which individuals acquire infection (Anderson & Nokes, 1991). Unfortunately, the force of infection will rarely be independent of age or time. Methods are available to deal with the problem of age dependence (Grenfell & Anderson, 1985; Farrington, 1990), provided either age-specific case notification records or age-specific serological records are available. If age-specific serological records are available at several times, it may even be possible to account simultaneously for both time and age-specific variation in the force of infection (Ades & Nokes, 1993). Age-specific serology is more likely than age-specific case notifications to be available for non-human hosts, but it will not give an accurate picture of age at infection if there is substantial disease-induced mortality. Use of age-specific serology in wildlife also presupposes the age of the animals can be estimated accurately.

For macroparasites, the equation equivalent to eqn (10.14) is

$$\frac{dM}{da} = \Lambda - \mu M, \tag{10.16}$$

where M is the mean parasite burden at age a and μ is the death rate of parasites on the host. Equation (10.16) has a solution

$$M(a) = \frac{\Lambda}{\mu}(1 - \exp[-\mu a]). \tag{10.17}$$

The force of infection can therefore be estimated from the asymptote of the relationship between mean parasite burden and age, provided an independent estimate of the parasite death rate is available. It is possible to generalize these results to cases in which either the force of infection or parasite death rate varies with the age of the host (Bundy *et al.*, 1987).

There are few cases in which age-specific prevalence or intensity of infection has been used to estimate the force of infection in wildlife. Two notable examples are the study of transmission of the nematode *Trichostrongylus tenuis* in the red grouse *Lagopus lagopus scoticus* by Hudson and Dobson (1997), and the study of force of infection of schistosome infections in snails by Woolhouse and Chandiwana (1992). These examples demonstrate that age-prevalence or age-intensity data are potentially powerful tools to estimate the force of infection in natural populations.

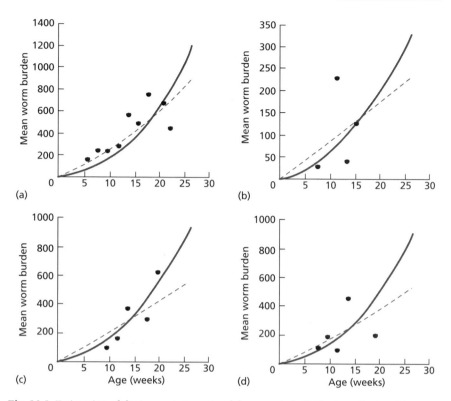

Fig. 10.1 Estimation of the transmission rate of the nematode *Trichostrongylus tenuis* in red grouse (*Lagopus lagopus scotius*). Dots show the observed relationship between age and intensity. For each of the four years, the dashed line is the fit of a constant-transmission-rate model, whereas the solid line assumes that the worm intake is proportional to the body size and hence to feeding rate. (a) 1982; (b) 1984; (c) 1985; (d) 1987. From Hudson and Dobson (1997).

Hudson and Dobson (1997) caught and marked grouse chicks between 2 and 15 days old. When tagged birds were shot during subsequent shooting seasons, their viscera were collected and the intensity of worm infection was estimated. In this way, parasite infection levels could be determined as a function of an accurately estimated age. The force of infection was assumed either to be constant, or to be directly proportional to food consumption, which was in turn assumed to be proportional to body mass. The results of this study are shown in Fig. 10.1.

Woolhouse and Chandiwana (1992) recorded the prevalence of *Schistosoma mansoni* infection in *Biomphalaria pfeifferi*, a species of freshwater snail, in Zimbabwe. The data they used were prevalence of infection, stratified by size, from snails collected in January 1987. The force of infection is likely to depend on water temperature, but, at the particular study site in January, this would be approximately constant over at least the preceding eight weeks.

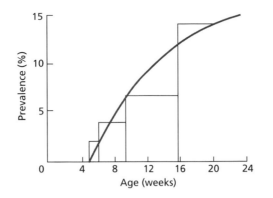

Fig. 10.2 Estimation of the force of infection of schistosomes in freshwater snails in Zimbabwe. The histogram shows observed prevalence from field data. The force of infection was estimated by fitting an equation similar to eqn (10.15), but with the addition of both natural and parasite-induced host mortality. All parameters other than the infection rate were estimated from independent data. From Woolhouse and Chandiwana (1992).

Snail size was converted to age, using laboratory growth curves and a correction for the effect of infection on growth. The fitted relationship is shown in Fig. 10.2.

Transmission rate

The force of infection depends on the number of infected hosts or on the number of infective stages in a population. This means that estimating force of infection is not, in itself, sufficient for situations in which the level of infection in the population varies through time. It is necessary to determine how the force of infection varies with the density of infective stages. If infection occurs through a simple mass-action process, as assumed by eqns (10.2) and (10.7), then the force of infection Λ is simply $\Lambda = \beta Y$ for a microparasitic infection or $\Lambda = \beta W$ for a macroparasitic infection. Such transmission is sometimes called density-dependent transmission, as it depends on the density of infected individuals or of infective stages (De Leo & Dobson, 1996). In sexually transmitted and some vectorially transmitted pathogens, transmission may depend on the proportion of individuals in the population that are infected, rather than on their absolute density. This is called frequency-dependent transmission, and $\Lambda = \beta Y/N$.

Provided it can be assumed that the appropriate mode of transmission has been correctly identified, and an estimate of the number of infected hosts or infective stages is available, it is straightforward to obtain an estimate of the transmission rate β from an estimated force of infection. This estimate will, however, apply only to the particular environment from which it was obtained.

Experimental estimation of transmission

Several studies have attempted to quantify transmission in macroparasites under experimental conditions (Anderson, 1978; Anderson *et al.*, 1978; McCallum, 1982a). The process occurring in an experimental infection arena can be represented by the following two equations, for the infective-stage population size W and parasite population on hosts P (McCallum, 1982a):

$$\frac{dW}{dt} = -\mu W - \beta WH, \tag{10.18}$$

$$\frac{dP}{dt} = \beta s WH. \tag{10.19}$$

Here, μ is the death rate of the infective stages, s is the proportion of infections that successfully develops into parasites, and H is the number of hosts present. The remainder of infection events remove the infective stage from the system, but a parasite infection does not result.

Equations (10.18) and (10.19) have a time-dependent solution,

$$P(t) = \frac{HsW(0)}{(\mu/\beta) + H}(1 - \exp(-[\mu + \beta H]t)). \tag{10.20}$$

Simple infection experiments thus do not produce a straightforward estimate of β, unless the exposure time is very short, in which case eqn (10.20) is approximately

$$P(t) = s\beta W(0)Ht. \tag{10.21}$$

If a constant number of hosts is exposed to varying numbers of infective stages until all infective stages are dead, eqn (10.20) also predicts that a relationship of the form $P = kHW$ will result, but the constant k is $s/(\mu/\beta + H)$, not β itself. In general, to estimate β, it is necessary to use several exposure times and to have an independent estimate of μ. An example of an experiment of this type (from McCallum, 1982a) on the ciliate *Ichthyophthyrius multifiliis* infecting black mollies (*Poecilia latipinna*) is shown in Fig. 10.3.

Estimating transmission rates from an observed epidemic

When a disease is introduced into a naïve population (a population without recent prior exposure to the disease) which exceeds the threshold population size, an epidemic results, in which the number of new cases per unit of time will typically follow a humped curve. If there are no new entrants into the susceptible class (either by recruitment or loss of immunity), the epidemic will cease. If a satisfactory model is developed to describe the host–pathogen

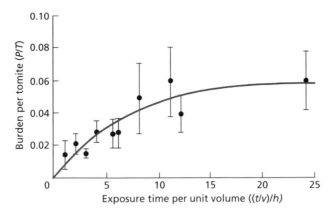

Fig. 10.3 Experimental infection of black mollies by the ciliate *Ichthyophthyrius multifiliis*. The horizontal axis shows the exposure time of the fish per unit of volume, and the vertical axis the number of parasites established per infective stage added. The curve shown is eqn (10.20), with β and s estimated by nonlinear least squares. The mortality rate of the infective stages μ was obtained in a separate experiment. From McCallum (1982a).

interaction, the shape of the curve will be a function of the various parameters in the model. In principle, then, it should be possible to estimate the model parameters by fitting the model to the observed epidemic curve. In practice, all but the most trivial epidemic models have so many parameters that estimating all of them from a single epidemic curve is impractical. However, if independent estimates of all the other parameters are available, it may be possible to estimate the transmission rate from the course of the epidemic. In fact, estimation of the infection rate is possible with as few as two data points: the number of animals present at the commencement of the epidemic, and the number left unaffected at its conclusion. That said, with only two points it is not possible to get any idea of the precision with which the transmission rate has been estimated. Suitable data are rarely available for free-ranging populations, but there are some examples of these approaches being used.

Hone *et al.* (1992) used data from an experimental introduction of classical swine fever into a wild boar population in Pakistan. The data available were an estimate of the initial population size, the total area of the forest occupied by the pigs, and the cumulative number of deaths as a function of time since the disease was introduced (Fig. 10.4). Unfortunately, data were available only for the first 69 days of the epidemic. The epidemic continued for some time further. Given independent estimates of the incubation period of the disease, the recovery rate and the mortality rate, the transmission rate coefficient was estimated as 0.000 99 d^{-1}. As is discussed in later in the chapter, there are potential problems in applying and scaling such an estimate for use in other environments.

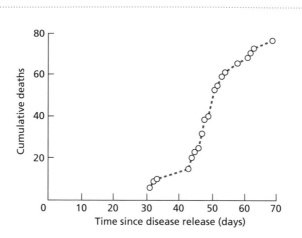

Fig. 10.4 Cumulative deaths in an epidemic of classic swine fever in Pakistan. Classic swine fever was introduced into a population of wild boar *Sus scrofa* in a forest plantation of 44.6 km² in Pakistan. The initial population was estimated to be 465 individuals, and cumulative deaths for the first 69 days are shown. Independent estimates of parameters other than the transmission rate are: mortality rate, 0.2 d⁻¹; recovery rate 0.067 d⁻¹; incubation period 4.75 d. From Hone *et al.* (1992).

Grenfell *et al.* (1992) used a broadly similar approach to estimate the transmission rate of phocine distemper in several northern European seal colonies, following epizootics in 1988. Using a simple epidemic model in the general form of eqns (10.1) to (10.3), but without births, disease-independent deaths or loss of immunity, they derived the following relationship between S, the numbers of animals unaffected by the disease at the end of the epidemic, and the transmission rate, β:

$$0 = N_0 - S + \frac{\alpha'}{\beta}\ln(N_0/S). \tag{10.22}$$

Here, N_0 is the number of animals present intially, and α' is the rate of loss from the infective class. In the notation of eqns (10.1) to (10.3), $\alpha' = \alpha + \gamma$, as it includes animals leaving the infective class from both death and recovery. The data available to Grenfell *et al.* (1992) were N_0, the number of seals present initially, and N_S, the number surviving at the end of the epidemic. Seals may survive by either not being infected, or by surviving infection. Hence,

$$N_S = S + (1 - \delta)(N_0 - S), \tag{10.23}$$

where δ is the probability of a seal surviving infection. Given an estimate of α of around 26 y⁻¹, and 0.63 for δ, the estimate of β obtained for a colony in Kattegut, Denmark, was 81 y⁻¹. Very different levels of total mortality were observed in other seal colonies in northern Europe, suggesting that either the transmission coefficient, β, or the proportion of infected seals dying, δ, varied

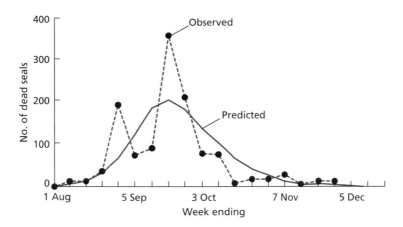

Fig. 10.5 Observed and predicted epidemic curves for phocine distemper in common seals (*Phoca vitulia*) in The Wash, UK, in 1988. From Grenfell *et al.* (1992).

considerably between colonies. Serological evidence suggested that it was the transmission rate that varied. This example emphasizes that transmission rates estimated for a given host–pathogen interaction in one place cannot easily be applied to the same interaction in another place. An example of observed and predicted epidemic curves obtained for phocine distemper is shown in Fig. 10.5.

Allometric approaches

It should be clear that direct estimation of the transmission rate is difficult. De Leo and Dobson (1996) have suggested a method based on allometry (scaling with size), which can be used in the absence of experimental data. As with any allometric approach to parameter estimation (see also Chapter 5), it should be considered a method of last resort. It does, however, permit a minimum plausible transmission rate to be estimated to the correct order of magnitude.

For any given epidemic model, it is possible to calculate a minimum transmission rate, which must be exceeded if a disease is able to invade a population of susceptible hosts at a particular carrying capacity. For example, De Leo and Dobson (1996) show that, for a model of the general form of eqns (10.1)–(10.3),

$$\beta > \frac{\alpha + b + \gamma}{K}, \tag{10.24}$$

where α is the disease-induced death rate, b is the disease-free host death rate, γ is the recovery rate and K is the carrying capacity in the absence of infection. There are well-established relationships between body size and both carrying capacity and lifespan in vertebrates (see Peters, 1983). As any disease that exists in a given species must be able to invade a fully susceptible population, eqn (10.24) can be combined with these allometric relationships to generate the minimum transmission rate for a disease with a given mortality rate to persist in a host species with a mean body weight w:

$$\beta_{min} = 2.47 \times 10^{-2}(m + \gamma/b)w^{-0.44}. \tag{10.25}$$

Here, m is the factor by which the disease-induced death rate exceeds the disease-free death rate (i.e. $m = \alpha/b$).

Similar relationships can be derived for other host–microparasite models (see De Leo & Dobson, 1996).

Scaling, spatial structure and the transmission rate

Applying an estimate of a transmission rate obtained in one situation (perhaps an experimental arena) to another situation (perhaps to a field population) is not straightforward. The binary collision model assumed in eqn (10.2) or eqn (10.7) is an idealization borrowed from the physical sciences (de Jong *et al.*, 1995), and is at best a very crude caricature of interactions between biological organisms. As I have said earlier, transmission of infection depends on at least two processes: contacts between susceptible hosts and infectious hosts or stages, and the successful establishment of infection following contact. Equation (10.20) shows how these processes can be separated in some experimental situations. In natural or semi-natural populations, this will rarely be possible. It is the contact rate that causes problems in scaling.

The binary collision model assumes that infected and susceptible animals contact each other randomly and instantaneously, as if they were randomly bouncing balls in a closed chamber. This means that the probability of a contact between a particular susceptible individual and any infected individual is proportional to the *density* of infected individuals per unit area or volume. De Jong *et al.* (1995) describe this assumption as the 'true mass-action assumption'. If transmission is indeed being considered within an arena of fixed size, and all individuals move randomly throughout the arena, then the probability of contact between a susceptible individual and any infected individual is also proportional to the *number* of infected individuals. It does not matter whether the equations are formulated in terms of density per unit area or number of individuals. Real populations do not live in arenas, and herein lies a problem.

Equations (10.1)–(10.3) are usually presented with X, Y and Z representing population sizes, not densities per unit area. De Jong *et al.* (1995) argue that, if density remains constant, independent of population size, the term βXY in eqn (10.2) does not represent 'true mass action'. If population density is constant, but the actual population size varies, the total number of contacts per unit of time made by a susceptible individual will be constant, and the probability of infection will depend on Y/N, the proportion of those contacts which are with infected individuals. Thus, they argue that, for 'true mass action' βXY should be replaced by $\beta XY/N$, and that βXY represents 'pseudo-mass action'. Reality probably lies somewhere between these two. Individuals do not interact equally with all members of their population, but do so mostly with their near-neighbours. This means that local population density is the key variable to which transmission is probably proportional. For most animals, an increase in total population size will increase local population density, but territorial or other spacing behaviour is likely to mean that the increase in density is not proportional to overall population size, unless the population is artificially constrained. A possible empirical solution might be to use a transmission term of the form $\beta XY/N^{\alpha}$, where $0 < \alpha < 1$ is a parameter to be estimated.

There is little empirical evidence available to distinguish between these two models of transmission. Begon *et al.* (1998) used two years of time-series data on cowpox infection in bank voles (*Clethrionomys glareolus*) in an attempt to resolve whether pseudo- or true mass action better predicted the dynamics of infection. Models of transmission based on pseudo-mass action performed marginally better, but the results were inconclusive. One conclusion was, however, clear. The study used two separate populations, and whether pseudo- or mass action was assumed, the estimated transmission parameter differed substantially between the two populations. This emphasizes the difficulty of applying a transmission rate estimated in one place to another.

The terms 'pseudo-' and 'true' mass action are potentially confusing. De Leo and Dobson (1996) consider the same two models, but use the term 'density-dependent transmission' for a transmission term βXY, and 'frequency-dependent transmission' for a transmission term $\beta XY/N$. This terminology is less ambiguous.

The rather unsatisfactory conclusion to the scaling problem is that there is really no way that an estimate based on experiments in a simple, homogeneous arena can be applied to a heterogeneous, natural environment. There is simply too much biological complexity compressed into a single transmission term. The best suggestion that can be made is to deal with densities per unit area, rather than actual population size. If this is done, then the coefficient β is not dimensionless: it will have units of area per unit of time.

Parasite-induced mortality

As discussed in Chapter 4, mortality or survival in captive situations is rarely a good indication of mortality in the field, but mortality in the field is not an easy thing to measure. These problems also, of course, apply to attempts to estimate parasite-induced mortality, but are exacerbated because parasites frequently affect host behaviour, possibly leading to increased exposure to predation or decreased foraging efficiency (Dobson, 1988; Hudson *et al.*, 1992a). Experimental infections of animals in laboratory conditions will almost always underestimate disease-induced mortality rates in the field, but field data will not be easy to obtain.

The 'ideal experiment'

In principle, the best way to obtain such data would be to capture a sample of animals from a population in which the parasite or pathogen does not occur, to infect some with known doses of the pathogen or parasite, to sham-infect others to serve as controls and then to release them and monitor survival. The resulting data could then be analysed using standard survival analysis techniques, treating infection as a covariate (see Chapter 4). This approach has rarely been used, for several reasons. If the pathogen is highly transmissible, the controls may become infected. There are obvious ethical problems in experimentally infecting wild animals, particularly if the study species is endangered. Even if this could be done, there are also difficulties in relating the parasite burden established on the treated animals to the infective dose given to them. All statistical models that attempt to estimate a parameter as a function of a predictor variable produce a biased parameter estimate, if the predictor is measured with error. Some methods to deal with this problem are discussed in Chapter 11.

Comparing survival of hosts treated for infection with controls

Rather than infecting uninfected hosts, an alternative approach is to treat some hosts for the infection, and then to compare survival between these and control hosts. This is obviously an easier approach to justify ethically. A number of studies have used this approach (e.g. Munger & Karasov, 1991; Gulland, 1992; Hudson *et al.*, 1992b), but usually with the objective of determining whether the parasite or pathogen has any impact, rather than quantifying the extent of impact.

There are several design issues that should be addressed if this approach is to be successful. Some of these are standard principles of experimental design, such as appropriate replication and controls. The process of capture and administration of the treatment may well have impacts on survival and behaviour, and, as far as possible, these impacts should be the same for both treatments and controls. Others are specific to experiments of this type. For example, it is important to have some idea of the infection status of each experimental animal before the manipulation takes place.

In a microparasite infection, if only a fraction of the untreated animals are infected in the first place, then the impact of the disease on survival will be

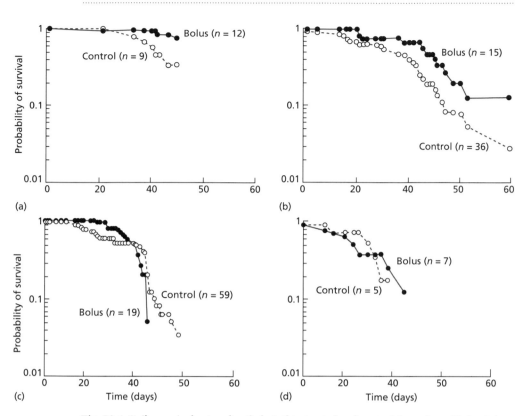

Fig. 10.6 Daily survival rates of anthelminthic-treated and control Soay sheep (*Ovis aries*) on St Kilda, Scotland, during a population crash in 1989. The *x* axis shows the time in days from when the first dead sheep was found, and the *y* axis shows the probability of survival on a logarithmic scale: (a) ewes, χ^2 for difference in daily survival 4.26, 1 df, $p < 0.05$; (b) male lambs, χ^2 for difference in daily survival 4.57, 1 df, $p < 0.05$; (c) yearling males, χ^2 for difference in daily survival 0.0, 1 df, not significant; (d) two-year-old males, χ^2 for difference in daily survival 0.1, 1 df, not significant. From Gulland (1992).

underestimated by the experiment. Ideally, one would want to perform a two-way experiment, with infection status before the experiment crossed with treatment. Less obviously, if there is a treatment effect, survival will be more variable in the untreated group than in the treated group, because only some of the untreated group will be infected. This will produce problems for most statistical techniques.

Quantifying the impact of a macroparasitic infection with an experiment of this type is even less straightforward. During a population crash, Gulland (1992) compared the survival of Soay sheep treated with an anthelminthic bolus with control sheep. There was a significant difference in survival between treated and control ewes, and for male lambs, but not for yearling and adult males (see Fig. 10.6). This population was very closely monitored so that accurate post-mortem worm counts could be obtained from all animals that died during the 60 days of the crash. Even so, these results do not allow a straightforward estimation of α, the increase in mortality rate per parasite. The difference in survival between the groups could be used as an estimate of $\alpha(M_C - M_T)$, where M_C is the mean parasite burden of the control animals and M_T that of the treated animals, but only if the relationship between the parasite burden and mortality was linear.

An alternative approach would be to determine parasite burdens in a random sample from a target population, to divide the hosts into several classes according to severity of infection, and then to treat half of each class for infection and monitor survival. Such an experiment is technically awkward. It is difficult to census most macroparasites in a way that is non-destructive of both host and parasite populations. Egg counts are often used as an index of the level of helminth infection, and whilst there is almost invariably a statistically significant correlation between worm burden and egg count, the relationship is often nonlinear and also has so much variation about it that it is a poor predictor. I know of no example of an experiment of this type being carried out to quantify the form of a parasite-induced mortality function in a macroparasite in the field.

Laboratory experiments

Most attempts to estimate the effect of parasitic infection on survival have been based on laboratory studies. For example, the classic experiments on the changes in mortality of wild rabbits exposed to the myxoma virus (Fenner & Ratcliffe, 1965) following the introduction of the disease into Australia were largely carried out by experimental inoculation of animals held in animal houses. Some experiments were also undertaken in enclosures in the natural environment. Surprisingly, the mortality rate was in fact greater in the controlled environment of the animal houses.

For macroparasites, a question that is difficult to assess other than by laboratory experiments is the functional form of the relationship between parasite burden and host death rate. The elementary model in eqn (10.6) assumes that the relationship is linear. In many cases, this seems to be a reasonable assumption, but there are other cases where the relationship between the parasite burden and mortality rate is nonlinear (see Anderson & May, 1978; May & Anderson, 1978). 'Thresholds' appear to be rare.

Observational methods

The extent to which it is possible either to determine that a parasite is having an impact on survival, or to quantify the extent of impact from entirely non-manipulative studies is a continuing area of debate. Some methods simply do not work. For example, it is quite incorrect to suppose that a high level of disease prevalence or high parasite burdens in morbid or recently dead animals indicate that the disease is having an impact on the host population. In fact, it is more likely that the opposite is the case, as non-pathogenic diseases are likely to occur at high prevalence (McCallum & Dobson, 1995).

It is possible to estimate pathogenicity from the reduction in population size following a short epidemic, provided serology is available after the epidemic to determine whether survivors have been exposed or not. For example, Fenner (1994) found that the rabbit count on a standard transect at Lake Urana, New South Wales, declined from 5000 in September 1951 to 50 in November 1951, following a myxomatosis epidemic. Of the survivors, serology showed that 75% had not been infected. From this information, the case mortality was estimated as 99.8%. The calculation would have run roughly as follows:

Rabbits dying between September and November:	$5000 - 50 = 4950$
Infected rabbits surviving:	$50 \times 0.25 = 12.5$
Total rabbits infected:	$4950 + 12.5 = 4962.5$
Case mortality:	$(4950/4962.5) \times 100$
	$= 99.75\%$

This calculation can be criticized on several grounds. It is assumed that all rabbit deaths over the two-month period were due to myxomatosis, it is assumed that there were no additions to the population, the spotlight counts are only an imprecise index, etc. Nevertheless, the data are perfectly adequate to show that the case mortality rate was very high indeed. One would need to be careful in applying the above logic to a disease with a much lower mortality rate, or to a case where there was more time between the initial and final population estimates.

Hudson *et al.* (1992b) used nine years of data on a grouse population in the north of England to obtain the relationship shown in Fig. 10.7 between mean worm burden and winter mortality. They fitted a linear regression to these data, and estimated α from the gradient. As Hudson *et al.* (1992b) discuss, this approach could be criticized on the grounds that this relationship is a correlation, rather than a demonstration of cause and effect. A more important issue limiting the application of this approach to other populations is that it could be used only because a long time series was available, in which the mean parasite burden varied over an order of magnitude. Unfortunately, such data are extremely unusual.

A much more controversial suggestion is that it is possible to infer macroparasite-induced mortality from the distribution of macroparasites within the host population. This idea was first proposed by Crofton (1971). The idea is that mortality amongst the most heavily infected individuals in a

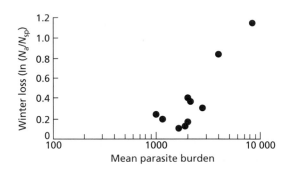

Fig. 10.7 Annual winter loss from a grouse population as a function of mean parasite burden. The annual winter loss for the years 1980–89 is shown as $\ln(N_a/N_{sp})$, where N_a is the number of breeding adults in autumn, and N_{sp} is the number in the following spring. The horizontal axis is the mean parasite burden, as determined by samples taken from the subsequent shooting season. The slope of a linear regression of winter loss on mean parasite burden was used as an estimate of α on an annual time scale. This assumes that almost all parasite-induced mortality occurred over the winter period. From Hudson *et al.* (1992b).

population will remove the upper tail of the frequency distribution of parasites amongst hosts. If this truncation can be detected and quantified, it can be used to detect parasite-induced mortality. Crofton (1971) noted that parasite distributions amongst hosts are usually negative binomial, and suggested that the 'missing tail' could be detected by fitting a truncated negative binomial.

There are two principal problems with this idea. First, the method relies on the assumption that the distribution of parasites on hosts actually is negative binomial before the parasite-induced mortality truncates it. The negative binomial certainly fits most such distributions quite well, but this does not mean that the distributions are in fact negative binomial. Second, there are many other factors that may truncate parasite distributions on hosts. Anderson and Gordon (1982) investigated this problem in considerable detail using simulation models. They concluded that parasite-induced mortality would cause the mean parasite burden to be highest in hosts of intermediate age, with both mean parasite burden and the degree of aggregation declining in older hosts, but found that age-related changes in the strength of host resistance, age-related changes in the rate of infection, or density-dependent death of parasites within the host could produce identical patterns. More recently, Rousset *et al.* (1996) have reproduced some of these results analytically, but as they set the death rate of parasites on hosts to zero, their results provide no more guidance on how to resolve truncation caused by parasite-induced mortality from the other factors listed above. In summary, no method for inferring parasite-induced mortality from the distribution of parasites amongst hosts can be recommended.

Nevertheless, the approach has been used several times, with increasing sophistication. For example, Lanciani and Boyett (1980) found that the negative binomial was a good fit to the distribution of mites on the aquatic insect *Sigara ornata*, on which they have little effect, but an untruncated distribution was a poor fit to the distribution of mites on *Anopheles crucians* larvae, on which laboratory experiments indicate that mites have a major effect. Royce and Rossignol (1990) used the method to infer mortality caused by tracheal mites on honey bees. Adjei *et al.* (1986) fitted a negative binomial to the first few terms of the distribution of a cestode parasite on lizard fish (*Saurida* spp.), and then extrapolated this to infer a distribution tail that could have been truncated by host mortality.

Effects of infection on host fecundity

Far more attention has been given to estimating parasite and pathogen effects on host survival than on host fecundity. This is somewhat surprising. Models suggest that a pathogen that decreases the fecundity of its hosts may have a greater impact on the host population than one that increases mortality (McCallum & Dobson, 1995). Applying estimates of fecundity from laboratory experiments to wild populations is also more likely to be valid than is applying laboratory estimates of survival to wild populations.

As well as the possibility of outright infertility, the fecundity of infected hosts may be decreased, relative to uninfected hosts, in several ways. There may be a decrease in litter or clutch size, the interval between breeding events may increase, or there may be an increase in the interval between birth and first breeding. Delay in first breeding has a particularly large effect on the intrinsic rate of increase of a population (see Chapter 5), and so parasites that delay the onset of breeding (perhaps by slowing growth) may have a profound effect on the population dynamics of their host. In many ecological models, 'fecundity' parameters include early juvenile survivorship (see Chapter 5), so a decrease in juvenile survival because of maternal infection may also be included as a component of parasite impact on fecundity.

A laboratory study by Feore *et al.* (1997) into the effect of the cowpox virus on the fecundity of bank voles (*Clethrionomys glareolus*) and wood mice (*Apodemus sylvaticus*) nicely illustrates some points about disease impact on fecundity. In the UK, cowpox is endemic in these small rodents, but has no demonstrable effect on survival. Feore *et al.* (1997) dosed pairs of the rodents shortly after weaning, and compared reproductive parameters with control, sham-treated pairs. There were no differences either in the proportion of pairs that reproduced over the 120 days of the experiment, or in the litter size produced. However, infected pairs of both species took significantly longer to produce their first litter than did uninfected pairs. It is interesting that it did not seem to matter whether both parents, or only one parent, was infected. As single-parent infections only occurred when the virus failed to take in one

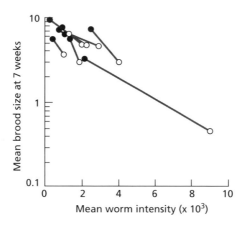

Fig. 10.8 Mean brood size of female grouse as a function of mean worm burden and treatment. The mean brood size (at seven weeks of age) of female grouse treated with anthelminthic (solid circles) is compared with the mean brood size of control females (open circles). Data points from the same year are connected with a line. The mean worm intensity was estimated from shot samples of both treated and control birds. From Hudson *et al.* (1992b).

parent, the corresponding sample sizes were very small, which would have limited the power of the experiment to detect any difference between dual-parent and single-parent infections.

Parasite reduction experiments can also be used to investigate parasite effects on fecundity. For example, Hudson *et al.* (1992b) treated red grouse females with anthelminthics over a period of eight years. In years of heavy infection, they found significant increases in both mean brood size and hatching success in treated females, compared with controls. The combined data (Fig. 10.8) allowed them to estimate the impact of the parasite on fecundity, as a function of parasite density. It is important to note that the intensity of infection in both treated and control birds could be estimated from birds shot as game.

Parasite parameters within the host

Most simple models of microparasitic infection do not attempt to model explicitly the dynamics of the parasite population within the host. However, it is still necessary to estimate the length of the latent period between a host being infected and becoming infectious, the duration of infection, and the rate at which immunity is lost. Some of these, of course, may be zero. It may also be necessary to estimate an incubation period, which is the time between the infection being acquired and it being detectable. This is not necessarily the same as the latent period. Each of these processes may potentially be quite complex and heterogeneous in some diseases (see Anderson & May, 1991). For example, some individuals may become asymptomatic carriers, which are infectious for a long period, but show few symptoms. The presence of carriers, even if rare, may profoundly alter the dynamics of a disease. In simple cases, however, latent periods, recovery rates, etc. can be estimated by direct observation of infected individuals.

The demographic parameters of parasites within their hosts are far more difficult to deal with in macroparasitic infections. In macroparasites, the rate of adult parasite mortality, parasite fecundity, maturation and growth rates often are all dependent on the intensity of infection in an individual host (Anderson & May, 1991). These density-dependent effects are frequently quite crucial to the overall dynamics of the macroparasite population. As the density dependence occurs within an individual host, its consequences for the population as a whole depend not only on the mean parasite burden, but also on the way in which the parasites are distributed between hosts.

Parasite survival within individual hosts can be measured in experimental situations by recording the decay in parasite infection levels within a host cohort that is no longer exposed to infection. The hosts may either be removed from a natural infective environment, or may be experimentally infected at a single time. Such experiments are rarely carried out in the field, because it is necessary to keep the hosts in an environment in which no further infection occurs. As there is evidence that the survival of parasites may depend strongly on the nutritional status of the host (Michael & Bundy, 1989), extrapolation of laboratory results to field populations should be done with caution. Host immune responses to macroparasites exist, but are poorly understood (Anderson & May, 1991; Grenfell *et al.*, 1995), introducing further complications.

Density-dependent fecundity in macroparasites is often detectable. It is probably the major factor regulating the level of macroparasitic infection (Anderson & May, 1991). Numerous studies have estimated the fecundity of gastrointestinal helminths by recording egg output in faeces as a function of worm burden, determined either by expulsion chemotherapy or by subsequent dissection of the hosts (Anderson & May, 1991; McCallum & Scott, 1994). Some examples of density-dependent fecundity in macroparasites are shown in Fig. 10.9. In general, there is a very high degree of variation in the egg output produced by hosts with a given parasite burden. There are several ways in which density dependence can be quantified. A regression of log(eggs per female worm per day) versus either worm burden itself (Fig. 10.9(a) and (c)) or log(worm burden) (Fig. 10.9(b)) can be used, with negative slope coefficients indicating density dependence. Alternatively, a simple regression of log(egg output) versus log(parasite burden) can be used (Fig. 10.9(e)), and the gradient of the log-log regression can be used to quantify density dependence (a slope less than 1 indicates direct density dependence). Some of the extreme

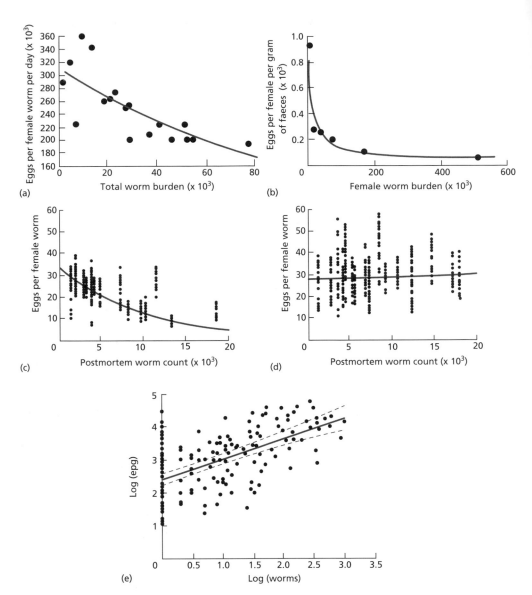

Fig. 10.9 Density-dependent fecundity in macroparasites. (a) *Ascaris lumbricoides* in humans in Malaysia. The solid line is an exponential model of the form $y = a\exp(-bx)$. Here, $a = 3.07 \times 10^5$ and $b = 0.007\,67$. From Anderson and May (1991). (b) Mixed hookworm infections (*Necator americanus* and *Ancylostoma duodenale*) in humans in India. The solid line is a power function of the form $y = ax^b$. Here, $a = 287.4$ and $b = -0.210$. From Anderson and May (1991). (c) Nematodes (*Ostertagia* spp.) in experimentally infected Soay sheep (*Ovis aries*). Worm counts were determined *post mortem*, when the number of eggs *in utero* of 20 randomly selected female worms was also counted. The solid line is an exponential model $y = a\exp(-bx)$. From Gulland and Fox (1992). (d) As for (c), but in wild sheep that had died during a population crash. Here, there is no evidence of density dependence in fecundity ($F_{1,398} = 0.33$, $p > 0.5$), despite both worm counts and eggs per female covering similar ranges as in (c). (e) *Opisthorchis viverrini* in humans in Thailand. The worm burden was determined after chemotherapy to expel worms, and is expressed here as $\log_{10}$. The vertical axis is eggs per gram of faeces, collected before chemotherapy, also expressed as $\log_{10}$. The linear regression shown has an intercept of 2.389, and a slope of 0.637. From McCallum and Scott (1994), using data from M. Haswell-Elkins.

variation in parasite fecundity at a given worm burden may be caused by short-term variation in egg output, but Figs 10.9(c) and (d) show that there is still a lot of variation in fecundity, even if the eggs per female worm can be counted directly.

Basic reproductive number R_0 of a parasite

The basic reproductive number is often the critical property of a host–parasite interaction that needs to be measured for predictive purposes. If your object-ive is to eliminate a pathogen or parasite from a population, the problem is one of reducing R_0 to below 1. Conversely, if a pathogen or parasite is con-templated for biological control, a necessary (but not sufficient) condition for success is that $R_0 > 1$. As can be seen for simple models in eqns (10.4) and (10.12), R_0 can be defined as a function of more elementary parameters of the host–parasite interaction, together with host density. Similar expressions can be derived for more complex models.

One way to estimate R_0 is thus to estimate all the parameters that con-tribute to R_0 in a model appropriate to the particular host–parasite interaction, and then to combine them. This approach has been taken in a number of studies (e.g. Lord *et al.*, 1995). The problem with this approach is not only that many separate parameters must be estimated, but also that each parameter contributing to the final R_0 is estimated with uncertainty, and probably is estimated by a different method. Attempting to place some sort of confidence bound around the final estimate is therefore difficult. Woolhouse *et al.* (1996) addressed this problem by conducting a sensitivity analysis, in which each parameter in turn was varied over its plausible range, and the effect on the estimated R_0 was investigated. Sanchez and Blower (1997) used a much more formal sensitivity analysis, with Latin hypercube sampling (see Chapter 2), to assess the uncertainty of an estimate of R_0 for tuberculosis in humans. This is the ideal way to approach the problem.

A second way to estimate R_0 is to do so directly. This may be possible, provided the system can be assumed to be in equilibrium: the approach is appropriate for endemic but not for epidemic infections. By definition, at equi-librium, R, the actual mean number of secondary cases per current case, is 1. In a microparasitic infection, if the population is homogeneously mixed, the number of secondary cases produced by an infective individual will be propor-tional to the probability that any contact is with a susceptible individual (Anderson & May, 1991). Thus, $R = R_0 x$, where x is the proportion of suscept-ible individuals in the population. Hence, at equilibrium.

$$R_0 = 1/x^*, \tag{10.26}$$

where x^* is the proportion of susceptible individuals in the population at equilibrium. This proportion can often be estimated using serological techniques, providing a quick and approximate estimate of R_0. The critical limitation is that the system must be at equilibrium. It is not easy to determine whether this is so, particularly in a wildlife population that may not have been monitored continuously for a long period.

A similar approach can be taken with macroparasites. At equilibrium, in the host population as a whole, each parasite must replace itself with one other parasite in the next generation. Hence,

$$R_0 f(M^*) = 1. \tag{10.27}$$

Here, f is some function describing the joint effect of all the density-dependent factors influencing the parasite population, and M^* is the equilibrium mean parasite burden. If all density dependence derives from density dependence in egg output, Anderson and May (1991) show that

$$f(M) = \left\{ 1 + \frac{M(1 - \exp[-\gamma])}{k} \right\}^{-(k+1)} \tag{10.28}$$

where k is the parameter of the negative binomial distribution that inversely describes the degree of aggregation and γ is the exponent in the following relationship between per capita mean parasite fecundity $\lambda(i)$ and parasite burden i:

$$\lambda(i) = \lambda_0 \exp(-\gamma(i - 1)). \tag{10.29}$$

Provided estimates of k, M^* and γ can be obtained, R_0 can then be estimated as

$$R_0 = 1/f(M^*). \tag{10.30}$$

Threshold population density necessary for disease transmission, N_T

N_T can be calculated from its component parameters, using eqns (10.5) or (10.13), or modified versions for more complex models. As with R_0, there are potential problems in combining estimates of a large number of parameters, each estimated with uncertainty. Appropriate sensitivity analysis should always be carried out. In some cases, it may be possible to estimate N_T, given an estimate of R_0. For example, in the simple, directly transmitted microparasite model described by eqns (10.1)–(10.3),

$$N_T = N_R/R_0, \tag{10.31}$$

where N_R is the number of animals in the population from which the estimate of R_0 was obtained. More generally, R_0 is a nonlinear function of population

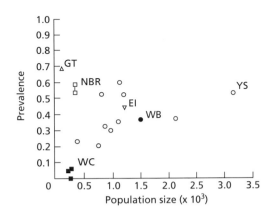

Fig. 10.10 The relationship between seroprevalence of *Brucella* and herd size in bison. The data were obtained from bison herds in the following North American national parks: Yellowstone (YS), the National Bison Range (NBR), Wind Cave (WC), Grand Teton (GT), Wood Buffalo (WB) and Elk Island (EI). From Dobson and Meagher (1996).

size (Anderson & May, 1991), and a model of the particular interaction would need to be examined to see what the predicted relationship between R_0 and N_T was, and parameters in addition to R_0 might be required to estimate N_T.

For example, for the simple macroparasite model described by eqns (10.6)–(10.8), the following relationship between N_T and R_0 can be obtained from eqns (10.12) and (10.13):

$$N_T = \frac{\frac{\mu}{\beta}}{R_0(1 + \frac{\mu}{\beta N_R}) - 1}. \tag{10.32}$$

So, to calculate N_T from R_0, you would need an estimate of μ/β, the ratio of the infective-stage death rate to the transmission rate, which is the proportion of infective stages that die before they are picked up by a host, and an estimate of N_R, which is the host population size from which the estimate of R_0 was obtained.

Sometimes, data on the presence or absence of the infection in populations of various sizes can be used to infer directly the size of N_T. For example, Dobson and Meagher (1996) obtained data on seroprevalence of the bacterial infection brucellosis amongst bison herds in Yellowstone National Park and other reserves. They observed that brucellosis tended not to be present when herds contained less than 200 individuals, and inferred that the threshold host density was therefore around 200. Clearly, this is a very approximate technique, and the data are not entirely convincing (Fig. 10.10), but it does provide a quick means of obtaining a rough estimate.

Summary and recommendations

1   The problems of parameter estimation are quite different for microparasites (viruses, bacteria and protozoa) and macroparasites (typically helminths).

2   For macroparasites, several of the key processes are functions of the parasite burden on individual hosts. This means that the way in which the parasites are distributed within the host population is very important.

3   The process of transmission is particularly difficult to parameterize. The relationship between the level of infection and host age is often useful in estimating the rate at which hosts acquire infection in a given environment.

4   There are substantial problems in applying an estimate of a transmission rate obtained in one location to another location, especially if there are major differences in scale or the extent of habitat heterogeneity. In particular, it is difficult to apply estimates of transmission rates obtained in the laboratory to field situations.

5   Parasite or pathogen effects on host death or fecundity are most appropriately estimated in the field. However, it is difficult to do so without manipulative experiments, particularly for endemic infections that do not change substantially in intensity through time. Observational methods based on the truncation of macroparasite frequency distributions cannot be recommended.

6   The basic reproductive number R_0 is a key parameter in both microparasite and macroparasite infections. It can be estimated from a combination of more basic parameters, but there are then difficulties in assessing the uncertainty in the resulting estimate. If infections, whether microparasitic or macroparasitic, can be assumed to be at equilibrium, there are simple, approximate ways of estimating R_0 directly. R_0 is a function of population density.

7   By far the best-understood macroparasite system in wildlife is the red grouse–nematode interaction studied by Hudson, Dobson and colleagues. This study is based on at least 20 years' work in a species for which large shot samples can be obtained. There are few short-cuts to producing precise parameter estimates for parasite models.

CHAPTER 11

The state of the art

Introduction

Reliable, statistically sound methods are now available to estimate population size, the survival rate and fecundity of individuals, and the growth rate of populations. I reviewed these methods in Chapters 3, 4 and 5. Given a particular ecological situation, it is possible to make clear recommendations as to which of these methods is the most appropriate to use. Some of these methods, those for mark–recapture data in particular, require quite sophisticated and elaborate computer programs for their analysis. These programs are readily available on the World Wide Web. There is no doubt that the current rapid development of methods better able to handle the complexities of real data will continue. It will always be worth checking the major link sites for statistical analysis of wildlife data for the most recent developments.

The critical issue in applying methods that estimate parameters associated with a particular population at a particular time in models is to ensure that these are the most appropriate parameters for the specific question the model is intended to address. For example, survival rates measured in a wild population of a particular species might well be a good description of the survival rates in that particular population for the time period in question. They would probably be entirely inappropriate to use in an attempt to estimate the maximum potential rate of increase of that species, even in a similar environment, as the measured survival rates would be likely to include density-dependent components.

Most of the second part of the book discusses the estimation of parameters to describe interactions. Chapter 6 describes the estimation of parameters concerned with density dependence, or interactions within species, and Chapters 8, 9 and 10 deal with interactions between species. In most cases, estimating interaction parameters is a matter of applying one of the basic parameter estimation techniques described in the first part of the book to one species, at differing values of some other variable, such as population size of the same species (for density dependence) or another species (for interspecific interactions). This process is not straightforward, for a number of reasons that are discussed throughout the book, and are reiterated below. In general, it is not possible to make prescriptive recommendations about the best methods for estimating particular interaction parameters. The principal problems are in experimental design, and the constraints on the design will depend on the use to which the parameters are to be put, the ecology of the organisms in question, and the resources available to the investigator. I have therefore taken a more descriptive approach in the second part of the book, critically reviewing methods that have been used in practice, and with which there have been varying degrees of success. Where possible, I have made suggestions and recommendations, but readers will have to develop their own experimental methods, for their own situation, in the light of the cases I have discussed.

Observation and process error

A recurring theme, particularly in the latter part of the book, is that most ecological data contain errors, or uncertainties, of two types. To reiterate, process error is uncertainty that occurs in the actual values of variables in ecological systems. For example, consider the type II functional response introduced in Table 9.1,

$$f(N) = \frac{\alpha N}{1 + \frac{\alpha}{\beta}N}. \tag{11.1}$$

This equation describes the number of prey eaten per unit of time, per predator, as a function of prey density N. At low densities the prey consumption rate increases linearly with prey density, with a slope α, but at high prey densities the consumption rate tends to β per unit of time. If the actual number of prey taken by a particular individual predator per unit of time was recorded at a variety of prey densities, the points would be most unlikely to fall precisely along the curve described by eqn (11.1), even if the equation was a good description of the functional response in that situation. There would be random variation in the number of prey caught at any prey density, the parameters themselves might vary between sampling times, or the functional response might vary somewhat from eqn (11.1). All these sources of variation contribute to process error: the number of prey actually eaten per unit of time will vary from that predicted by the equation. There may also be observation error: the recorded number of prey eaten per unit of time may be only an estimate, with uncertainty, of the actual number of prey consumed. There may also be observation error in the independent (or predictor) variable, in this case, the population size of the prey N.

As I discuss in Chapters 5, 6, 8 and 9, the distinction between process and observation error is very important when dealing with time series data. Quite different estimation procedures need to be applied in cases in which process errors predominate than where estimation errors predominate. Process errors propagate in a time series: the variables themselves are affected by the error, so that process error at one time will influence the rest of the series. Observation errors do not propagate: the error is in the recorded value, and has no impact on the actual value of the time series, and thus no influence on the remainder of the series. In general terms, methods designed to handle process errors predict changes in the variables over one time step, whereas those for observation error aim to predict the actual values of the variables at each time they are recorded. More detail can be found in the relevant chapters.

Unfortunately, almost all ecological data will contain both process and observation error. If the probability distribution of the observation errors is known, or their variance can be estimated separately, it is technically possible to fit a model with both observation and process errors. The papers by Carpenter *et al.* (1994), discussed in Chapter 9, and Pascual and Kareiva (1996), discussed in Chapter 8, show how this might be done, although the details are beyond the scope of this book. Replicated observations of the same time series at each point where the system is measured can provide an idea of the size of the observation error, independent of the process error. This is always useful information. Failing this, it is worth comparing the estimates from methods assuming only process error with estimates from methods assuming only observation error. As a crude generalization, for most ecological systems, and most estimation methods, process error is likely to be more important than observation error. Development of easy-to-use methods for the analysis of time series containing both process and observation error is another area in which there should be progress over the next few years.

Observation error is also important when it occurs in predictor variables. It is rare that the value of any ecological variable, including population size or density, can be measured without considerable error. I discussed the problem, but not possible solutions, in Chapters 6, 8, 9 and 10. There is a large statistical literature on the subject, where the issue is usually called 'measurement error' or 'errors-in-variables'. Fuller (1987) summarizes the literature for linear models, and Carroll *et al.* (1995) extend the discussion to nonlinear models.

In most cases, if the objective is to predict the value of a variable Y, given a set of predictor variables $X_1, \ldots, X_k$, observation error does not change the method of analysis or its interpretation (Carroll *et al.*, 1995). However, this book is concerned with parameter estimation, in which case there is a real problem. The simplest case is a univariate linear regression, where observation errors in the predictor are unbiased and are uncorrelated with the error in the response. Observation error then causes 'bias towards the null': whatever the true relationship is between the predictor and response, it tends to be lessened and obscured by measurement error. In more complicated cases, the true relationship may be distorted in other ways. For example, in multiple regression problems where the predictor variables are correlated, observation error in one or more of the predictor variables may result in overestimation of the effect of the predictor, or may even cause the sign of its effect to be estimated incorrectly. Nonlinear models with observation errors behave in a broadly similar way, but the effects of observation error are more complex still (Carroll *et al.*, 1995). Some generalizations can be made about the likely effects of observation error for some classes of models. The best advice is to follow the suggestion in Hilborn and Mangel (1997), and to simulate data with properties like those of your system, thus exploring the consequences of observation error empirically.

There are several possible ways to correct for observation error bias. These are discussed in Fuller (1987) and Carroll *et al.* (1995). One approach, which can be applied to a wide range of generalized linear models, including linear and logistic regression, is regression calibration (Carroll *et al.*, 1995). The simplest version is applicable in the following situation:

(i)   There is one predictor variable X measured with error, such that what is recorded is the random variable $W = X + U$, where U is the measurement error. U has a variance σ_u^2, which is estimated by $\hat{\sigma}_u^2$.

(ii)   There may be several covariates (other predictor variables) Z, but these are measured without error.

(iii)   The objective is to predict a response Y, via a linear predictor of the form $\beta_0 + \beta_x X + \boldsymbol{\beta}_z^t \mathbf{Z}$. Here, X is a scalar variable, $\mathbf{Z}$ is a column vector of covariates, β_0 and β_x are scalar parameters, and $\boldsymbol{\beta}_z^t$ is a row vector of regression coefficients. This linear predictor may generate the predicted response via one of a range of link functions (see Chapter 2). The problem is to estimate β_x.

(iv) X, $\mathbf{Z}$ and W are normally distributed.

The regression calibration estimator, $\hat{\beta}_x$, can then be obtained by the following steps:

1   Obtain the 'naïve' estimator $\hat{\beta}_x^*$, by fitting the model, ignoring the measurement error.

2   Regress W, the predictor with measurement error, on the other covariates $\mathbf{Z}$. From this regression, obtain the mean square error. Call this $\hat{\sigma}_{w|z}^2$. If there are no covariates $\mathbf{Z}$, this will just be the sample variance of W.

3   Calculate the corrected parameter estimate

$$\hat{\beta}_x = \frac{\hat{\beta}_x^* \hat{\sigma}_{w|z}^2}{\hat{\sigma}_{w|z}^2 - \hat{\sigma}_u^2}.$$

This correction requires an estimate of the observation error variance $\hat{\sigma}_u^2$. In some cases, it may readily be available. For example, if the problem is to estimate the effect of population size on a parameter, all the methods described in Chapter 3 will return an estimate of the standard error of the estimated population size. This can simply be squared to give an estimate of the error variance. However, a possible problem in this situation is that it is quite likely that the error will not be additive with constant variance, and may well be non-normal. Carroll *et al.* (1995) discuss some possible solutions, but these are beyond the scope of this book.

Spatial structure and scale

There is no doubt that models dealing explicitly with spatial structure and dynamics will increasingly be of importance both to theoretical ecology and for answering practical problems of ecological management (Levin *et al.*, 1997). I discuss methods for estimation of spatial parameters in Chapter 7, but this is again an area in which rapid developments can be expected over the next few years.

Even if spatial structure is not built explicitly into models, it introduces substantial problems into estimation of parameters, particularly the interaction parameters discussed in Chapters 9 and 10. The problem is straightforward: interactions are more likely to occur between close neighbours than between individuals that are far apart. If spatial separation is not included explicitly in a model, an estimate of the mean value of an interaction parameter is a function of the probability distribution of separation distances (or contact frequencies) of the interacting entities. This causes major problems if the spatial scale of the system in which the parameters are estimated differs from the spatial scale at which the parameter estimates are to be used. Further, even at the same spatial scale, the way in which contact frequencies scale with population size is not easy to determine (see the discussion of 'true' and 'pseudo-' mass action in Chapter 10).

A solution to this problem is to build interaction models in which the spatial relationships between the interacting entities are dealt with explicitly. I briefly discuss some host–pathogen models of this type in Chapter 10. The obvious difficulty is that the model itself becomes considerably more complex.

Levels of abstraction: the mismatch between data and models

The problem I have just discussed is an example of a much wider issue in model parameterization. The level of abstraction at which many models are built is different from the level of abstraction at which parameters can easily be estimated. A simple example occurs in Chapter 4. Models that do not include age structure use a single parameter to describe the average birth rate of individuals in a population. However, in no species can individuals reproduce immediately after birth. The average fecundity in a population is a function of at least four separate parameters: the average time from birth until individuals become reproductive; the proportion of individuals that survive to become reproductive; the proportion of individuals that reproduce once that age is reached; and the number of offspring produced per reproductive individual. These are the quantities that potentially can be estimated from an actual population, but they must be combined into a single parameter for use in a model without age structure. To do so, it is necessary to make simplifying assumptions (for example, that the population has reached a stable age distribution). These assumptions inevitably involve loss of information. It is also necessary to use one model (in this case, an age-structured model) to generate the summary parameter from the data. You need to ask whether it is better to use an age-structured model throughout the modelling process, rather than to use an age-structured model to generate a summary parameter for use in a non-age-structured model. The answer to this question depends on the intended use of the model. It is far easier to examine the qualitative behaviour of a non-age-structured model algebraically than it is to do so for an age-structured model. However, the age-structured model, with parameters directly related to the data, will probably better predict the quantitative behaviour of the system. If you only intend to analyse the model numerically, modern computing power is such that there is little point in using an over-simplified model structure.

The role of brute-force computing

The rate at which computer processing speed and memory capacity have increased over the last few years will be well known to every reader. Modelling approaches that were not feasible 10 years ago, because of the computer resources required, can now be used on a laptop computer. There is little doubt that methods that are impractical now will become commonplace in the near future. We can expect numerically intensive methods to be increasingly practical, both for parameter estimation (see bootstraps and jackknives in Chapter 2), and for analysing models themselves (for example, individual-based models, see Chapter 2; and spatially explicit models, see Chapter 7). In both cases, it will be increasingly unnecessary to make simplifying assumptions for the sake of mathematical tractability alone. Paradoxically, in many cases this means that the models can be conceptually much simpler. Rather than needing to assume a complex functional form for the probability distribution of some property in a population, one can model individuals or sampled units directly, and allow the computing power to take care of working out empirically what the probability distribution should be.

The challenge for the future is to use increased computer power wisely. As I stated at the outset of the book, the skill in model building is to produce a model that is as simple as possible, but which still performs its task. The ability to analyse highly complex models may tempt ecologists into building over-elaborate models that require a plethora of parameter estimates, and from which it is impossible to obtain general insights. For example, Ruckelshaus *et al.* (1997) simulated spatially explicit models, and concluded that they were 'staggeringly sensitive to the details of dispersal behaviour'. Alternatively, we should be able to develop models that are powerful, elegant and general, but are freed from the constraints of computational tractability. The best strategy for progress in ecological model building is to model a system at several different levels of abstraction and aggregation, and then to compare the results. In this way, you can determine the appropriate level of complexity for solution of a given problem.

References

Abrams, P.A. (1994) The fallacies of 'ratio-dependent' predation. *Ecology* **75**, 1842–50.

Abramsky, Z., Bowers, A. & Rosenzweig, M.L. (1986) Detecting interspecific competition in the field: testing the regression model. *Oikos* **47**, 19–209.

Abramsky, Z., Rosenzweig, M.L. & Pinshow, B. (1991) The shape of a gerbil isocline measured using principles of optimal habitat selection. *Ecology* **72**, 329–40.

Abramsky, Z., Rosenzweig, M.L. & Zubach, A. (1992) The shape of a gerbil isocline: an experimental field study. *Oikos* **63**, 193–9.

Abramsky, Z., Ovadia, O. & Rosenzweig, M.L. (1994) The shape of a *Gerbillus pyramidium* (Rodentia: Gerbillinae) isocline: an experimental field study. *Oikos* **69**, 318–26.

Ades, A.E. & Nokes, D.J. (1993) Modeling age- and time-specific incidence from seroprevalence: toxoplasmosis. *American Journal of Epidemiology* **137**, 1022–34.

Adjei, E.L., Barnes, A. & Lester, R.J.G. (1986) A method for estimating possible parasite-related host mortality, illustrated using data from *Callitetrarhynchus gracilis* (Cestoda: Trypanorhyncha) in lizardfish (*Saurida* spp.). *Parasitology* **92**, 227–43.

Agresti, A. (1994) Simple capture–recapture models permitting unequal catchability and variable sampling effort. *Biometrics* **50**, 494–500.

Akaike, H. (1985) Prediction and entropy. In: *A Celebration of Statistics: The ISI Centenary Volume* (eds Atkinson, A.C. & Fienberg, S.E.), pp. 1–24. Springer-Verlag, Heidelberg.

Akçakaya, H.R. (1992) Population cycles of mammals: evidence for a ratio-dependent predation hypothesis. *Ecological Monographs* **62**, 119–42.

Akçakaya, H.R. & Ferson, S. (1990) *RAMAS/space User Manual: Spatially Structured Population Models for Conservation Biology.* Applied Biomathematics/Exeter Software, New York.

Akçakaya, H.R., Arditi, R. & Ginsberg, L.R. (1995) Ratio dependent predation: an abstraction that works. *Ecology* **76**, 995–1004.

Allee, W.C., Emerson, A.E., Park, O., Park, T. & Schmidt, K.P. (1949) *Principles of Animal Ecology.* Saunders, Philadelphia.

Allen, L., Engeman, R. & Krupa, H. (1996) Evaluation of three relative abundance indices for assessing dingo populations. *Wildlife Research* **23**, 197–206.

Anderson, R.M. (1978) Population dynamics of snail infection by miracidia. *Parasitology* **77**, 201–24.

Anderson, R.M. (1979) Parasite pathogenicity and the depression of host population equilibria. *Nature* **279**, 150–2.

Anderson, R.M. (1980) Depression of host population abundance by direct life cycle macroparasites. *Journal of Theoretical Biology* **82**, 283–311.

Anderson, R.M. & Gordon, D.M. (1982) Processes influencing the distribution of parasite numbers within host populations, with special emphasis on parasite-induced host mortalities. *Parasitology* **85**, 373–98.

Anderson, R.M. & May, R.M. (1978) Regulation and stability of host–parasite interactions. I. Regulatory processes. *Journal of Animal Ecology* **47**, 219–47.

Anderson, R.M. & May, R.M. (1979) Population biology of infectious diseases. Part I. *Nature* **280**, 361–7.

Anderson, R.M. & May, R.M. (1985) Helminth infections of humans: mathematical models, population dynamics, and control. *Advances in Parasitology* **24**, 1–99.

Anderson, R.M. & May, R.M. (1991) *Infectious diseases of humans.* Oxford University Press, Oxford.

Anderson, R.M. & Nokes, D.J. (1991) Mathematical models of transmission and control. In: *Oxford Textbook of Public Health*, Vol. 2 (eds Holland, W.W., Detels, R. & Knox, G.), 2nd edn, pp. 225–52. Oxford University Press, Oxford.

Anderson, R.M., Whitfield, P.J. & Dobson, A.P. (1978) Experimental studies of infection dynamics: infection of the definitive host by the cercariae of *Transversotrema patialense*. *Parasitology* **77**, 189–200.

Anderson, R.M., Jackson, H.C., May, R.M. & Smith, A.M. (1981) Population dynamics of fox rabies in Europe. *Nature* **289**, 765–71.

Andreasen, V. & Christiansen, F.B. (1995) Slow coevolution of a viral pathogen and its diploid host. *Philosophical Transactions of the Royal Society of London Series B* **348**, 341–54.

Andrew, N.L. & Mapstone, B.D. (1987) Sampling and the description of spatial pattern in marine ecology. *Oceanography and Marine Biology Annual Review* **25**, 39–90.

Arditi, R. & Saïah, H. (1992) Empirical evidence of the role of heterogeneity in ratio-dependent consumption. *Ecology* **73**, 1544–51.

Bailey, N.T.J. (1975) *The Mathematical Theory of Infectious Diseases and Its Applications*, 2nd edn. Griffin, London.

Baltensweiler, W. & Fischlin, A. (1987) The larch budmoth in the Alps. In: *Dynamics of Forest Insect Populations: Patterns, Causes, Implications* (ed. Berryman, A.A.), pp. 331–51. Plenum, New York.

Banse, K. & Mosher, S. (1980) Adult body mass and annual production/biomass relationships of field populations. *Ecological Monographs* **50**, 355–79.

Barker, R.J. & Sauer, J.R. (1992) Modelling population change from time series data. In: *Wildlife 2001: Populations* (eds McCullough, D.R. & Barrett, R.H.), pp. 182–94. Elsevier Applied Science, London.

Barlow, N.D. (1991) Control of endemic bovine TB in New Zealand possum populations: results from a simple model. *Journal of Applied Ecology* **28**, 794–809.

Barlow, N.D. (1997) Modelling immunocontraception in disseminating systems. *Reproduction, Fertility and Development* **9**, 51–60.

Barrowclough, G.F. (1978) Sampling bias in dispersal studies based on finite area. *Bird-banding* **49**, 333–41.

Bartlett, M.S. (1961) Monte Carlo studies in ecology and epidemiology. In: *Proceedings of the Fourth Berkeley Symposium on Mathematical Statistics and Probability*, Vol. 4 (ed. Neyman, J.), pp. 39–55. University of California Press, Berkeley.

Bates, D.M. & Watts, D.G. (1988) *Nonlinear Regression Analysis and Its Applications*. Wiley, New York.

Bayliss, P. (1987) Kangaroo dynamics. In: *Kangaroos: Their Ecology and Management in the Sheep Rangelands of Australia* (eds Caughley, G., Shepherd, N. & Short, J.), pp. 119–34. Cambridge University Press: Cambridge.

Beck, K. (1984) Coevolution: Mathematical analysis of host–parasite interactions. *Journal of Mathematical Biology* **19**, 63–77.

Beddington, J.R. & Basson, M. (1994) The limits to exploitation on land and sea. *Philosophical Transactions of the Royal Society of London Series B* **343**, 87–92.

Begon, M., Harper, J.L. & Townsend, C.R. (1990) *Ecology: Individuals, Populations and Communities*, 2nd edn. Blackwell Scientific, Boston.

Begon, M., Harper, J.L. & Townsend, C.R. (1996a) *Ecology*, 3rd edn. Blackwell Science, Oxford.

Begon, M., Mortimer, M. & Thompson, D.J. (1996b) *Population Ecology: A Unified Study of Animals and Plants*, 3rd edn. Blackwell Science, Oxford.

Begon, M., Feore, S.M., Brown, K., Chantrey, J., Jones, T. & Bennett, M. (1998) Population and transmission dynamics of cowpox in bank voles: testing fundamental assumptions. *Ecology Letters* **1**, 82–6.

Bellhouse, D.R. (1988) Systematic sampling. In: *Handbook of Statistics*, Vol. 6. *Sampling* (eds Krishnaiah, P.R. & Rao, C.R.), pp. 125–45. Elsevier, Amsterdam.

Bellows, T.S. (1981) The descriptive properties of some models for density dependence. *Journal of Animal Ecology* **50**, 139–56.

Bender, E.A., Case, T.J. & Gilpin, M.E. (1984) Perturbation experiments in community ecology: theory and practice. *Ecology* **65**, 1–13.

Berger, J.O. & Berry, D.A. (1988) Statistical analysis and the illusion of objectivity. *American Scientist* **76**, 159–65.

Bernoulli, D. (1760) Essai d'une nouvelle analyse de la mortalité causée par la petite Vérole, et des advantages de l'Inoculation pour la prévenir. *Mémoires de mathématique et de physique, tirés des registres de l'Académie Royale des Sciences, Paris,* 1–45.

Berryman, A.A. (1992) The origins and evolution of predator–prey theory. *Ecology* **73**, 1530–5.

Berryman, A.A., Gutierrez, A.P. & Arduti, R. (1995) Credible, parsimonious and useful predator–prey models – a reply to Abrams, Gleeson, and Sarnelle. *Ecology* **76**, 1980–5.

Bissell, A.F. & Ferguson, R.A. (1974) The jackknife – toy, tool or two-edged weapon? *The Statistician* **24**, 79–100.

Blueweiss, L., Fox, H., Kudzma, V., Nakashima, D., Peters, R. & Sams, S. (1978) Relationships between body size and some life history parameters. *Oecologia* **37**, 257–72.

Bolker, B.M. & Grenfell, B.T. (1993) Chaos and biological complexity in measles dynamics. *Proceedings of the Royal Society of London Series B* **251**, 75–81.

Bossart, J.L. & Prowell, D.P. (1998) Genetic estimates of population structure and gene flow: limitations, lessons and new directions. *Trends in Ecology and Evolution* **13**, 202–6.

Boudreau, P.R. & Dickie, L.M. (1989) Biological model of fisheries production based on physiological and ecological scalings of body size. *Canadian Journal of Fisheries and Aquatic Sciences* **46**, 614–23.

Boutin, S. (1995) Testing predator–prey theory by studying fluctuating populations of small mammals. *Wildlife Research* **22**, 89–100.

Box, G.E.P. & Cox, D.R. (1964) A analysis of transformation. *Journal of the Royal Statistical Society Series B* **26**, 211–52.

Boyce, M.S. (1992) Population viability analysis. *Annual Review of Ecology and Systematics* **23**, 481–506.

Brownie, C., Hines, J.E., Nichols, J.D., Pollock, K.H. & Hestbeck, J.B. (1993) Capture–recapture studies for multiple strata including non-Markovian transitions. *Biometrics* **49**, 1173–87.

Buckland, S.T., Anderson, D.R., Burnham, K.P. & Laake, J.L. (1993) *Distance Sampling: Estimating Abundance of Biological Populations.* Chapman & Hall, London.

Buckland, S.T., Burnham, K.P. & Augustin, N.H. (1997) Model selection: an integral part of inference. *Biometrics* **53**, 603–18.

Bulmer, M.G. (1975) The statistical analysis of density dependence. *Biometrics* **31**, 901–11.

Bundy, D.A.P., Cooper, E.S., Thompson, D.E., Didier, J.M. & Simmons, I. (1987) Epidemiology and population dynamics of *Ascaris lumbricoides* and *Trichuris trichiura* infection in the same community. *Transactions of the Royal Society of Tropical Medicine and Hygiene* **81**, 987–93.

Burgman, M.A., Ferson, S. & Akçakaya, H.R. (1993) *Risk Assessment in Conservation Biology.* Chapman & Hall, London.

Burnett, T. (1958) A model of a host–parasite interaction. In: *Proceedings of the 10th International Congress of Entomology,* Vol. 2 (ed. Becker, E.C.), pp. 679–86. Mortimer, Ottawa.

Burnham, K.P. & Anderson, D.R. (1992) Data-based selection of an appropriate biological model: the key to modern to data analysis. In: *Wildlife 2001: Populations* (eds McCullough, D.R. & Barrett, R.H.), pp. 16–30. Elsevier Applied Science, London.

Burnham, K.P., Anderson, D.R. & Laake, J.L. (1980) *Estimation of Density from Line Transect Sampling of Biological Populations.* Wildlife Monographs Vol. 72. The Wildlife Society, Bethesda, MD.

Burnham, K.P., Anderson, D.R. & White, G.C. (1995) Selection among open population models when capture probabilities are heterogeneous. *Journal of Applied Statistics* **22**, 611–24.

Burt, W.H. (1943) Territoriality and home range concepts as applied to mammals. *Journal of Mammalogy* **24**, 346–52.

Cairns, S.C. (1989) Models of macropodid populations. In: *Kangaroos, Wallabies and Rat Kangaroos* (eds Grigg, G.C., Jarman, P. & Hume, I.), pp. 695–704. Surrey Beatty & Sons, Chipping Norton, New South Wales.

Carey, J.R. (1989) The multiple decrement life table: a unifying framework for cause-of-death analysis in ecology. *Oecologia* **78**, 131–7.

Carey, J.R. (1993) *Applied Demography for Biologists with Special Emphasis on Insects.* Oxford University Press, New York.

Carpenter, S.R., Cottingham, K.L. & Stow, C.A. (1994) Fitting predator–prey models to time series with observational errors. *Ecology* **75**, 1254–64.

Carroll, R.J., Ruppert, D. & Stefanski, L.A. (1995) *Measurement Error in Nonlinear Models.* Chapman & Hall, London.

Caswell, H. (1989) *Matrix Population Models: Construction, Analysis, and Interpretation.* Sinauer Associates Inc., Sunderland, MA.

Caswell, H. & Etter, R.J. (1993) Ecological interactions in patchy environments: from patch-occupancy models to cellular automata. In: *Patch Dynamics* (eds Levin, S.A., Powell, T.M. & Steele, J.H.), pp. 93–109. Springer-Verlag, Berlin.

Caswell, H. & Twombly, S. (1989) Estimation of stage-specific demographic parameters for zooplankton populations: methods based on stage-classified matrix projection methods. In: *Estimation and Analysis of Insect Populations* (eds McDonald, L., Manly, B., Lockwood, J. & Logan, J.), pp. 93–107. Springer-Verlag, Berlin.

Caughley, G. (1977a) *Analysis of Vertebrate Populations.* Wiley, London.

Caughley, G. (1977b) Sampling in aerial survey. *Journal of Wildlife Management* **41**, 605–15.

Caughley, G. (1987) Ecological relationships. In: *Kangaroos: Their Ecology and Management in the Sheep Rangelands of Australia* (eds Caughley, G., Shepherd, N. & Short, J.), pp. 159–87. Cambridge University Press, Cambridge.

Caughley, G. (1993) Elephants and economics. *Conservation Biology* **7**, 943–5.

Caughley, G. & Gunn, A. (1993) Dynamics of large herbivores in deserts: kangaroos and caribou. *Oikos* **67**, 47–55.

Caughley, G. & Krebs, C.J. (1983) Are big mammals simply little mammals writ large? *Oecologia* **59**, 7–17.

Caughley, G. & Sinclair, A.R.E. (1994) *Wildlife Ecology and Management.* Blackwell Science, Boston.

Chao, A., Lee, S.M. & Jeng, S.L. (1992) Estimating population size for capture–recapture data when capture probabilities vary by time and individual animal. *Biometrics* **48**, 201–16.

Chase, J.M. (1996) Differential competitive interactions and the included niche: an experimental analysis with grasshoppers. *Oikos* **76**, 103–12.

Choquenot, D. (1993) Growth, condition and demography of wild banteng (*Bos javanicus*) on Cobourg Peninsula, northern Australia. *Journal of Zoology, London* **231**, 533–42.

Clark, C.W. (1973a) The economics of overexploitation. *Science* **181**, 630–4.

Clark, C.W. (1973b) Profit maximization and the extinction of animal species. *Journal of Political Economy* **81**, 950–61.

Clobert, J. & Lebreton, J.D. (1991) Estimation of demographic parameters in bird populations. In: *Bird Population Studies: Relevance to Conservation and Management* (eds Perrins, C.M., Lebreton, J.D. & Hirons, G.J.M.), pp. 75–104. Oxford University Press, New York.

Clobert, J., Lebreton, J.D. & Allaine, D. (1994) The estimation of age-specific breeding probabilities from recaptures or resightings in vertebrate populations: II. Longitudinal models. *Biometrics* **50**, 375–87.

Cochran, W.G. (1977) *Sampling Techniques,* 3rd edn. Wiley, New York.

Cochran, W.G. & Cox, G.M. (1957) *Experimental Designs,* 2nd edn. Wiley, New York.

Cock, M.J.W. (1977) *Searching behaviour of polyphagous predators.* PhD thesis, Imperial College, London.

Cole, L.C. (1954) The population consequences of life history phenomena. *Quarterly Review of Biology* **29**, 103–37.

Collett, D. (1991) *Modelling Binary Data.* Chapman & Hall, London.

Connell, J.H. (1983) On the prevalence and relative importance of interspecific competition: evidence from field experiments. *American Naturalist* **122**, 661–96.

Connor, E.F. & Simberloff, D. (1986) Competition, scientific method, and null models in ecology. *American Scientist* **74**, 155–62.

Conroy, M.J., Cohen, Y., James, F.C., Matsinos, Y.G. & Maurer, B.A. (1995) Parameter estimation, reliability, and model improvement for spatially explicit models of animal populations. *Ecological Applications* **5**, 17–19.

Cooch, E. & White, G. (1998) Using MARK – a gentle introduction. http://www.biol.sfu.ca/cmr/mark/.

Cormack, R.M. (1964) Estimates of survival from the sighting of marked animals. *Biometrika* **51**, 429–38.

Cormack, R.M. (1985) Examples of the use of GLIM to analyse capture–recapture studies. In: *Statistics in Ornithology* (eds Morgan, B.J.T. & North, P.M.), pp. 243–73. Springer-Verlag, New York.

Cormack, R.M. (1989) Log-linear models for capture–recapture. *Biometrics* **45**, 395–413.

Cormack, R.M. (1993) Variances of mark–recapture estimates. *Biometrics* **49**, 1188–93.

Cormack, R.M. (1994) Unification of mark–recapture analyses by loglinear modelling. In: *Statistics in Ecology and Environmental Monitoring* (eds Fletcher, D.J. & Manly, B.F.J.), pp. 19–32. University of Otago Press, Dunedin.

Cowlishaw, G. (1997) Trade-offs between foraging and predation risks determine habitat use in a desert baboon population. *Animal Behaviour* **53**, 667–86.

Cox, D.R. & Oakes, D. (1984) *Analysis of Survival Data.* Chapman & Hall, London.

Crawford, H.S. & Jennings, D.T. (1989) Predation by birds on spruce budworm *Choristoneura fumiferana*: functional, numerical and total responses. *Ecology* **70**, 152–63.

Crawley, M.J. (1983) *Herbivory: The Dynamics of Animal–Plant Interactions.* Blackwell Scientific, Oxford.

Crawley, M.J. (1992) Population dynamics of natural enemies and their prey. In: *Natural Enemies: The Population Biology of Predators, Parasites and Diseases* (ed. Crawley, M.J.), pp. 40–89. Blackwell Science, Oxford.

Crawley, M.J. (1993) *GLIM for Ecologists.* Blackwell Science, Oxford.

Crawley, M.J. (ed.) (1997) *Plant Ecology,* 2nd edn. Blackwell Science, Oxford.

Crofton, H.D. (1971) A quantitative approach to parasitism. *Parasitology* **62**, 179–94.

Crowell, K. & Pimm, S.L. (1976) Competition and niche shifts of mice introduced onto small islands. *Oikos* **27**, 251–8.

Crowley, P.H. (1992). Resampling methods for computation-intensive data analysis in ecology and evolution. *Annual Review of Ecology and Systematics* **23**, 405–47.

de Jong, M.C.M., Diekmann, O. & Heesterbeck, H. (1995) How does transmission of infection depend on population size? In: *Epidemic Models: Their Structure and Relation to Data* (ed. Mollison, D.), pp. 84–94. Cambridge University Press, Cambridge.

De Leo, G.A. & Dobson, A.P. (1996) Allometry and simple epidemic models for microparasites. *Nature* **379**, 720–2.

DeAngelis, D.L., Gross, L.J., Huston, M.A., Wolff, W.F., Fleming, D.M., Comiskey, E.J. & Sylvester, S.M. (1998) Landscape modelling for Everglades ecosystem restoration. *Ecosystems* **1**, 64–75.

Dennis, B. (1996) Discussion: should ecologists become Bayesians? *Ecological Applications* **6**, 1095–103.

Dennis, B. & Taper, M.L. (1994) Density dependence in time series observations of natural populations: estimation and testing. *Ecological Monographs* **64**, 205–24.

Dennis, B., Munholland, P.L. & Scott, J.M. (1991) Estimation of growth and extinction parameters for endangered species. *Ecological Monographs* **61**, 115–43.

Diaconis, P. & Efron, B. (1983) Computer-intensive methods in statistics. *Scientific American* **248**, 96–109.

Diamond, J. (1986) Overview. Laboratory experiments, field experiments and natural experiments. In: *Community Ecology* (eds Diamond, J.M. & Case, T.J.), pp. 3–22. Harper & Row, New York.

Dixon, K.R. & Chapman, J.A. (1980) Harmonic mean measure of animal activity areas. *Ecology* **61**, 1040–4.

Dobson, A.P. (1988) The population biology of parasite-induced changes in host behaviour. *Quarterly Review of Biology* **63**, 139–65.

Dobson, A.P. & Meagher, M. (1996) The population dynamics of brucellosis in the Yellowstone National Park. *Ecology* **77**, 1026–36.

Dunning, J.B.J., Stewart, D.J., Danielson, B.R., Noon, B.R., Root, T.L., Lamberson, R.H. & Stevens, E.E. (1995) Spatially explicit population models: current forms and future uses. *Ecological Applications* **5**, 3–11.

Edwards, A.W.F. (1972) *Likelihood*. Cambridge University Press, Cambridge.

Efron, B. & Tibshirani, R.J. (1993) *An Introduction to the Bootstrap*. Chapman & Hall, New York.

Elliott, J.M. & Persson, L. (1978) The estimation of daily rates of food consumption for fish. *Journal of Animal Ecology* **47**, 977–91.

Ellison, A.M. (1996) An introduction to Bayesian inference for ecological research and environmental decision-making. *Ecological Applications* **6**, 1036–46.

Ellner, S. & Turchin, P. (1995) Chaos in a noisy world: new methods and evidence from time-series analysis. *American Naturalist* **145**, 343–75.

Elton, C. (1933) *The Ecology of Animals*. Methuen, London.

Endean, R. & Stablum, W. (1975) Population explosions of *Acanthaster planci* and associated destruction of the hard-coral cover of reefs of the Great Barrier Reef, Australia. *Environmental Conservation* **2**, 247–56.

Erlinge, S., Göransson, G., Hansson, L., Högstedt, G., Liberg, O., Nilsson, I.N., Nilsson, T., von Schantz, T. & Sylvén, M. (1983) Predation as a regulating factor on small rodent populations in southern Sweden. *Oikos* **40**, 36–52.

Ewens, W.J., Brockwell, P.J., Gani, J.M. & Resinick, S.I. (1987) Minimum viable population size in the presence of catastrophes. In: *Viable Populations for Conservation* (ed. Soulé, M.E.), pp. 59–68. Cambridge University Press, Cambridge.

Fan, Y. & Petitt, F.L. (1994) Parameter estimation of the functional response. *Environmental Entomology* **23**, 785–94.

Farrington, C.P. (1990) Modelling forces of infection for measles, mumps and rubella. *Statistics in Medicine* **9**, 953–67.

Favre, L., Balloux, F., Goudet, J. & Perrin, N. (1997) Female-biased dispersal in the monogamous mammal *Crocidura russula*: evidence from field data and microsatellite patterns. *Proceedings of the Royal Society of London Series B* **264**, 127–32.

Fenchel, T. (1974) Intrinsic rate of natural increase: the relationship with body size. *Oecologia* **14**, 317–26.

Fenner, F. (1994) Myxomatosis. In: *Parasitic and Infectious Diseases: Epidemiology and Ecology* (eds Scott, M.E. & Smith, G.), pp. 337–46. Academic Press, San Diego, CA.

Fenner, F. & Ratcliffe, F.N. (1965) *Myxomatosis*. Cambridge University Press, Cambridge.

Feore, S.M., Bennett, M., Chantrey, J., Jones, T., Baxby, D. & Begon, M. (1997) The effect of cowpox virus infection on fecundity in bank voles and wood mice. *Proceedings of the Royal Society of London Series B* **264**, 1457–61.

Ferson, S. (1991) *RAMAS/stage: Generalized Stage-Based Modeling for Population Dynamics*. Applied Biomathematics: Setauket, NY.

Ferson, S. & Akçakaya, H.R. (1988) *RAMAS/age User Manual; Modeling Fluctuations in Age-Structured Populations*. Applied Biomathematics/Exeter Software: Setauket, New York.

Ferson, S., Ginzburg, L. & Silvers, A. (1989) Extreme event risk analysis for age-structured populations. *Ecological Modelling* **47**, 175–87.

Fitt, B.D.L. & McCartney, H.A. (1986) Spore dispersal in relation to epidemic models. In: *Plant Disease Epidemiology*, Vol. 1 (eds Leonard, K.J. & Fry, W.E.), pp. 311–45. Macmillan, New York.

Fowler, C.W. & Baker, J.D. (1991) A review of animal population dynamics at extremely reduced population levels. *Report of the International Whaling Commission* **41**, 545–54.

Fox, B.J. & Luo, J. (1996) Estimating competition coefficients from census data: a revisit of regression technique. *Oikos* **77**, 291–300.

Fox, B.J. & Pople, A.R. (1984) Experimental confirmation of interspecific competition between native and introduced mice. *Australian Journal of Ecology* **9**, 323–34.

Fox, D.R. & Ridsdillsmith, J. (1995) Tests for density dependence revisited. *Oecologia* **103**, 435–43.

Francis, R.I.C.C. (1988) Are growth parameters estimated from tagging and age-length data comparable? *Canadian Journal of Fisheries and Aquatic Sciences* **45**, 936–42.

Fretwell, S.D. (1972) *Populations in a Seasonal Environment.* Princeton University Press, Princeton, NJ.

Fuller, W.A. (1987) *Measurement Error Models.* Wiley, New York.

Gaston, K.J. (1988) The intrinsic rates of increase of insects of different sizes. *Ecological Entomology* **13**, 399–409.

Getz, W.M. & Swartzman, G.L. (1981) A probability transition matrix model for yield estimation in fisheries with highly variable recruitment. *Canadian Journal of Fisheries and Aquatic Sciences* **38**, 847–55.

Ginzburg, L.R., Ferson, S. & Akcakaya, H.R. (1990) Reconstructability of density dependence and the conservative assessment of extinction risks. *Conservation Biology* **4**, 63–70.

Gleeson, S.K. (1994) Density dependence is better than ratio dependence. *Ecology* **75**, 1834–5.

Godfray, H.C.J. (1994) *Parasitoids: Behavioral and Evolutionary Ecology.* Princeton University Press, Princeton, NJ.

Godfray, H.C.J. & Pacala, S.W. (1992) Aggregation and the population dynamics of parasitoids and predators. *American Naturalist* **140**, 30–40.

Godfray, H.C.J. & Waage, J.K. (1991) Predictive modelling in biological control: the mango mealy bug (*Rastrococcus invadens*) and its parasitoids. *Journal of Applied Ecology* **28**, 434–53.

Goodman, D. (1987) The demography of chance extinction. In: *Viable Populations for Conservation* (ed. M.E. Soulé), pp. 11–35. Cambridge University Press, Cambridge.

Goudet, J. (1995) FSTAT (Version 1.2): a computer program to calculate *F* statistics. *Journal of Heredity* **86**, 485–6.

Grenfell, B.T. & Anderson, R.M. (1985) The estimation of age-related rates of infection from case notifications and serological data. *Journal of Hygiene* **95**, 419–36.

Grenfell, B.T., Lonergan, M.E. & Harwood, J. (1992) Quantitative investigations of the epidemiology of phocine distemper virus (PDV) in European common seal populations. *Science of the Total Environment* **115**, 15–29.

Grenfell, B.T., Dietz, K. & Roberts, M.G. (1995) Modelling the immuno-epidemiology of macroparasites in naturally-fluctuating host populations. In: *Ecology of Infectious Diseases in Natural Populations* (eds Grenfell, B.T. & Dobson, A.P.), pp. 362–83. Cambridge University Press, Cambridge.

Grigg, G.C., Pople, A.R. & Beard, L.A. (1995) Movements of feral camels in central Australia determined by satellite telemetry. *Journal of Arid Environments* **31**, 459–69.

Guillemette, M. & Himmelman, J.H. (1996) Distribution of wintering common eiders over mussel beds: Does the ideal free distribution apply? *Oikos* **76**, 435–42.

Guinet, C., Khoudil, M., Bost, C.A., Durbec, J.P., Georges, J.Y., Mouchot, M.C. & Jouventin, P. (1997) Foraging behaviour of satellite-tracked king penguins in relation to sea-surface temperatures obtained by satellite telemetry at Crozet Archipelago, a study during three austral summers. *Marine Ecology – Progress Series* **150**, 11–20.

Gulland, F.M. (1992) The role of nematode parasites in Soay sheep (*Ovis aries* L.) mortality during a population crash. *Parasitology* **105**, 493–503.

Gulland, F.M. & Fox, M. (1992) Epidemiology of nematode infections of Soay sheep (*Ovis aries* L.) on St Kilda. *Parasitology* **105**, 481–92.

Gurney, W.S.C., Nisbet, R.M. & Lawton, J.H. (1983) The systematic formulation of tractable single-species population models incorporating age structure. *Journal of Animal Ecology* **52**, 479–95.

Hairston, N.G. & Twombly, S. (1985) Obtaining life table data from cohort analyses: a critique of current methods. *Limnology and Oceanography* **30**, 886–93.

Halley, J.M. (1996) Ecology, evolution and 1/f noise. *Trends in Ecology and Evolution* **11**, 33–8.

Hanski, I. (1994) A practical model of metapopulation dynamics. *Journal of Animal Ecology* **63**, 151–62.

Hanski, I. (1997) Metapopulation dynamics: from concepts and observations to predictive models. In: *Metapopulation Biology: Ecology, Genetics and Evolution* (eds Hanski, I. & Gilpin, M.E.), pp. 69–92. Academic Press: London.

Hanski, I. (1997) Predictive and practical metapopulation models: the incidence function approach In *Spatial Ecology* (eds Tilman, D. & Kareiva, P.), pp. 21–45. Princeton University Press, Princeton, NJ.

Hanski, I. & Korpimäki, E. (1995) Microtine rodent dynamics in northern Europe: parameterized models for the predator–prey interaction. *Ecology* **76**, 840–50.

Hanski, I., Hansson, L. & Henttonen, H. (1991) Specialist predators, generalist predators and the microtine rodent cycle. *Journal of Animal Ecology* **60**, 353–67.

Hanski, I., Turchin, P., Korpimäki, E. & Henttonen, H. (1993a) Population oscillations of boreal rodents: regulation by mustelid predators leads to chaos. *Nature* **364**, 232–5.

Hanski, I., Woiwood, I. & Perry, J. (1993b) Density dependence, population persistence, and largely futile arguments. *Oecologia* **95**, 595–8.

Hardman, J.R.P. (1996) *The wild harvest and marketing of kangaroos: a case study of the profitability of kangaroos compared with sheep/beef in Queensland.* Queensland Department of Primary Industries, Brisbane.

Harper, J.L. (1977) *Population Biology of Plants.* Academic Press, London.

Harrison, S. (1994) Metapopulations and conservation. In: *Large-Scale Ecology and Conservation Biology* (eds May, R.M. & Webb, N.R.), pp. 111–28. Blackwell Science, Oxford.

Harrison, S. & Cappuccino, N. (1995) Using density manipulation methods to study population regulation. In: *Population Dynamics: New Approaches and Synthesis* (eds Cappuccino, N. & Price, P.W.), pp. 131–47. Academic Press, San Diego, CA.

Hartl, D.L. & Clark, A.G. (1997) *Principles of Population Genetics*, 3rd edn. Sinauer Associates, Sunderland, MA.

Hassell, M.P. (1978) *The Dynamics of Arthropod Predator–Prey Systems.* Princeton University Press, Princeton, NJ.

Hassell, M.P. & Godfray, H.C.J. (1992) The population biology of insect parasitoids. In: *Natural Enemies: The Population Biology of Predators, Parasites and Diseases* (ed. Crawley, M.J.), pp. 265–92. Blackwell Science, Oxford.

Hassell, M.P., Lawton, J.H. & May, R.M. (1976) Patterns of dynamical behaviour in single-species populations. *Journal of Animal Ecology* **45**, 471–86.

Hassell, M.P., Latto, J. & May, R.M. (1989) Seeing the wood for the trees: detecting density dependence from existing life-table studies. *Journal of Animal Ecology* **58**, 883–92.

Hassell, M.P., Comins, H.N. & May, R.M. (1991a) Spatial structure and chaos in insect population dynamics. *Nature* **353**, 255–8.

Hassell, M.P., May, R.M., Pacala, S.W. & Chesson, P.L. (1991b) The persistence of host–parasitoid associations in patchy environments: I. A general criterion. *American Naturalist* **138**, 568–83.

Hassell, M.P., Comins, H.N. & May, R.M. (1994) Species coexistence and self-organizing spatial dynamics. *Nature* **370**, 290–2.

Hastings, A. (1996) Models of spatial spread: is the theory complete? *Ecology* **77**, 1675–9.

Henderson-Sellers, B. & Henderson-Sellers, A. (1993) Factorial techniques for testing environmental model sensitivity. In: *Modelling Change in Environmental Systems* (eds Jakeman, A.J., Beck, M.B. & McAleer, M.J.), pp. 59–76. Wiley, Chichester.

Hengeveld, R. (1994) Small-step invasion research. *Trends in Ecology and Evolution* **9**, 339–42.

Hennemann, W.W. (1983) Relationship among body mass, metabolic rate and the intrinsic rate of natural increase in mammals. *Oecologia* **56**, 104–8.

Heppell, S.S., Walters, J.R. & Crowder, L.B. (1994) Evaluating management alternatives for red-cockaded woodpeckers: a modeling approach. *Journal of Wildlife Management* **58**, 479–87.

Hilborn, R. & Mangel, M. (1997) *The Ecological Detective*. Princeton University Press, Princeton, NJ.

Hilborn, R. & Walters, C.J. (1992) *Quantitative Fisheries Stock Assessment: Choice Dynamics and Uncertainty*. Chapman & Hall, New York.

Hill, J.K., Thomas, C.D. & Lewis, O.T. (1996) Effects of habitat patch size and isolation on dispersal by *Hesperia comma* butterflies: implications for metapopulation structure. *Journal of Animal Ecology* **65**, 725–35.

Hillis, D.M., Mortiz, C. & Mable, B.K. (eds) (1996) *Molecular Systematics*, 2nd edn. Sinauer Associates, Sunderland, MA.

Hines, J.E. (1993) *MSSURVIV User's Manual*. National Biological Resources Unit, Patuxent Wildlife Research Centre, Laurel, MD.

Holling, C.S. (1959) The components of predation as revealed by a study of small mammal predation on the European pine sawfly. *Canadian Entomology* **91**, 293–320.

Holling, C.S. (1966) The strategy of building models of complex ecological systems. In: *Systems Analysis in Ecology* (ed. Watt, K.E.F.), pp. 195–214. Academic Press, New York.

Holt, R.D. (1987) On the relationship between niche overlap and competition: the effect of incommensurable niche dimensions. *Oikos* **48**, 110–14.

Holyoak, M. & Lawton, J.H. (1993) Comment arising from a paper by Wolda and Dennis: using and interpreting the results of tests for density dependence. *Oecologia* **95**, 592–4.

Hone, J., Pech, R. & Yip, P. (1992) Estimation of the dynamics and rate of transmission of classical swine fever (hog cholera) in wild pigs. *Epidemiology and Infection* **108**, 377–86.

Höss, M., Kohn, M., Pääbo, S., Knauer, F. & Schröder, W. (1992) Excrement analysis by PCR. *Nature* **359**, 199.

Hoyle, S.D., Horsup, A.B., Johnson, C.N., Crossman, D.G. & McCallum, H.I. (1995) Live trapping of the northern hairy-nosed wombat (*Lasiorhinus krefftii*): population size estimates and effects on individuals. *Wildlife Research* **22**, 741–55.

Hudson, P.J. & Dobson, A.P. (1989) Population biology of *Trichostrongylus tenuis*, a parasite of economic importance for red grouse management. *Parasitology Today* **5**, 283–91.

Hudson, P.J. & Dobson, A.P. (1997) Transmission dynamics and host–parasite interactions of *Trichostrongylus tenuis* in red grouse (*Lagopus lagopus scoticus*). *Journal of Parasitology* **83**, 194–202.

Hudson, P.J., Dobson, A.P. & Newborn, D. (1992a) Do parasites make prey vulnerable to predation? Red grouse and parasites. *Journal of Animal Ecology* **61**, 681–92.

Hudson, P.J., Newborn, D. & Dobson, A.P. (1992b) Regulation and stability of a free-living host–parasite system: *Trichostrongylus tenuis* in red grouse. I. Monitoring and parasite reduction experiments. *Journal of Animal Ecology* **61**, 477–86.

Hutchings, J.A. & Myers, R.A. (1994) What can be learned from the collapse of a renewable resource? Atlantic cod, *Gadus morhua*, of Newfoundland and Labrador. *Canadian Journal of Fisheries and Aquatic Sciences* **51**, 2126–46.

Iachan, R. (1985) Optimum stratum boundaries for shellfish surveys. *Biometrics* **41**, 1053–62.

Iman, R.L. & Conover, W.J. (1980) Small sample sensitivity analysis techniques for computer models, with an application to risk assessment. *Communications in Statistics: Theory and Methods* **A9**, 1749–842.

Ives, A.R. (1992) Density-dependent and density-independent parasitoid aggregation in model host–parasitoid systems. *American Naturalist* **140**, 912–37.

Ivlev, V.S. (1961) *Experimental Ecology of the Feeding of Fishes*. Yale University Press, New Haven, CT.

James, I.R. (1991) Estimation of von Bertalanffy growth curve parameters from recapture data. *Biometrics* **47**, 1519–30.

Jarne, P. & Lagoda, P.J.L. (1996) Microsatellites, from molecules to populations and back. *Trends in Ecology and Evolution* **11**, 424–9.

Jenkins, S.H. (1988) Use and abuse of demographic models of population growth. *Bulletin of the Ecological Society of America* **69**, 201–7.

Jennrich, R.I. & Turner, F.B. (1969) Measurement of non-circular home range. *Journal of Theoretical Biology* **22**, 227–37.

Jones, T.H., Hassell, M.P. & Pacala, S.W. (1993) Spatial heterogeneity and the population dynamics of a host–parasitoid system. *Journal of Animal Ecology* **62**, 252–62.

Judson, O.P. (1994) The rise of the individual-based model in ecology. *Trends in Ecology and Evolution* **9**, 9–14.

Juliano, S.A. & Williams, F.M. (1987) A comparison of methods for estimating the functional response parameters of the random predator equation. *Journal of Animal Ecology* **56**, 641–53.

Kaufmann, K.W. (1981) Fitting and using growth curves. *Oecologia* **49**, 293–9.

Kendall, S. (1976) *Time Series*, 2nd edn. Griffin, London.

Kendall, W.L., Pollock, K.H. & Brownie, C. (1995) A likelihood-based approach to capture–recapture estimation of demographic parameters under the robust design. *Biometrics* **51**, 293–308.

Kendall, W.L., Nichols, J.D. & Hines, J.E. (1997) Estimating temporary emigration using capture–recapture data with Pollock's robust design. *Ecology* **78**, 563–78.

Kennedy, M. & Gray, R.D. (1993). Can ecological theory predict the distribution of foraging animals? A critical analysis of experiments on the ideal free distribution. *Oikos* **68**, 158–66.

Key, G.E. & Woods, R.D. (1996) Spool-and-line studies on the behavioural ecology of rats (*Rattus* spp.) in the Galapagos Islands. *Canadian Journal of Zoology* **74**, 733–7.

Keyfitz, N. (1977) *Applied Mathematical Demography*. Wiley-Interscience, New York.

Kie, J.G., Baldwin, J.A. & Evans, C.J. (1996) CALHOME: a program for estimating animal home ranges. *Wildlife Society Bulletin* **24**, 342–4.

Kimura, D.K. (1980) Likelihood methods for the von Bertalanffy growth curve. *Fishery Bulletin* **77**, 765–76.

Kimura, D.K. & Lemberg, N.A. (1981) Variability of line intercept density estimates (a simulation study of the variance of hydroacoustic biomass estimates). *Canadian Journal of Fisheries and Aquatic Sciences* **38**, 1141–52.

Koenig, W.D., van Duren, D. & Hooge, P.N. (1996) Detectability, philopatry and the distribution of dispersal distances in vertebrates. *Trends in Ecology and Evolution* **11**, 514–17.

Kohlmann, S.G. & Risenhoover, K.L. (1997) White-tailed deer in a patchy environment: a test of the ideal-free-distribution theory. *Journal of Mammalogy* **78**, 1261–72.

Kohn, M.H. & Wayne, R.K. (1997) Facts from feces revisited. *Trends in Ecology and Evolution* **12**, 223–7.

Kot, M., Lewis, M.A. & van den Driessche, P. (1996) Dispersal data and the spread of invading organisms. *Ecology* **77**, 2027–42.

Krebs, C.J. (1985) *Ecology*, 2nd edn. Harper & Row, New York.

Krebs, C.J. (1989) *Ecological Methodology*. HarperCollins, New York.

Krebs, C.J., Singleton, G.R. & Kenney, A.J. (1994) Six reasons why feral house mouse populations might have low recapture rates. *Wildlife Research* **21**, 559–67.

Krebs, C.J., Boutin, S., Boonstra, R., Sinclair, A.R.E., Smith, J.N.M., Dale, M.R.T., Martin, K. & Turkington, R. (1995) Impact of food and predation on the snowshoe hare cycle. *Science* **269**, 1112–15.

Kuusaari, M., Saccheri, I., Camara, M. & Hanski, I. (1998) Allee effect and population dynamics in the Glanville fritillary butterfly. *Oikos* **82**, 384–92.

Lacy, R.C. (1993) Vortex: a computer simulation model for population viability analysis. *Wildlife Research* **20**, 45–65.

Lahaye, W.S., Gutierrez, R. & Akçakaya, H.R. (1994) Spotted owl metapopulation dynamics in Southern California. *Journal of Animal Ecology* **63**, 775–85.

Lamberson, R.H., McKelvey, R., Noon, B.R. & Voss, C. (1992) A dynamic analysis of northern spotted owl viability in a fragmented forest landscape. *Conservation Biology* **6**, 505–12.

Lanciani, C.A. & Boyett, J.M. (1980) Demonstrating parasitic water mite induced mortality in natural host populations. *Parasitology* **81**, 465–75.

Lande, R. (1988) Demographic models of the northern spotted owl (*Strix occidentalis caurina*). *Oecologia* **75**, 601–7.

Lande, R. (1993) Risks of population extinction from demographic and environmental stochasticity and random catastrophes. *American Naturalist* **142**, 911–27.

Laska, M.S. & Wootton, J.T. (1998) Theoretical concepts and empirical approaches to measuring interaction strength. *Ecology* **79**, 461–76.

Lawton, J.H. & Hassell, M.P. (1981) Asymmetrical competition in insects. *Nature* **289**, 793–5.

Lebreton, J.D., Burnham, K.P., Clobert, J. & Anderson, D.R. (1992) Modeling survival and testing biological hypotheses using marked animals: a unified approach with case studies. *Ecological Monographs* **62**, 67–118.

Leslie, P.H. (1945) On the use of matrices in certain population mathematics. *Biometrika* **33**, 183–212.

Levin, S.A. (1992) The problem of pattern and scale in ecology. *Ecology* **73**, 1943–67.

Levin, S.A., Grenfell, B., Hastings, A. & Perelson, A.S. (1997) Mathematical and computational challenges in population biology and ecosystems science. *Science* **275**, 334–43.

Levins, R. (1969) Some demographic and genetic consequences of environmental heterogeneity for biological control. *Bulletin of the Entomological Society of America* **15**, 237–40.

Levins, R. (1970) Extinction. In: *Some Mathematical Questions in Biology*. Lectures on Mathematics in the Life Sciences, Vol. 2, pp. 75–107. American Mathematical Society, Providence, RI.

Liermann, M. & Hilborn, R. (1997) Depensation in fish stocks: a hierarchic Bayesian meta-analysis. *Canadian Journal of Fisheries and Aquatic Sciences* **54**, 1976–84.

Lindenmayer, D.B. & Lacy, R.C. (1995a) Metapopulation viability of arboreal marsupials in fragmented old growth forests: comparison among species. *Ecological Applications* **5**, 183–99.

Lindenmayer, D.B. & Lacy, R.C. (1995b) Metapopulation viability of Leadbeater's possum *Gymnobelideus leadbeateri*, in fragmented old growth forests. *Ecological Applications* **5**, 164–82.

Link, W.A. & Hahn, D.C. (1996) Empirical Bayes estimation of proportions with application to cowbird parasitism rates. *Ecology* **77**, 2528–37.

Lipsitz, S.R., Dear, K.B.G. & Zhao, L. (1994) Jackknife estimators of variance for parameter estimates from estimating equations with applications to clustered survival data. *Biometrics* **50**, 842–6.

Lomnicki, A. (1988) *Population Ecology of Individuals*. Princeton University Press, Princeton, NJ.

Lord, C.C., Woolhouse, M.E.J., Heesterbeek, J.A.P. & Mellor, P.S. (1995) Vector-borne diseases and the basic reproduction number: a case study of African horse sickness. *Medical and Veterinary Entomology* **10**, 19–28.

Ludwig, D. (1996) Uncertainty and the assessment of extinction probabilities. *Ecological Applications* **6**, 1067–76.

Ludwig, D. & Walters, C.J. (1981) Measurement errors and uncertainty in parameter estimates for stock and recruitment. *Canadian Journal of Fisheries and Aquatic Sciences* **38**, 711–20.

Manly, B.F.J. (1989) A review of methods for the analysis of stage–frequency data. In: *Estimation and Analysis of Insect Populations* (eds McDonald, L., Manly, B., Lockwood, J. & Logan, J.), pp. 3–69. Springer-Verlag, Berlin.

Manly, B.F.J. (1990) *Stage-Structured Populations: Sampling, Analysis, and Simulation*. Chapman & Hall: London.

May, R.M. (1974a) Biological populations with nonoverlapping generations: stable points, stable cycles and chaos. *Science* **186**, 645–7.

May, R.M. (1974b) *Stability and Complexity in Model Ecosystems*, 2nd edn. Princeton University Press, Princeton, NJ.

May, R.M. (1976) Models for two interacting populations. In: *Theoretical Ecology* (ed. R.M. May), pp. 49–70. Blackwell Science, Oxford.

May, R.M. (1978) Host–parasitoid models in patchy environments: a phenomenological model. *Journal of Animal Ecology* **47**, 833–44.

May, R.M. & Anderson, R.M. (1978) Regulation and stability of host–parasite interactions. II. Destabilizing processes. *Journal of Animal Ecology* **47**, 249–67.

May, R.M. & Anderson, R.M. (1979) Population biology of infectious diseases. Part II. *Nature* **280**, 455–61.

May, R.M. & Hassell, M.P. (1988) Population dynamics and biological control. *Philosophical Transactions of the Royal Society of London Series B* **318**, 129–69.

McArthus, R.H. & Levins, R. (1968) The limiting similarity, convergence and divergence of coexisting species. *American Naturalist* **101**, 377–85.

McCallum, H.I. (1982a) Infection dynamics of *Ichthyophthirius multifiliis*. *Parasitology* **85**, 475–88.

McCallum, H.I. (1982b) *Population dynamics of Ichthyophthirius multifiliis*. PhD thesis, University of London.

McCallum, H.I. (1985) Population effects of parasite survival of host death: experimental studies of the interaction of *Ichthyophthirius multifiliis* and its fish host. *Parasitology* **90**, 529–47.

McCallum, H.I. (1987) Predator regulation of *Acanthaster planci*. *Journal of Theoretical Biology* **127**, 207–20.

McCallum, H.I. (1993) Evaluation of a nematode (*Capillaria hepatica* Bancroft, 1893) as a control agent for populations of house mice (*Mus musculus domesticus* Schwartz and Schwartz, 1943). *Revue Scientifique et Technique, Office International des Epizooties* **12**, 83–93.

McCallum, H.I. (1994) Quantifying the impact of disease on threatened species. *Pacific Conservation Biology* **1**, 107–17.

McCallum, H.I. (1995a) Modelling translocation strategies for the bridled nailtail wallaby (*Onychogalea fraenata* Gould, 1840): effects of reintroduction size. In: *Reintroduction Biology* (ed. Serena, M.), pp. 7–14. Surrey Beatty, Chipping Norton, New South Wales.

McCallum, H.I. (1995b) Would property rights over kangaroos necessarily lead to their conservation? Implications of fisheries models. In: *Conservation through Sustainable Use of Wildlife* (eds Grigg, G.C., Hale, P.T. & Lunney, D.), pp. 215–23. Centre for Conservation Biology, University of Queensland, Brisbane.

McCallum, H.I. & Dobson, A.P. (1995) Detecting disease and parasite threats to endangered species and ecosystems. *Trends in Ecology and Evolution* **10**, 190–4.

McCallum, H.I. & Scott, M.E. (1994) Quantifying population processes: experimental and theoretical approaches. In: *Parasitic and Infectious Diseases: Epidemiology and Ecology* (eds Scott, M.E. & Smith, G.), pp. 29–45. Academic Press, San Diego, CA.

McCallum, H.I. & Singleton, G.R. (1989) Models to assess the potential of *Capillaria hepatica* to control population outbreaks of house mice. *Parasitology* **98**, 425–37.

McCallum, H.I., Timmers, P. & Hoyle, S. (1995) Modelling the impact of predation on reintroductions of bridled nailtail wallabies. *Wildlife Research* **22**, 163–71.

McCarthy, M.A., Burgman, M.A. & Ferson, S. (1995) Sensitivity analysis for models of population viability. *Biological Conservation* **73**, 93–100.

McCullagh, P. & Nelder, J.A. (1989) *Generalized Linear Models*, 2nd edn. Chapman & Hall, London.

McDonald, D.B. & Caswell, H. (1993) Matrix methods for avian demography. In: *Current Ornithology*, Vol. 10. (ed. Power, D.M.), pp. 139–85. Plenum Press, New York.

Menkens, G.E. & Anderson, S.H. (1988) Estimation of small mammal population size. *Ecology* **69**, 1952–9.

Messier, F. (1994) Ungulate population models with predation: a case study with the North American moose. *Ecology* **75**, 478–88.

Messier, F. & Crête, M. (1985) Moose–wolf dynamics and the natural regulation of moose populations. *Oecologia* **65**, 503–12.

Michael, E. & Bundy, D.A.P. (1989) Density dependence in establishment, growth and worm fecundity in intestinal helminthiasis: The population biology of *Trichuris muris* (Nematoda) infection in CBA/Ca mice. *Parasitology* **98**, 451–8.

Mills, N.J. & Getz, W.M. (1996) Modelling the biological control of insect pests: a review of host–parasitoid models. *Ecological Modelling* **92**, 121–43.

Millistein, J.A. & Turchin, P. (1994) *RAMAS/time, Ecological Time Series Modeling and Forecasting*. Applied Biomathematics, Setauket, NY.

Moran, P.J. (1986) The Acanthaster phenomenon. *Oceanography and Marine Biology Annual Review* **24**, 379–480.

Moritz, C. & Lavery, S. (1996) Molecular ecology: contributions from molecular genetics to population ecology. In: *Frontiers of Population Ecology* (eds Floyd, R.B., Sheppard, A.W. & De Barro, P.J.), pp. 433–50. CSIRO, Melbourne.

Munger, J.C. & Karasov, W.H. (1991) Sublethal parasites in white-footed mice: impact on survival and reproduction. *Canadian Journal of Zoology* **69**, 398–404.

Murdoch, W.W. & Briggs, C.J. (1996) Theory for biological control: recent developments. *Ecology* **77**, 2001–13.

Murdoch, W.W. & Oaten, A. (1975) Predation and population stability. *Advances in Ecological Research* **9**, 1–131.

Murdoch, W.W. & Stewart-Oaten, A. (1989) Aggregation by parasitoids and predators: effects on equilibrium and stability. *American Naturalist* **134**, 288–310.

Murdoch, W.W. & Walde, S.J. (1989) Analysis of insect population dynamics. In: *Towards a More Exact Ecology* (eds Grubb, P.J. & Whittaker, J.B.), pp. 113–40. Blackwell Scientific Publications, Oxford.

Murdoch, W.W., Reeve, J.D., Huffaker, C.B. & Kennett, C.E. (1984) Biological control of olive scale and its relevance to ecological theory. *American Naturalist* **123**, 371–92.

Murdoch, W.W., Chesson, J. & Chesson, P.L. (1985) Biological control in theory and practice. *American Naturalist* **125**, 344–66.

Murdoch, W.W., Nisbet, R.M., Blythe, S.P., Gurney, W.S.C. & Reeve, J.D. (1987) An invulnerable age class and stability in delay-differential parasitoid–host models. *American Naturalist* **129**, 263–82.

Murray, J.D. (1989) *Mathematical Biology*. Springer-Verlag, Berlin.

Myers, R.A., Barrowman, N.J., Hutchings, J.A. & Rosenberg, A.A. (1995). Population dynamics of exploited fish stocks at low population levels. *Science* **269**, 1106–8.

Newsome, A.E., Catling, P.C. & Corbett, L.K. (1983) The feeding ecology of the dingo. II. Dietary and numerical relationships with fluctuating prey populations in south-eastern Australia. *Australian Journal of Ecology* **8**, 345–66.

Newsome, A.E., Parer, I. & Catling, P.C. (1989) Prolonged prey suppression by carnivores – predator-removal experiments. *Oecologia* **78**, 458–67.

Nisbet, R.M. & Gurney, W.S.C. (1983) The systematic formulation of population models for insects with dynamically varying instar duration. *Theoretical Population Biology* **23**, 114–35.

Nisbet, R.M., Gurney, W.S.C., Murdoch, W.W. & McCauley, E. (1989) Structured population models: a tool for linking effects at individual and population level. *Biological Journal of the Linnean Society* **37**, 79–99.

Odum, E.P. (1971) *Fundamentals of Ecology*. W.B. Saunders, Philadelphia.

Okubo, A. & Levin, S.A. (1989) A theoretical framework for data analysis of wind dispersal of seeds and pollen. *Ecology* **70**, 329–38.

Otis, D.L., Burnham, K.P., White, G.C. & Anderson, D.R. (1978) *Statistical Inference from Capture Data on Closed Animal Populations*. Wildlife Monographs Vol. 62. The Wildlife Society, Bethesda, MD.

Pacala, S.W. & Hassell, M.P. (1991) The persistence of host–parasitoid associations in patchy environments. II. Evaluation of field data. *American Naturalist* **138**, 584–605.

Pacala, S.W., Hassell, M.P. & May, R.M. (1990) Host–parasitoid associations in patchy environments. *Nature* **344**, 150–3.

Paetkau, D. & Strobeck, C. (1998) Ecological genetic studies of bears using microsatellite analysis. *Ursus* **10**, 299–306.

Paetkau, D., Calvert, W., Stirling , I. & Strobeck, C. (1995) Microsatellite analysis of population structure in Canadian polar bears. *Molecular Ecology* **4**, 347–54.

Paine, R.T. (1992) Food-web analysis through field measurement of per capita interaction strength. *Nature* **355**, 73–5.

Paloheimo, J.E. (1988) Estimation of marine production from size spectrum. *Ecological Modelling* **42**, 33–44.

Parker, P.G., Snow, A.A., Schug, M.D., Booton, G.C. & Fuerst, P.A. (1998) What molecules can tell us about populations: choosing and using a molecular marker. *Ecology* **79**, 361–82.

Pascual, M.A. & Kareiva, P. (1996) Predicting the outcome of competition using experimental data: maximum likelihood and Bayesian approaches. *Ecology* **77**, 337–49.

Pech, R.P., Sinclair, A.R.E. & Newsome, A.E. (1992) Limits to predator regulation of rabbits in Australia: evidence from predator-removal experiments. *Oecologia* **89**, 102–12.

Pech, R.P., Sinclair, A.R.E. & Newsome, A.E. (1995) Predation models for primary and secondary prey species. *Wildlife Research* **22**, 55–64.

Peters, R.H. (1983) *The Ecological Implications of Body Size*. Cambridge University Press, Cambridge.

Petersen, R.O. & Page, R.E. (1988) The rise and fall of Isle Royale wolves, 1975–1986. *Journal of Mammalogy* **69**, 89–99.

Pfister, C.A. (1995) Estimating competition coefficients from census data: a test with field manipulations of tide pool fishes. *American Naturalist* **146**, 271–91.

Pimm, S.L. (1985) Estimating competition coefficients from census data. *Oecologia* **67**, 588–90.

Pollard, E., Lakhani, K.H. & Rothery, P. (1987) The detection of density dependence from a series of annual censuses. *Ecology* **68**, 2046–55.

Pollock, K.H. (1991) Modeling capture, recapture and removal statistics for estimation of demographic parameters for fish and wildlife populations: past, present and future. *Journal of the American Statistical Association* **86**, 225–38.

Pollock, K.H. (1995) The challenges of measuring change in wildlife populations: a biometrician's perspective. In: *Conservation through the Sustainable Use of Wildlife* (eds Grigg,G.C., Hale, P.T. & Lunney, D.), pp. 117–21. Centre for Conservation Biology, University of Queensland, Brisbane.

Pollock, K.H. & Otto, M.C. (1983) Robust estimation of population size in closed animal populations from capture–recapture experiments. *Biometrics* **39**, 1035–49.

Pollock, K.H., Winterstein, S.R., Bunck, C.M. & Curtis, P.D. (1989) Survival analysis in telemetry studies: the staggered entry design. *Journal of Wildlife Management* **53**, 7–15.

Pollock, K.H., Nichols, J.D., Brownie, C. & Hines, J.E. (1990) *Statistical Inference for Capture–Recapture Experiments*. Wildlife Monographs Vol. 107. The Wildlife Society, Bethesda, MD.

Possingham, H., Davies, I. & Noble, I. (1992a) *Alex: An Operating Manual*. Dept of Applied Mathematics, University of Adelaide, Adelaide.

Possingham, H.P., Davies, I., Noble, I.R. & Norton, T.W. (1992b) A metapopulation simulation model for assessing the likelihood of plant and animal extinctions. *Mathematics and Computers in Simulation* **33**, 367–72.

Possingham, H.P., Lindenmayer, D.B. & Davies, I. (1994) Metapopulation viability analysis of the greater glider *Petauroides volans* in a wood production area. *Biological Conservation* **70**, 227–36.

Pradel, R. & Lebreton, J.D. (1993) *User's Manual for Program SURGE Version 4.2*. CEPE/CNRS, Montpellier, France.

Press, W.H., Teukolsky, S.A., Vetterling, W.T. & Flannery, B.P. (1994) *Numerical Recipes in C*, 2nd edn. Cambridge University Press, Cambridge.

Quinnell, R.J. (1992) The population dynamics of *Heligmosomoides polygyrus* in an enclosure population of wood mice. *Journal of Animal Ecology* **61**, 669–79.

Rannala, B. & Mountain, J.L. (1997) Detecting immigration by using multilocus genotypes. *Proceedings of the National Academy of Science, USA* **94**, 9197–201.

Real, L.A. (1977) The kinetics of functional response. *American Naturalist* **111**, 289–300.

Rexstad, E. & Burnham, K. (1991) *'User's Guide for Interactive Program CAPTURE'*. Colorado Cooperative Fish and Wildlife Research Unit, Colorado State University, Fort Collins.

Reynolds, J.C. & Aebischer, N.J. (1991) Comparison and quantification of carnivore diet by faecal analysis: a critique, with recommendations, based on a study of the fox *Vulpes vulpes*. *Mammal Review* **21**, 97–122.

Rhodes, C.J. & Anderson, R.M. (1997) Epidemic thresholds and vaccination in a lattice model of disease spread. *Journal of Theoretical Biology* **52**, 101–18.

Ricker, W.E. (1975) Computation and interpretation of biological statistics of fish populations. *Fisheries Research Board of Canada, Bulletin* **191**.

Ricker, W.E. (1979) Growth rates and models. In: *Fish Physiology*, Vol. 8. (eds Hoar, W.S., Randall, D.J. & Brett, J.R.), pp. 677–743. Academic Press, New York.

Rigler, F.H. & Cooley, J.M. (1974) The use of field data to derive population statistics of multivoltine copepods. *Limnology and Oceanography* **19**, 636–55.

Ripa, J. & Lundberg, P. (1996) Noise colour and the risk of population extinctions. *Proceedings of the Royal Society of London Series B* **263**, 1751–3.

Roberts, M.G., Smith, G. & Grenfell, B.T. (1995) Mathematical models for macroparasites of wildlife. In: *Ecology of Infectious Diseases in Natural Populations* (eds Grenfell, B.T. & Dobson, A.P.), pp. 177–208. Cambridge University Press, Cambridge.

Robertson, G. (1987) Plant dynamics. In: *Kangaroos: Their Ecology and Management in the Sheep Rangelands of Australia* (eds Caughley, G., Shepherd, N. & Short, J.), pp. 50–68. Cambridge University Press, Cambridge.

Robson, D.S. & Regier, H.A. (1964) Sample size in Petersen mark–recapture experiments. *Transactions of the American Fisheries Society* **93**, 215–26.

Rogers, D. (1972) Random search and insect population models. *Journal of Animal Ecology* **41**, 369–83.

Rohner, C. (1995) Great horned owls and snowshoe hares: what causes the time lag in the numerical response of predators to cyclic prey? *Oikos* **74**, 61–8.

Rose, K.A., Christensen, S.W. & DeAngelis, D.L. (1993) Individual-based modeling of populations with high mortality: a new method based on following a fixed number of model individuals. *Ecological Modelling* **68**, 293–302.

Rosenzweig, M.L. & MacArthur, R.H. (1963) Graphical representation and stability conditions of predator–prey interactions. *American Naturalist* **97**, 209–23.

Rosenzweig, M.L., Abramsky, Z., Kotter, B. & Mitchel, W. (1985) Can interaction coefficients be determined from census data? *Oecologia* **66**, 194–8.

Ross, R. (1916) An application of the theory of probabilities to the study of a priori pathometry – Part I. *Proceedings of the Royal Society of London Series B* **92**, 204–30.

Ross, R. (1917) An application of the theory of probabilities to the study of a priori pathometry – Part II. *Proceedings of the Royal Society of London Series B* **93**, 212–40.

Roughgarden, J. (1983) Competition and theory in community ecology. *American Naturalist* **122**, 583–601.

Rousset, F., Thomas, F., de Meeûs, T. & Renaud, F. (1996) Influence of parasite induced host mortality from distributions of parasite loads. *Ecology* **77**, 2203–11.

Royama, T. (1971) A comparative study of models for predation and parasitism. *Researches on Population Ecology, Supplement* **1**, 1–99.

Royce, L.A. & Rossignol, P.A. (1990) Honey bee mortality due to tracheal mite parasitism. *Parasitology* **100**, 147–51.

Ruckelshaus, M., Hartway, C. & Kareiva, P. (1997) Assessing the data requirements of spatially explicit dispersal models. *Conservation Biology* **11**, 1298–1306.

Ryan, T.P. (1997) *Modern Regression Analysis*. Wiley-Interscience, New York.

Sabelis, M.W. (1992) Predatory arthropods. In: *Natural Enemies: The Population Biology of Predators, Parasites and Diseases* (ed. Crawley, M.J.), pp. 225–64. Blackwell Scientific, Oxford.

Sæther, B.-E., Ringsby, T.H. & Røskaft, E. (1996) Life history variation, population processes and priorities in species conservation: towards a reunion of research paradigms. *Oikos* **77**, 217–26.

Sainsbury, K.J. (1980) Effect of individual variability of the von Bertalanffy growth curve. *Canadian Journal of Fisheries and Aquatic Sciences* **37**, 241–7.

Sainsbury, K.J. (1986) Estimation of food consumption from field observations of fish feeding cycles. *Journal of Fish Biology* **29**, 23–36.

Sanchez, M.A. & Blower, S.M. (1997) Uncertainty and sensitivity analysis of the basic reproductive rate: tuberculosis as an example. *American Journal of Epidemiology* **145**, 1127–37.

Sauer, J.R., Barker, R.J. & Geissler, P.H. (1994) Statistical aspects of modeling population change from population size data. In: *Wildlife Toxicology and Population Modeling: Integrated Studies of Agroecosytems* (eds Kendall, R.J. & Lacher, L.E.), pp. 451–66. CRC Press, Boca Raton, FL.

Scandol, J.P. & James, M.K. (1992) Hydrodynamics and larval dispersal: a population model of *Acanthaster planci* on the Great Barrier Reef. *Australian Journal of Marine and Freshwater Research* **43**, 583–96.

Scheffer, V.B. (1951) The rise and fall of a reindeer herd. *Science Monthly* **73**, 356–62.

Schnute, J. (1981) A versatile growth model with statistically stable parameters. *Canadian Journal of Fisheries and Aquatic Sciences* **38**, 1128–40.

Schoener, T.W. (1974a) Competition and the form of habitat shift. *Theoretical Population Biology* **6**, 265–307.

Schoener, T.W. (1974b) Some methods for calculating competition coefficients from resource utilization spectra. *American Naturalist* **108**, 332–40.

Schoener, T.W. (1985) On the degree of consistency expected when different methods are used to estimate competition coefficients from census data. *Oecologia* **67**, 591–2.

Schwartz, C.J., Schweigert, J.F. & Aronson, A.N. (1993) Estimating migration rates using tag-recovery data. *Biometrics* **49**, 177–93.

Scott, M.E. (1987) Regulation of mouse colony abundance by *Heligmosomoides polygyrus*. *Parasitology* **95**, 111–24.

Scott, M.E. & Tanguay, G.V. (1994) *Heligmosomoides polygyrus:* a laboratory model for direct life cycle nematodes of humans and livestock. In: *Parasitic and Infectious Diseases* (eds Scott, M.E. & Smith, G.), pp. 279–300. Academic Press, San Diego, CA.

Seaman, D.E. & Powell, R.A. (1996) An evaluation of the accuracy of kernel density estimators for home range analysis. *Ecology* **77**, 2075–85.

Seber, G.A.F. (1982) *The Estimation of Animal Abundance and Related Parameters*, 2nd edn. Griffin: London.

Seber, G.A.F. & Wild, C.J. (1989) *Nonlinear Regression*. Wiley, New York.

Seifert, R.P. & Seifert, F.M. (1976) A community matrix analysis of *Heliconia* insect communities. *American Naturalist* **110**, 461–83.

Seifert, R.P. & Seifert, F.M. (1979) A *Heliconia* insect community in a Venezuelan cloud forest. *Ecology* **60**, 462–7.

Shao, J. & Tu, D. (1995) *The Jackknife and the Bootstrap*. Springer-Verlag, New York.

Shine, R. & Madsen, T. (1997) Prey abundance and predator reproduction: rats and pythons on a tropical Australian floodplain. *Ecology* **78**, 1078–86.

Short, J. (1987) Factors affecting the food intake of rangelands herbivores. In: *Kangaroos: Their Ecology and Management in the Sheep Rangelands of Australia* (eds Caughley, G., Shepherd, N. & Short, J.), pp. 84–99. Cambridge University Press, Cambridge.

Shostak, A.W. & Scott, M.E. (1993) Detection of density-dependent growth and fecundity of helminths in natural infections. *Parasitology* **106**, 527–39.

Simberloff, D. (1983) Competition theory, hypothesis testing, and other community ecological buzzwords. *American Naturalist* **122**, 626–35.

Simmons, R.E., Avery, D.M. & Avery, G. (1991) Biases in diets determined from pellets and remains: correction factors for a mammal and bird eating raptor. *Raptor Research* **25**, 63–7.

Sinclair, A.R.E., Olsen, P.D. & Redhead, T.D. (1990) Can predators regulate small mammal populations? Evidence from house mouse outbreaks in Australia. *Oikos* **59**, 382–92.

Singleton, G.R. (1989) Population dynamics of an outbreak of house mice (*Mus domesticus*) in the mallee wheatlands of Australia – hypothesis of plague formation. *Journal of Zoology, London* **219**, 495–515.

Sjögren Gulve, P. (1994) Distribution and extinction patterns within a northern metapopulation of the pool frog *Rana lessonae*. *Ecology* **75**, 1357–67.

Skalski, J.R. (1991) Using sign counts to quantify animal abundance. *Journal of Wildlife Management* **55**, 705–15.

Skellam, J.G. (1951) Random dispersal in theoretical populations. *Biometrika* **38**, 196–218.

Skogland, T. (1985) The effects of density-dependent resource limitation on the demography of wild reindeer. *Journal of Animal Ecology* **54**, 359–74.

Slatkin, M. (1985) Gene flow in natural populations. *Annual Review of Ecology and Systematics* **16**, 393–430.

Slatkin, M. (1987) Gene flow and the geographical structure of natural populations. *Science* **236**, 787–92.

Slatkin, M. & Barton, N.H. (1989) A comparison of three indirect methods for estimating average levels of gene flow. *Evolution* **43**, 1349–68.

Smith, A.D.M. & Walters, C.J. (1981) Adaptive management of stock-recruitment systems. *Canadian Journal of Fisheries and Aquatic Sciences* **38**, 690–703.

Smith, D.R., Conroy, M.J. & Brakhage, D.H. (1995) Efficiency of adaptive cluster sampling for estimating density of wintering waterfowl. *Biometrics* **51**, 777–88.

Smith, E.B., Williams, F.M. & Fisher, C.R. (1997) Effects of intrapopulation variability on von Bertalanffy growth parameter estimates from equal mark–recapture intervals. *Canadian Journal of Fisheries and Aquatic Sciences* **54**, 2025–32.

Smith, F.E. (1954) Quantitative aspects of population growth. In: *Dynamics of Growth Processes* (ed. Boell, E.), pp. 277–94. Princeton University Press, Princeton.

Smith, G.C. & Trout, R.C. (1994) Using Leslie matrices to determine wild rabbit population growth and the potential for control. *Journal of Applied Ecology* **31**, 223–30.

Sokal, R.R. & Rohlf, F.J. (1995) *Biometry*, 3rd edn. W.H. Freeman, New York.

Spalinger, D.E. & Hobbs, N.T. (1992) Mechanisms of foraging in mammalian herbivores: new models of functional response. *American Naturalist* **140**, 325–48.

Starfield, A.M. (1997) A pragmatic approach to modeling for wildlife management. *Journal of Wildlife Management* **61**, 261–70.

Stenseth, N.C. & Lidicker, W.Z.J. (1992) Appendix 3. The use of radioisotopes in the study of dispersal: with a case study. In: *Animal Dispersal* (eds Stenseth, N.C. & Lidicker, W.Z.J.), pp. 333–52. Chapman & Hall, London.

Stenseth, N.C., Falck, W., Bjørnstad, O.N. & Krebs, C.J. (1997) Population regulation in snowshoe hare and the Canadian lynx: asymmetric food web configurations between hare and lynx. *Proceedings of the National Academy of Science, USA* **94**, 5147–52.

Strong, D.R. (1986) Density-vague population change. *Trends in Ecology and Evolution* **1**, 39–42.

Sutherland, W.J. (ed.) (1996) *Ecological Census Techniques: A Handbook*. Cambridge University Press, Cambridge.

Swartzman, G.L. & Kaluzny, S.P. (1987) *Ecological Simulation Primer*. Macmillan, New York.

Tanner, J.T. (1975) The stability and the intrinsic growth rates of prey and predator populations. *Ecology* **56**, 855–67.

Taylor, L.R. (1961) Aggregation, variance and the mean. *Nature* **189**, 732–5.

Tchamba, M.N., Bauer, H. & De-Iongh, H.H. (1995) Application of VHF-radio and satellite telemetry techniques on elephants in northern Cameroon. *African Journal of Ecology* **33**, 335–46.

Thompson, S.D. (1987) Body size, duration of parental care, and the intrinsic rate of natural increase in eutherian and metatherian mammals. *Oecologia* **71**, 201–9.

Thompson, S.K. (1990) Adaptive cluster sampling. *Journal of the American Statistical Association* **85**, 1050–9.

Thompson, S.K. (1991a) Adaptive cluster sampling: designs with primary and secondary units. *Biometrics* **47**, 1103–15.

Thompson, S.K. (1991b) Stratified adaptive cluster sampling. *Biometrika* **78**, 389–97.

Thompson, S.K. (1992) *Sampling*. Wiley-Interscience, New York.

Tilman, D. (1982) *Resource Competition and Community Structure*. Princeton University Press, Princeton, NJ.

Tilman, D. (1987) The importance of the mechanisms of interspecific competition. *American Naturalist* **129**, 769–74.

Tilman, D. (1990) Mechanisms of plant competition for nutrients: the elements of a predictive theory for competition. In: *Perspectives on Plant Competition* (eds Grace, J.B. & Tilman, D.), pp. 117–41. Academic Press, San Diego, CA.

Tilman, D., Lehman, C.L. & Kareiva, P. (1997) Population dynamics in spatial habitats. In: *Spatial Ecology: The Role of Space in Population Dynamics and Interspecific Interactions* (eds Tilman, D. & Kareiva, P.M.), pp. 3–20. Princeton University Press, Princeton, NJ.

Tong, II. (1990) *Non-linear Time Series: A Dynamical System Approach*. Oxford University Press, Oxford.

Trostel, K., Sinclair, A.R.E., Walters, C.J. & Krebs, C.J. (1987) Can predation cause the 10-year hare cycle? *Oecologia* **74**, 185–92.

Tuljapurkar, S. (1989) An uncertain life: Demography in random environments. *Theoretical Population Biology* **35**, 227–94.

Turchin, P. (1990) Rarity of density dependence or population regulation with lags? *Nature* **344**, 660–3.

Turchin, P. (1995) Population regulation: old arguments and a new synthesis. In: *Population Dynamics: New Approaches and Synthesis* (eds Cappuccino, N. & Price, P.W.), pp. 19–40. Academic Press, San Diego, CA.

Turchin, P. (1996) Nonlinear time series modeling of vole population fluctuations. *Researches on Population Ecology* **38**, 121–32.

Turchin, P. & Hanski, I. (1997) An empirically based model for latitudinal gradient in vole population dynamics. *American Naturalist* **149**, 842–74.

Turchin, P. & Ostfeld, R.S. (1997) Effects of density and season on the population rate of change in the meadow vole. *Oikos* **78**, 355–61.

Turchin, P. & Taylor, A.D. (1992) Complex dynamics in ecological time series. *Ecology* **73**, 289–305.

Underwood, A.J. (1986) The analysis of competition by field experiments. In: *Community Ecology: Pattern and Process* (eds J. Kikkawa & D.J. Anderson), pp. 240–68. Blackwell, Scientific Publications, Melbourne.

van den Bosch, F., Hengeveld, F.R. & Metz, J.A.J. (1992) Analysing the velocity of animal range expansion. *Journal of Biogeography* **19**, 135–50.

Veit, R.R. & Lewis, M.A. (1996) Dispersal, population growth, and the Allee effect: dynamics of the house finch invasion of eastern North America. *American Naturalist* **148**, 255–74.

Venzon, D.J. & Moolgavkar, S.H. (1988) A method for computing profile-likelihood-based confidence intervals. *Applied Statistics* **37**, 87–94.

von Bertalanffy, L. (1938) A quantitative theory of organic growth. *Human Biology* **10**, 181–213.

Wahlberg, N., Moilanen, A. & Hanski, I. (1996) Predicting occurrence of endangered species in fragmented landscapes. *Science* **273**, 1536–8.

Walters, C.J. & Ludwig, D. (1981) Effects of measurement errors on the assessment of stock–recruitment relationships. *Canadian Journal of Fisheries and Aquatic Science* **38**, 704–10.

Warner, A.C.I. (1981) Rate of passage of digesta through the gut of mammals and birds. *Nutrition Abstracts and Reviews B* **51**, 789–820.

Waster, P.M. & Strobeck, C. (1998) Genetic signatures of interpopulation dispersal. *Trends in Ecology and Evolution* **13**, 43–4.

Weir, B.S. (1996) Interspecific differentiation. In: *Molecular Systematics* (eds Hillis, D.M., Moritz, C. & Mable, B.K.), 2nd edn., pp. 385–406. Sinauer Associates, Sunderland, MA.

White, G.C. & Garrott, R.A. (1990) *Analysis of Wildlife Radio-Tracking Data*. Academic Press, San Diego, CA.

White, G.C., Anderson, D.R., Burnham, K.P. & Otis, D.L. (1982) Capture, recapture and removal methods for sampling closed populations. Los Alamos National Laboratory, Los Alamos, NM.

Williams, F.M. & Juliano, S.A. (1996) Functional responses revisited. *Environmental Entomology* **25**, 549–50.

Wilson, K. & Grenfell, B.T. (1997) Generalized linear modelling for parasitologists. *Parasitology Today* **13**, 33–8.

Woiwod, I.P. & Hanski, I. (1992) Patterns of density dependence in moths and aphids. *Journal of Animal Ecology* **61**, 619–29.

Wolda, H. & Dennis, B. (1993) Density dependence tests, are they? *Oecologia* **95**, 581–91.

Wolfram, S. (1991) *Mathematica: A System of Doing Mathematics by Computer*, 2nd edn. Addison-Wesley, Redwood City, CA.

Wood, S.N. (1993) How to estimate life history durations from stage structured population data. *Journal of Theoretical Biology* **163**, 61–76.

Wood, S.N. (1994) Obtaining birth and mortality patterns from structured population trajectories. *Ecological Monographs* **64**, 23–44.

Woolhouse, M.E.J. & Chandiwana, S.K. (1992) A further model for temporal patterns in the epidemiology of schistosome infections of snails. *Parasitology* **104**, 443–9.

Woolhouse, M.E.J., Hasibeder, G. & Chandiwana, S.K. (1996) On estimating the basic reproduction number for *Schistosoma haematobium*. *Tropical Medicine and International Health* **1**, 456–63.

Woolhouse, M.E., Dye, C., Etard, J.F., Smith, T., Charlwood, J.D., Garnett, G.P., Hagan, P., Hill, J.L., Ndhlovu, P.D., Quinnell, R.J., Watts, C.H., Chandiwana, S.K. & Anderson, R.M. (1997) Heterogeneities in the transmission of infectious agents: implications for the design of control programs. *Proceedings of the National Academy of Sciences, USA* **94**, 338–42.

Wright, S. (1951) The genetical structure of populations. *Annals of Eugenics* **15**, 323–54.

Zar, J.H. (1999) *Biostatistical Analysis*, 4th edn. Prentice Hall, Upper Saddler River, NJ.

Index

Page numbers in **bold** refer to tables and those in *italic* refer to figures